Building Controls

The United Association of Journeymen and Apprentices of the Plumbing and Pipe Fitting Industry of the United States and Canada

Three Park Place
Annapolis, MD 21401
Phone: 410-269-2000
Fax: 410-267-0261

International Pipe Trades Joint Training Committee, Inc.

AMERICAN TECHNICAL PUBLISHERS
Orland Park, Illinois 60467-5756

International Pipe Trades Joint Training Committee, Inc.

Composed of representatives of the

United Association of Journeymen and Apprentices of the Plumbing and Pipe Fitting Industry of the United States and Canada (UA), the Mechanical Contractors Association of America, Inc. (MCAA), the Union-Affiliated Contractors (UAC) affiliated with the Plumbing-Heating-Cooling Contractors–National Association (PHCC-NA) and the National Fire Sprinkler Association Inc. (NFSA)

Under the Direction
of the
United Association Training Department

ISBN 978-0-8269-2024-9

First Edition 2014

Printed in the United States of America on environmentally friendly, Recycled Paper with Soy Inks

Contents

STUDENT RESOURCES

- **Quick Quizzes®**
- **Illustrated Glossary**
- **Flash Cards**
- **Media Library**
- **Link to ATPeResources.com**

UA WEBSITES

United Association (UA) Information:

For more information about the UA and its training programs in plumbing, pipefitting, sprinklerfitting, welding, and HVAC service, please visit http://www.ua.org.

Ordering UA Training Materials:

UA local unions, their JATCs, and individual members may order this manual and other educational materials from the online bookstore by visiting the members-only section at http://www.ua.org.

Introduction

Building Controls provides a comprehensive approach to building automation skill development with an overview of building control systems and control concepts, network data communication, and system control devices and applications.

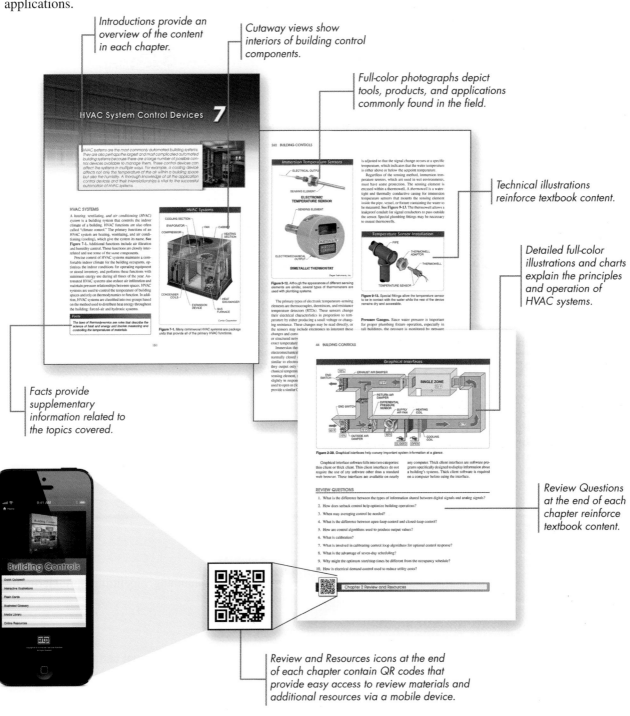

Introductions provide an overview of the content in each chapter.

Cutaway views show interiors of building control components.

Full-color photographs depict tools, products, and applications commonly found in the field.

Technical illustrations reinforce textbook content.

Detailed full-color illustrations and charts explain the principles and operation of HVAC systems.

Facts provide supplementary information related to the topics covered.

Review Questions at the end of each chapter reinforce textbook content.

Review and Resources icons at the end of each chapter contain QR codes that provide easy access to review materials and additional resources via a mobile device.

Interactive Student Resources

The *Building Controls* Interactive Student Resources are a self-study aid designed to supplement content and learning activities in the book. The Interactive Student Resources include the Building Controls Quick Quizzes®, an Illustrated Glossary, Flash Cards, a Media Library, and links to Online Resources.

The Interactive Student Resources enhance content in the textbook with the following features:

- **Quick Quizzes**® provide an interactive review of key topics covered in each chapter.
- **Illustrated Glossary** defines terms and includes links to selected illustrations and media.
- **Flash Cards** reinforce understanding of common industry terms and definitions.
- **Media Library** reinforces and expands upon book content with videos and animated graphics.
- **Internet Resources** provide access to online reference materials that support continued learning.

Acknowledgments

The publisher is grateful to the following companies, organizations, and individuals for providing information, photographs, and technical assistance.

Carrier Corporation
Cleaver-Brooks
Cooper Bussmann
Datastream Systems, Inc.
Dwyer Instruments, Inc.
Fireye, Inc.
Fluke Corporation
The Foxboro Company
GE Thermometrics

Jackson Systems, LLC
Leviton Manufacturing Co., Inc.
Siemens
Trane
U.S. Energy Information Administration
U.S. Green Building Council
Vico Software, Inc.
Watts Regulator Company

The publisher sincerely appreciates the technical information and assistance provided by the following content experts in the development of this textbook.

Shane McCarthy
Training Director
JTAC/UA Local 787/ORAC
Brampton, Ontario

Craig MacDonald
BAS Program Coordinator
JTAC/UA Local 787/ORAC
Brampton, Ontario

Introduction to Building Control Systems 1

Automated buildings are often called "smart" or "intelligent" buildings. This is because the building includes the controls to operate its systems optimally with little or no intervention from building personnel. The benefits of automating a building include significant energy savings, improved control, comfort, security, and convenience. Commercial building HVAC control systems include self-contained, electric, pneumatic, system-powered, electronic, direct digital, automated, and hybrid control systems. These systems differ in the technology used to provide control.

BENEFITS OF BUILDING AUTOMATION

Building automation is the control of the energy-using and resource-using devices in a building for optimization of building system operations. Building automation creates an intelligent building that improves occupant comfort and reduces energy use and maintenance. This can result in significant cost savings and improvements in safety and productivity. While building automation can be a significant initial investment, the return usually justifies the expense.

Energy Efficiency

Energy efficiency is probably the most significant benefit of a building automation system. The energy efficiency of a building is improved by controlling the electrical loads in such a way as to reduce their use without adversely affecting their purposes. For example, lighting is only used when rooms or areas are occupied, reducing electrical costs and increasing overall efficiency. Also, with the automated controls, this reduction in lighting does not negatively impact the use of the space. As the largest consumers of energy in a building, the lighting and heating, ventilating, and air conditioning (HVAC) systems are the primary targets for improving energy efficiency through automation.

Building automation also helps conserve other resources, such as water and fuel. Plumbing controls manage water use and HVAC controls reduce fuel consumption by regulating the boiler or furnace use.

Improved Control over Manual Operation

Most building system functions can be controlled manually, such as turning off lights and adjusting HVAC setpoints, but manual control is not as efficient as automated control. Automated controls can make control decisions much faster than a person and implement much smaller corrections to a system to optimize performance and efficiency.

Automated controls work continuously and, if properly implemented, are not subject to common human errors such as inaccurate calculations, forgetting a control step, or missing an important input. This results in much more consistent and predictable control over the systems. For example, indoor air temperature remains consistent throughout the day because the automated controller is constantly monitoring all of the variables that affect this parameter, making small adjustments to the system to maintain the desired temperature. **See Figure 1-1.** Manual control of the same variables would not likely be as effective at providing such consistent results.

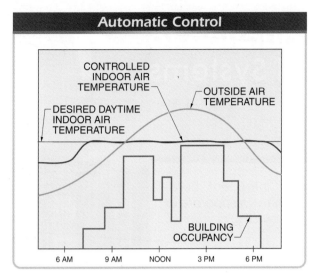

Figure 1-1. Automatic control provides control of building systems even while other parameters change.

Automated controls also allow greater flexibility in implementing control applications. For example, an output from one system can become an input to another, effectively integrating systems in new and unique ways. The building systems may also be more secure from accidental or intentional tampering since the automated controls can limit manual intervention.

Convenience

Building automation provides convenience for building owners, personnel, and occupants. After a system is commissioned and tested, it can operate with little human intervention. It reduces the workload of maintenance and security personnel by providing predictive maintenance information, automatically activating and deactivating some loads, and monitoring the building and its systems for problems. Building occupants find convenience in the automation of routine tasks, such as lights turning on when a person enters a room.

BUILDING CONTROLS HISTORY

Individuals have attempted to improve their living environment for thousands of years, whether by gathering fuel for a fire, experimenting with new fuels, or improving their shelter from the elements. Over time, innovations in technology have increased the level of comfort in living environments. For example, simple huts were followed by permanent structures made of wood or brick. Similar improvements have occurred in the area of HVAC control.

Early control of the level of comfort in a living environment was done by hand. For example, an individual may have controlled the temperature in the environment by adding fuel to a fire or allowing the fire to die down. As structures grew in size and contained a greater number of rooms, hot air from a central fire was regulated manually by opening or closing a diffuser (damper) through the use of an adjusting pulley.

In the early 1900s, electricity came into wide distribution in many cities. This new power source was adapted to provide automatic building environment control. Electricity allowed increased control and required less time and attention than a manual control system. Early electric controls were large, bulky, unreliable, inaccurate by modern standards, and dangerous, partly because early electrical power quality and distribution were poor by today's standards.

In the first quarter of the 20th century, pneumatic control systems came into use in commercial buildings. In a pneumatic control system, an electrical power supply is used to turn on the motor of an air compressor. The compressed air supply is piped through the building to power the HVAC control system.

In the 1960s, solid-state, low-voltage direct current (DC) devices came into common use. This technology was adapted for use in commercial HVAC control systems. Early solid-state systems were known as electronic control systems. More recently, with the advent of microchips that can hold thousands of solid-state devices on a $\frac{1}{4}''$ square chip, automated control systems were created that enabled improved control of the environment in commercial buildings.

The control of the environment in commercial buildings is based on energy efficiency and comfort. During the energy crisis of the 1970s, the cost of energy escalated dramatically. This spurred developments in advanced control systems and techniques. Many commercial buildings were not designed with energy-saving control systems. As a result, there has been a tremendous effort to equip old buildings with systems that can significantly reduce energy expenditures.

Self-Contained Control Systems

A *self-contained control system* is a control system that does not require an external power supply. Self-contained control systems are widely used basic control systems. Self-contained control systems have been successfully applied in certain applications for almost 100 years.

The power to a self-contained control system is supplied by a sealed, fluid-filled element (bulb). The fluid may be a gas, liquid, or mixture of both. The fluid-filled element is known as a power head or power element. The element may be attached to a pipe containing water or refrigerant. The element may also be attached to a wall and used as a building space thermostat.

As the temperature of the fluid in a pipe or air in a building space changes, heat is transferred to or from the fluid-filled element. The heat transfer changes the pressure of the fluid in the element. The pressure acts against a diaphragm that moves a valve body to regulate the flow of refrigerant, steam, hot water, or chilled water through the valve. Adjustments are often provided to change the temperature setpoint of the system. **See Figure 1-2.**

Heating and cooling anticipators work with an ON/OFF control to improve control. Line-voltage control systems are rarely used in commercial building spaces because of the danger to the occupants from electrical shock, fire, and possible explosion due to arcing of the circuit in an area that may contain flammable substances.

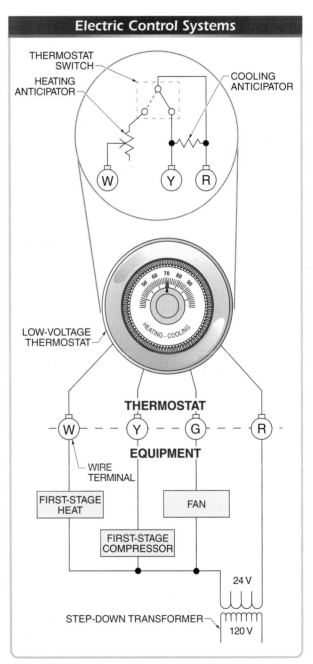

Figure 1-3. Electric control systems may combine heating and cooling controls into a single unit. Heating and cooling anticipators work with an ON/OFF control to improve control.

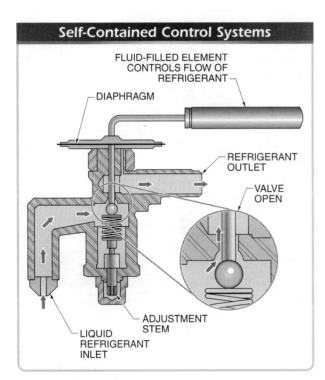

Figure 1-2. Self-contained control systems commonly use fluid-filled elements that do not require an external power supply for operation.

Electric Control Systems

An *electric control system* is a control system in which the power supply is line voltage (120 VAC or 220 VAC) or 24 VAC from a step-down transformer that is wired into the building power supply. **See Figure 1-3.**

Electric control systems commonly use a mechanical device to control the flow of electricity in a circuit. The mechanical device varies from system to system. For example, thermostats may use a bimetallic element to make or break contacts to control building space temperature. A *bimetallic element* is a sensing device that consists of two different metals jointed together that expand and contract at different rates with temperature change. The different metals bend at different rates when heated or cooled. **See Figure 1-4.** On a change in temperature, the bimetallic coil changes position, causing contacts to open or close or a mercury switch to tilt. The contacts or mercury switch make or break a circuit, which energizes or deenergizes the heating equipment. Pressurestats use a bellows for pressure control. Humidistats use a synthetic moisture-sensing element for humidity control.

Pneumatic Control Systems

A *pneumatic control system* is a control system in which compressed air is used to provide power and control signals for the control system. Pneumatic control systems were developed in the early 1900s as a primary method of controlling the environment in commercial buildings. Pneumatic control systems were developed because the existing electric control systems were dangerous or could not provide the flexibility and control necessary in commercial buildings.

Pneumatic control systems can be separated into four main groups of components based on their functions. These groups are the air compressor station, controlled devices, transmitters and controllers, and auxiliary devices. **See Figure 1-5.**

Pneumatic damper actuators are used to control the flow of air by opening and closing dampers.

Air Compressor Stations. The power source for a pneumatic control system is compressed air. The air compressor station consists of the air compressor and other devices that ensure that the air has the qualities needed by the control system. The pneumatic control system functions improperly if the air compressor does not deliver clean, dry, oil-free air at the correct pressure. The air compressor station dryers, filters, and pressure-reducing valves help ensure these qualities. **See Figure 1-6.**

Electrical Switches

MOVABLE CONTACT — STATIONARY CONTACT

MAGNET USED FOR SNAP ACTION

BIMETALLIC COIL

ELECTRICAL LEADS

OPEN CONTACT

GLASS BULB

MERCURY

ELECTRICAL LEADS

BIMETALLIC COIL

MERCURY BULB

Figure 1-4. A bimetallic element or a mercury bulb can be used to switch power in an electric control system.

Pneumatic Control Systems

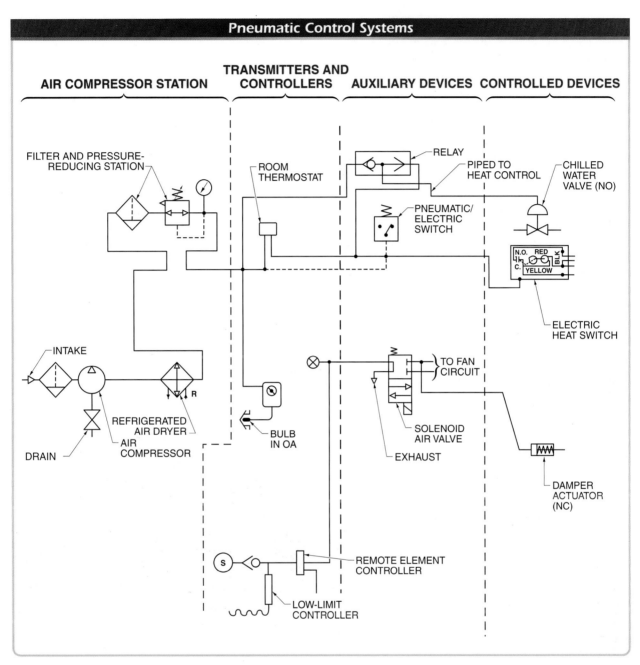

Figure 1-5. Pneumatic control systems include the air compressor station, transmitters and controllers, auxiliary devices, and controlled devices.

Controlled Devices. Controlled devices include dampers, valves, actuators, and switches that deliver the heating or cooling into the building spaces or system. Controlled devices are driven by the compressed air that is supplied to a controller and that causes the controlled devices to open or close properly. **See Figure 1-7.**

Facts

Pneumatic actuators are used frequently in commercial and light industrial applications because they are easy to troubleshoot and repair in the field, are relatively inexpensive, and create a large force using compressed air. The ability to troubleshoot pneumatic control systems is one of the most sought-after technical skills in the HVAC industry.

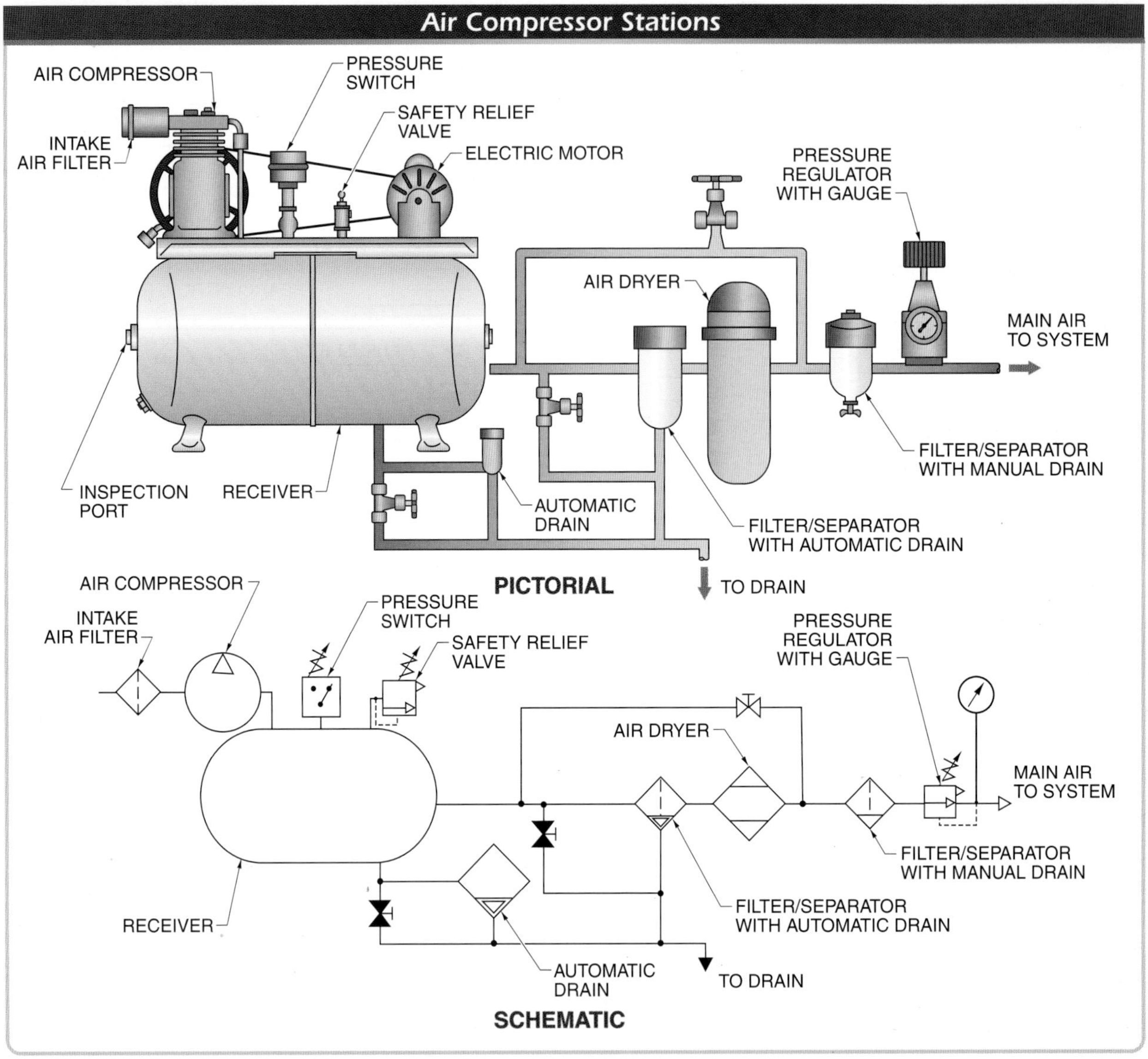

Air Compressor Stations

AIR COMPRESSOR

PRESSURE
SWITCH

SAFETY RELIEF
VALVE

ELECTRIC MOTOR

INTAKE
AIR FILTER

PRESSURE
REGULATOR
WITH GAUGE

AIR DRYER

MAIN AIR
TO SYSTEM

INSPECTION
PORT

RECEIVER

AUTOMATIC
DRAIN

FILTER/SEPARATOR
WITH AUTOMATIC DRAIN

FILTER/SEPARATOR
WITH MANUAL DRAIN

TO DRAIN

PICTORIAL

AIR COMPRESSOR

INTAKE
AIR FILTER

PRESSURE
SWITCH

SAFETY RELIEF
VALVE

PRESSURE
REGULATOR
WITH GAUGE

AIR DRYER

MAIN AIR
TO SYSTEM

RECEIVER

FILTER/SEPARATOR
WITH MANUAL DRAIN

FILTER/SEPARATOR
WITH AUTOMATIC DRAIN

AUTOMATIC
DRAIN

TO DRAIN

SCHEMATIC

Figure 1-6. Air compressor stations provide clean, dry, compressed air to pneumatic systems.

Three-Way Valve Controlled Devices

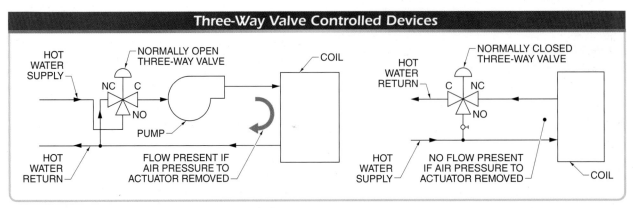

HOT
WATER
SUPPLY

NORMALLY OPEN
THREE-WAY VALVE

COIL

NC   C

NO

PUMP

HOT
WATER
RETURN

FLOW PRESENT IF
AIR PRESSURE TO
ACTUATOR REMOVED

NORMALLY CLOSED
THREE-WAY VALVE

HOT
WATER
RETURN

C   NC

NO

HOT
WATER
SUPPLY

NO FLOW PRESENT
IF AIR PRESSURE TO
ACTUATOR REMOVED

COIL

Figure 1-7. A normally open three-way valve is a controlled device that allows flow if the air pressure to the actuator is removed. A normally closed three-way valve is a controlled device that does not allow flow if the air pressure to the actuator is removed.

Many pneumatic controllers use bimetallic elements to cover or uncover a bleedport. A bleedport is an orifice that allows a small volume of air to be expelled to the atmosphere. The air pressure delivered by the controller to the valve or damper changes as the bimetallic element covers or uncovers the bleedport. **See Figure 1-8.** The air pressure in the line changes proportionally to the temperature change at the bimetallic element. The air pressure signal is piped to an actuator, which causes a valve or damper to open or close to control the temperature of the air in a building space.

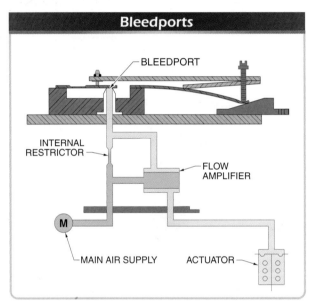

Figure 1-8. The air pressure delivered by the controller to the valve or damper changes as the bimetallic element covers or uncovers the bleedport.

Transmitters and Controllers. Transmitters and controllers sense and control the temperature, pressure, or humidity in a building space or system. Transmitters and controllers are connected to the air supply from the air compressor station. Transmitters and controllers change the pressure to the controlled devices, which causes the controlled devices to regulate the flow of the heating, cooling, or other medium into the building spaces. **See Figure 1-9.**

Auxiliary Devices. Auxiliary devices are devices that are normally located between the transmitters and controllers and the controlled devices. Auxiliary devices change or reroute the air supply from the transmitter or controller before it reaches the controlled devices.

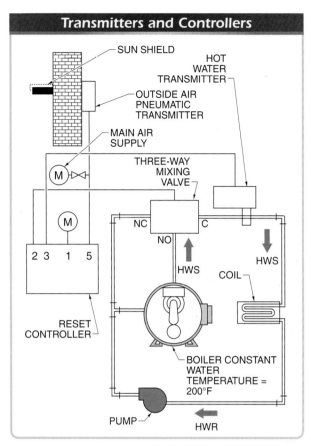

Figure 1-9. Transmitters and controllers change the pressure to the controlled devices, which causes the controlled devices to regulate the flow of the heating, cooling, or other medium into the building spaces.

System-Powered Control Systems

A *system-powered control system* is a control system in which the duct pressure developed by the fan system is used as the power supply. System-powered control systems are used sparingly. System-powered control systems were developed to avoid the installation costs of pneumatic piping runs from an air compressor station.

The power supply of a system-powered control system is the HVAC system itself. In a system-powered control system, the control system is connected to a duct that supplies air to a building space. This small amount of pressure, about 1″ WC, powers the control system. The air duct has a pressure tap that directs the duct system pressure to a filter. The filter supplies the low-pressure air to a system-powered thermostat. **See Figure 1-10.**

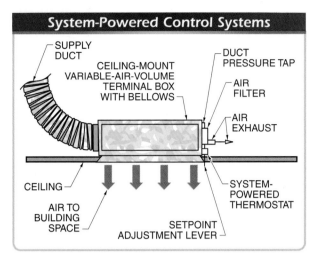

Figure 1-10. System-powered control systems are used in limited applications such as VAV terminal boxes.

The thermostat has a bimetallic element that senses building space temperature. The bimetallic element moves in response to changes in the building space temperature. As the bimetallic element changes position, the air pressure to an inflatable bellows changes. As the bellows inflates or deflates, it controls the primary airflow from the duct into the building space, causing a change in the building space temperature. This eliminates the need for pneumatic piping or wires from a transformer to power the control system.

Electronic Control Systems

An *electronic control system* is a control system in which the power supply is 24 VDC or less. Electronic control systems were originally developed to replace pneumatic control systems. Electronic control systems use solid-state components and are often confused with automated control systems.

Electronic control systems commonly have supply voltages of 10 VDC, 12 VDC, or 18 VDC. **See Figure 1-11.** The power supply of an electronic control system consists of a transformer that drops the supply voltage to 24 VAC and rectifiers that convert the AC to DC. The result is an output commonly between 10 VDC to 18 VDC. Filters may also be included to prevent voltage spikes from passing through the power supply and damaging the control system components.

In an electronic control system, a signal from a sensor is wired to a resistive bridge circuit. A resistive bridge circuit is a circuit containing four arms and four resistors. In a typical design, two of the resistors

are fixed and two are variable. The variable resistors may consist of a temperature sensor and setpoint potentiometer. The bridge circuit output value is zero when the system is at setpoint. When the sensor signal changes, the resistance bridge circuit becomes unbalanced, generating a response voltage in relation to the sensor input. The voltage signal is sent to an actuator, which opens or closes a valve or damper in response to the signal change. The signal may also be changed by adjusting the setpoint potentiometer.

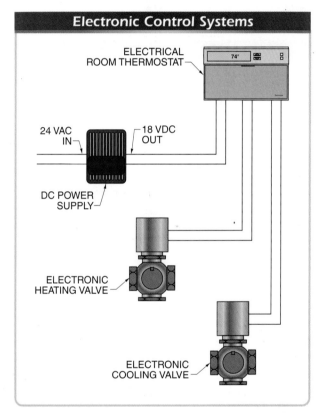

Figure 1-11. Electronic control systems commonly have supply voltages of 10 VDC, 12 VDC, or 18 VDC.

In an alternative design, a resistive bridge circuit consists of two variable resistors, each with a movable wiper. **See Figure 1-12.** The four resistors are the two halves of the two variable resistors. The sensing element moves the wiper on the controller potentiometer to create an unbalanced circuit. A relay is used to actuate a drive shaft that moves the wiper on the feedback potentiometer to balance the circuit. The output is connected to a valve or damper that adjusts the flow of a fluid that adjusts the temperature in the room.

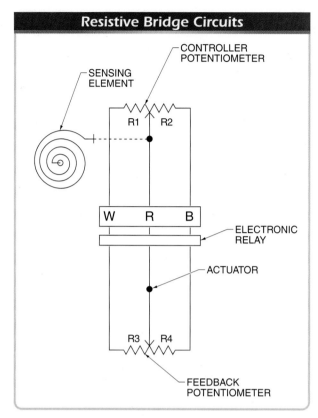

Figure 1-12. A resistive bridge circuit typically consists of two variable resistors, each with a movable wiper.

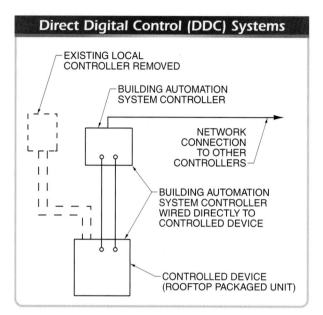

Figure 1-13. DDC systems use electronic controllers with their local sensors and actuators to control portions of a system.

Direct Digital Control Systems

Improvements in electronics allowed for a significant improvement over pneumatic control systems. Electronics technology was adapted for use in building control systems, though it is still primarily used for HVAC systems. A *direct digital control (DDC) system* is a control system in which the building automation system controller is wired directly to controlled devices and can turn them on or off or start a motion. **See Figure 1-13.**

Similar to pneumatic control systems, DDC systems include sensors, controllers, and actuators, although the electronic versions all operate from low-voltage DC power instead of air pressure. Sensors measure temperature, pressure, or humidity and provide the measured values as input to the controller. The controller compares the actual value to the programmed setpoint parameters. The controller also calculates the desired position of the controlled device and outputs a signal to it. The output of the controller is then sent to an electrically controlled actuator, such as a motor or solenoid. The controlled device changes position and supplies heating, ventilation, air conditioning, lighting, or some process variable.

DDC systems are reliable, accurate, and relatively inexpensive. DDC controllers allow far greater functionality in making optimal control decisions because each includes a microprocessor to quickly and accurately calculate the necessary output signal based on the information from the input signals. One disadvantage of DDC systems is that they may require special diagnostic tools and procedures. DDC controllers are also limited to communicating only with the few sensors and actuators to which they are directly connected.

Building Automation Systems

In order to share information between building systems, electronic controllers must have a way to communicate between themselves. A *building automation system (BAS)* is a system that uses a distributed system of microprocessor-based controllers to automate any combination of building systems. Building automation systems can control almost any type of building system, including HVAC equipment, lighting, and security systems.

Building automation system controllers are similar to DDC controllers in that each includes a microprocessor, memory, and a control program. The controller is connected to local sensors and actuators and manages a part of a building system, such as a variable-air-volume (VAV) terminal box. The difference is that the building automation system devices can all communicate with each other on the same shared network. **See Figure 1-14.**

Building Automation Systems

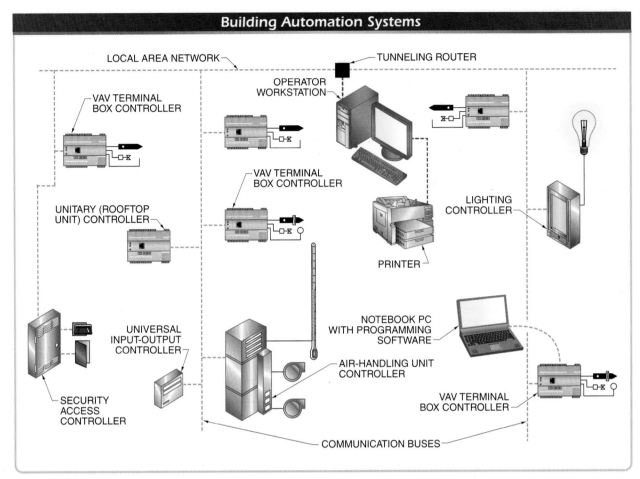

LOCAL AREA NETWORK

TUNNELING ROUTER

OPERATOR WORKSTATION

VAV TERMINAL BOX CONTROLLER

VAV TERMINAL BOX CONTROLLER

LIGHTING CONTROLLER

UNITARY (ROOFTOP UNIT) CONTROLLER

PRINTER

UNIVERSAL INPUT-OUTPUT CONTROLLER

NOTEBOOK PC WITH PROGRAMMING SOFTWARE

AIR-HANDLING UNIT CONTROLLER

SECURITY ACCESS CONTROLLER

VAV TERMINAL BOX CONTROLLER

COMMUNICATION BUSES

Figure 1-14. Building automation systems network electronic controllers together into a system that can share information between building systems.

Information is shared in the form of structured network messages that contain details of many control parameters. This information is encoded into a series of digital signals and sent to any other device that requires the information. For example, if one device reads from a sensor measuring outside air temperature, it can share that temperature reading with any other device, in any building system, that may have use for that information.

Building automation systems are extremely accurate, offer sophisticated features such as data acquisition and remote control, can integrate a variety of building systems, and are very flexible. A disadvantage of building automation systems is that the design and programming can be more involved than with other controls, requiring contractors with specialized knowledge of these systems.

Hybrid Control Systems

A *hybrid control system* is a control system that uses multiple control technologies. As control systems evolve, it is often necessary to use multiple control system technologies together in the same HVAC unit. **See Figure 1-15.** In hybrid control systems, transducers are often used as an interface between different control system technologies. Hybrid control systems are common in commercial building control.

Hybrid control systems contain multiple control technologies requiring different power supplies. Each power supply must be operating correctly in order for the entire system to work properly. For example, a hybrid control system that uses an automated control system to control pneumatic damper and valve actuators requires that the air compressor station and the electronic power supply are working to deliver the proper building space control.

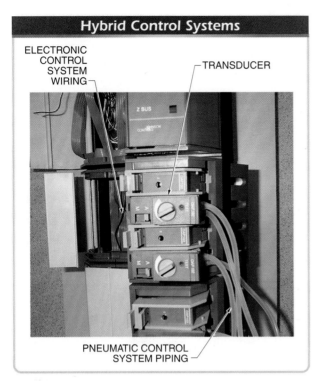

Figure 1-15. Hybrid control systems use different HVAC control technologies to control a single HVAC unit.

BUILDING AUTOMATION APPLICATIONS

Building automation systems can be installed in any residential, commercial, or industrial facility. The basic principles of automation are the same, regardless of the building use, though different types of facilities will likely have different requirements for system automation. Device manufacturers have developed different ways to implement automation and often selectively target certain markets based on building use.

Residential Buildings

Residential buildings are the least automated buildings as a group. Automation devices are becoming increasingly affordable for homeowners and landlords, and the installation and programming of the systems are becoming increasingly practical. One common automation technology for residential buildings is X10 technology.

X10 technology is a control protocol that uses power line signals to communicate between devices. The devices overlay control signals on the 60 Hz AC sine wave, which can be read by any other X10 device connected to the same electrical power conductors. The control signals are used to energize, deenergize, dim, or monitor common electrical loads. This technology is particularly attractive

for residential buildings because it uses the existing electrical infrastructure and requires no new wiring. The control devices simply plug into standard wall receptacles. The devices are also very simple to commission and program. **See Figure 1-16.**

For many of the same reasons, wireless devices are also becoming popular for residential control applications. No additional wiring is required and the risk of radio frequency interference in a residential environment is minimal. Also, residential control applications are not likely to be critical systems, so there is little risk to safety or property if there is a communication problem between control devices.

Commercial Buildings

Commercial buildings include offices, schools, hospitals, warehouses, and retail establishments. With growing emphasis on energy efficiency and "green" buildings, commercial buildings are probably the fastest growing market for implementing new automation systems. **See Figure 1-17.** New construction commercial buildings commonly include some level of automation and many existing buildings are being retrofitted with automation systems.

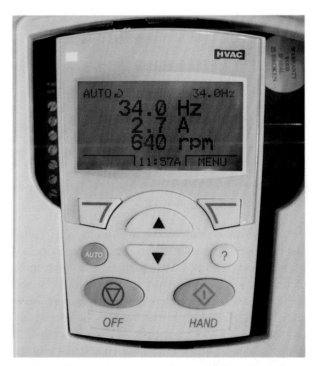

Variable-speed drives can be used to reduce energy use in air-handling units.

X10 Control Systems

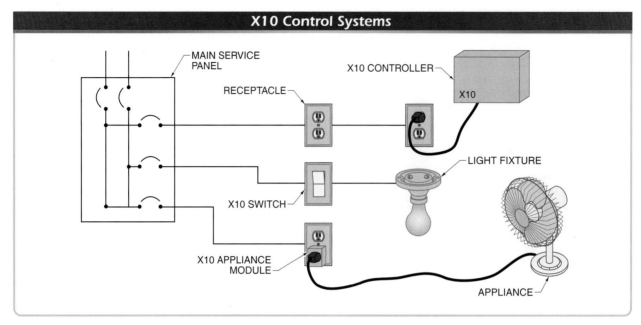

Figure 1-16. X10 is a common residential control system that uses the power line as a medium for sharing control information between devices.

Commercial Buildings

Figure 1-17. Commercial buildings are the most common applications of building automation systems.

Commercial building automation systems can control and integrate any or all of the building's systems, including electrical, lighting, HVAC, plumbing, fire protection, security, access control, voice-data-video (VDV), and elevator systems. Building automation systems are inherently flexible, so building owners can choose almost any level of sophistication, from one specific application to all of the building systems. Upgrades or additional applications can be added and integrated into the system at any time in the future.

Industrial/Process Facilities

Automation has a long history in industrial, manufacturing, and process facilities. In addition to the control of lighting, security, and sanitary water, industrial facilities typically automate manufacturing processes, though the building and manufacturing automation systems are typically separate. Manufacturing processes must be so tightly controlled in terms of time, safety, productivity, equipment efficiency, and product quality that automation is essential. **See Figure 1-18.**

Industrial automation systems operate in much the same way as residential or commercial building automation but for different purposes. The control devices in industrial environments must be more rugged. There may be more human interaction with the automation system in order to modify sequences for changing tasks. Industrial automation is commonly applied at the controller level rather than the device level. Also, the control sequences are primarily focused on safety and product manufacturing.

> **Facts**
>
> Pneumatic controls and communication systems are inherently safe for use in hazardous environments. Pneumatic instrumentation can be used in any area where the National Electrical Code® requires Hazardous Area Classifications.

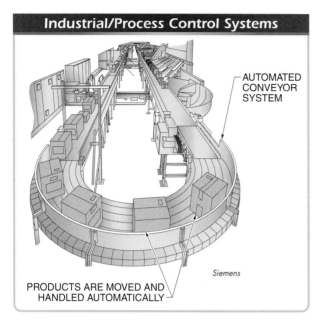

Figure 1-18. Industrial buildings include many control systems, but they are typically for manufacturing processes or material handling.

BUILDING AUTOMATION INDUSTRY

The building automation industry includes a wide range of individuals and groups that address the requirements, concerns, and ideas for new or existing buildings and use them to create state-of-the-art automated buildings. The participants in the building automation industry are the building owners, consulting-specifying engineers, controls contractors, and any authorities having jurisdiction.

Building Owners

The building owner decides which building automation strategies to implement based on the desired results. Owners may wish to build a fully automated showcase facility that is highly visible to the public, or they may wish to automate only the few building systems that are most important to the efficient operation of the building. The end use of the building may also affect the automation choices because the building's location, occupancy density, equipment and products, and special on-site processes dictate its automation priorities. For example, lighting and HVAC controls may be especially important for office buildings, while security and access controls may be important for warehouses full of valuable inventory.

The concerns of building owners also include initial costs, payback periods, maintenance, impact on occupants, system flexibility, system upgradeability, and environmental impacts. For example, automation systems typically return their extra costs within a few years but still require a greater initial investment. These factors are even more important in retrofitting existing buildings for automation, as the implementation possibilities are typically more limited than in new construction.

Programmable lighting controls are often used to reduce energy use in office areas.

Consulting-Specifying Engineers

For designing the most efficient system, the owner meets with consulting-specifying engineers to refine the automation plan. A *consulting-specifying engineer* is a building automation professional that designs the building automation system from the owner's list of desired features. This involves describing the necessary functionality of all the control devices and detailing the interactions between them. The consulting-specifying engineers work with the building owner, architect, and general contractor to make changes as needed to the planned automation system in order to maximize the benefits while working within any aesthetic, financial, or construction constraints.

Consulting-specifying engineers prepare contract specifications that include all of the control information. A contract specification is a document describing the desired performance of the purchased components and means and methods of installation. **See Figure 1-19.** Automation applications in specifications may be written out as detailed control sequence steps. These contract specifications are then made available to controls contractors to bid the project.

Contract Specifications

1.1 ELECTRONIC DATA INPUTS AND OUTPUTS

A. Input/output sensors and devices matched to requirements of remote panel for accurate, responsive, noise free signal output/input. Control input to be highly sensitive and matched to loop gain requirements for precise and responsive control.
  1. In no case shall computer inputs be derived from pneumatic sensors.

B. Temperature sensors:
  1. Except as indicated below, all space temperature sensors shall be provided with single sliding setpoint adjustment. Scale on adjustment shall indicate temperature. The following are exceptions to this:
     a. The following locations shall have sensor without setpoint adjustment:
        1) All electrical and communication rooms.
        2) All mechanical rooms.
        3) All unit heaters.
        4) All public elevator lobbies and entrance vestibules.
        5) Elevator equipment rooms.
  2. Duct temperature sensors to be averaging type. Averaging sensors shall be of sufficient length (a maximum of 1.8 sq ft of cross sectional area per 1 lineal foot of sensing element) to insure that the resistance represents an average over the cross section in which it is installed. The sensor shall have a bendable copper sheath. Water sensors provided with separable copper, monel, or stainless steel well. Outside air wall-mounted sensors provided with sun shield.

Figure 1-19. Contract specifications are produced by the consulting-specifying engineer and include the device requirements and control sequences for the building automation system.

Controls Contractors

Controls engineering has traditionally been considered a subset of mechanical work, especially since HVAC systems are the most commonly automated building systems. However, controls technology has evolved into a highly technical field that has allowed contractors to specialize in automation. This has also given controls contractors the freedom to work with every building system so that the systems can be seamlessly integrated.

The controls contractor uses the contract specification to choose the necessary components and infrastructure that can accomplish the requirements. For example, the specification may list a temperature sensor that must provide temperature measurements to an air-handling unit controller. The controls contractor chooses the manufacturer and model of the sensor that meets all the requirements for installation type, temperature range, environmental conditions, calibration needs, and signal type. The controls contractor submits a quote for the project based on the costs of the purchased equipment, installation labor, system commissioning, and tuning. If chosen, the controls contractor then performs the work.

Authorities Having Jurisdiction

The *authority having jurisdiction (AHJ)* is the organization, office, or individual responsible for approving the equipment and materials used for building automation installation. This includes all agencies and organizations that regulate, legislate, or create rules that affect the building industry including building codes, National Electrical Code® (NEC®) regulations, Occupational Safety and Health Administration (OSHA) regulations, local and municipal codes, fire protection standards, waste disposal regulations, and all levels of government or other industry agencies. These groups may overlap. Building automation is most affected by policies related to energy efficiency and indoor air quality regulations.

CONSTRUCTION DOCUMENTS

The planning and implementation of a building automation system requires detailed documentation. Controls contractors receive a set of construction documents that give a comprehensive view of the project from the consulting-specifying engineers. For example, controls contractors will receive sheet metal, plumbing, electrical, and piping drawings, as well as project schedules, scope, and other details.

Project Management

The project management process is based on *A Guide to the Project Management Body of Knowledge (PMBOK® Guide)*. The PMBOK provides an outline that can be used to analyze and manage a project from beginning to completion. It identifies the controls that are needed to complete a project within the specified time and financial constraints. Many project management professional (PMP) certificate holders use the PMBOK standard as management standards. ANSI has accepted the PMBOK as a standard. The Project Management Institute (PMI) periodically updates PMBOK, and ANSI periodically updates the corresponding standard. The most recent version of the standard is ANSI/PMI 99/001/2008.

Contract Documents

Contract documents are documents produced by the consulting-specifying engineer for use by a contractor to bid a project and consist of construction drawings and a book of contract specifications. The contract specifications describe the project requirements in language rather than graphics.

Contract documents cover the various design disciplines: architectural, civil, structural, mechanical, and electrical engineering. Contract documents are purposely written in a generic manner to accommodate a variety of available products and features. This allows the controls contractor the freedom to specify any manufacturers or models that fulfill the project requirements.

CSI MasterFormat® Division Specifications. The *Construction Specifications Institute (CSI)* is an organization that develops standardized construction specifications. The CSI, in cooperation with the American Institute of Architects (AIA), the Associated General Contractors of America (AGC), the Associated Specialty Contractors (ASC), and other industry groups, has developed the CSI MasterFormat® for Construction Specifications, which is a uniform system for construction specifications, data

filing, and cost accounting. CSI continually promotes the CSI MasterFormat and updates it periodically.

The MasterFormat is a master list of numbers and titles used for organizing information about construction requirements, products, and activities into a standard sequence. The MasterFormat consists of front end documents and 50 divisions. **See Figure 1-20.** The front end documents contain information on bidding requirements, contracting requirements, and conditions related to the construction project.

The CSI MasterFormat® is used to specify construction and installation requirements.

The 50 divisions of the body are numbered and designed to give complete written information about individual construction requirements for building and material needs. Not all sections of the CSI MasterFormat appear in every set of print specifications. Only the sections that are applicable to the construction project appear in the specifications.

Division 25 contains specifications for integrated automation that provide information on wiring, equipment, and controls for integrated automation systems. Division 25 information includes descriptions of operation and maintenance of integrated automation systems and lists of networking equipment. The specifications also contain descriptions of instrumentation and terminal devices, facility controls, and control sequences for multiple systems. **See Figure 1-21.**

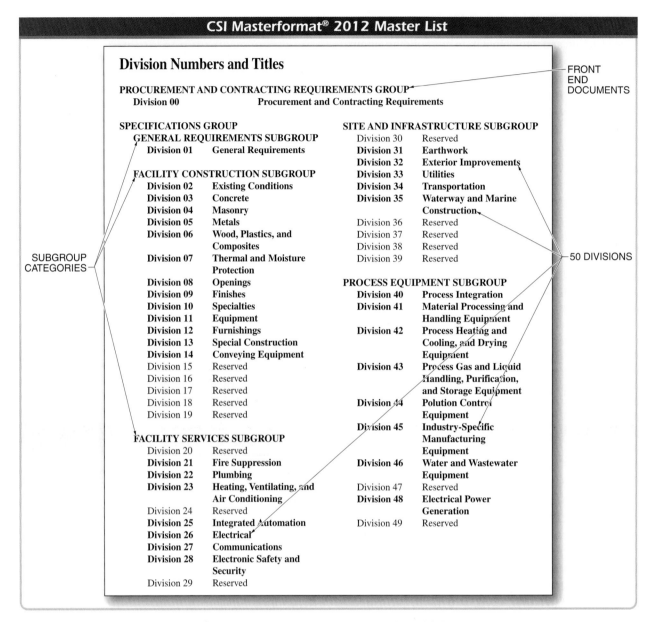

Figure 1-20. The CSI MasterFormat® consists of front end documents and 50 divisions.

Shop Drawings. The controls contractor's project bid will include shop drawings. A shop drawing is a document produced by the controls contractor with the details necessary for installation and includes two basic categories: components and software/wiring. The consulting-specifying engineer reviews the shop drawings to ensure that the intent of the contract documents has been met. Also, the shop drawings are used by the owner's building maintenance staff for troubleshooting after the building has been turned over.

Shop drawings show much more detail than the contract drawings because they are specific to a particular device manufacturer and model. For example, shop drawings include wiring details down to the individual device level. **See Figure 1-22.**

> **Facts**
>
> *Optimum start/stop control relies on scheduled periods of expected occupancy and cannot account for unexpected occupancy, such as early arrivals or visits to the building on the weekends. Access events can be used to trigger unscheduled operation at occupancy setpoints.*

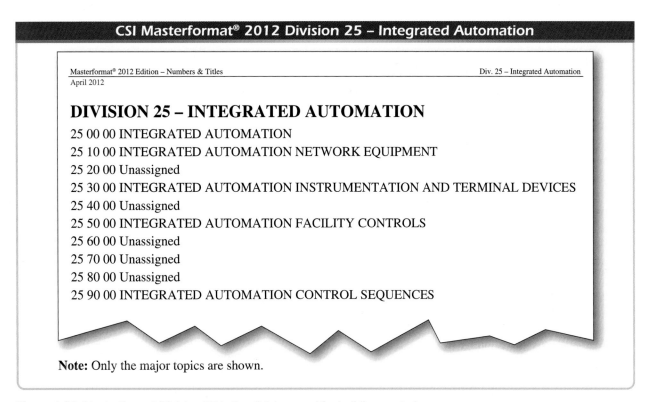

Figure 1-21. MasterFormat Division 25 is the division used for building controls.

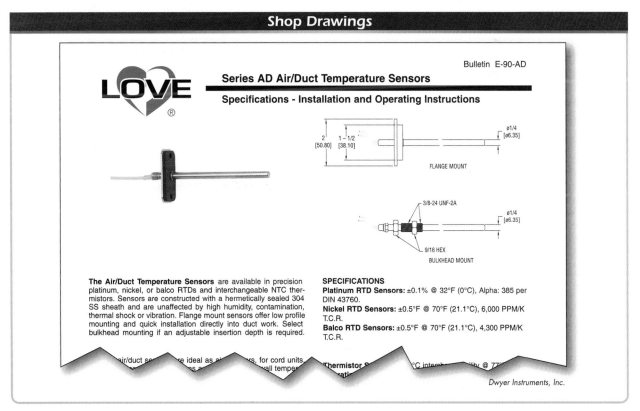

Figure 1-22. Shop drawings from a controls contractor include specific information on devices that satisfy the automation requirements.

The component portion of the shop drawing package includes information on every component and its size, configuration, and location. This allows the other tradesworkers to coordinate the installation of control components. For example, control valves are installed by pipefitters long before the controls contractor connects signal wiring to them. Manufacturer data sheets are included to describe the features of the component in detail.

The component shop drawing package is often developed and reviewed by the consulting-specifying engineers first so that purchases with long lead times or decisions that impact other contractors can be made in a timely manner. For example, the controls contractor may choose to have the air terminal unit controllers mounted by the manufacturer at the factory to save installation time.

The software/wiring portion of the shop drawing package addresses all of the control interactions between the components. A building-level overview explains the control system architecture. Wiring interconnection and block diagrams illustrate both signal and power wiring for each panel and device, including information on power sources. Flow charts for each control sequence show interrelationships between inputs, calculations (control loops), and outputs. Sequences of operation describe all flow chart functions in words. A detailed explanation of color conventions clarifies drawing symbology.

Project Closeout Information. The *project closeout information* is a set of documents produced by the controls contractor for the owner's use while operating the building. Project closeout information includes operating and maintenance manuals for all equipment, an owner instruction report, and a certified resolution of issues raised during the commissioning process.

AUTOMATED BUILDING SYSTEMS

Building automation may involve any or all of the building systems. The level of sophistication of a building automation system is affected by the number of integrated building systems. Common building systems controlled by building automation include electrical, lighting, HVAC, plumbing, fire protection, security, access control, voice-data-video (VDV), and elevator systems.

Electrical Systems

An *electrical system* is a combination of electrical devices and components connected by conductors that distributes and controls the flow of electricity from its source to a point of use. The control of the electrical system involves ensuring a constant and reliable power supply to all building loads by managing building loads, uninterruptible power supplies (UPSs), and back-up power supplies. Sophisticated switching systems are used to connect and disconnect power supplies as needed to maintain building operation and occupant productivity. Electrical systems affect many other systems, such as lighting, security, and VDV circuits. However, control of those systems is typically handled separately.

Lighting Systems

A *lighting system* is a building system that provides artificial light for indoor areas. Lighting is one of the single largest consumers of electricity in a commercial building. Therefore, improving energy efficiency and reducing electricity costs are the driving factors in lighting system control.

Lighting system control involves switching off or dimming lighting circuits as much as possible without adversely impacting the productivity and safety of the building occupants or the security of the building. Lighting systems are controlled based on schedules, occupancy, amount of available daylight, or timers. Specialty lighting control systems can also produce custom lighting scenes for special applications.

HVAC Systems

A *heating, ventilating, and air conditioning (HVAC) system* is a building system that controls the indoor climate of a building. Properly conditioned air improves the comfort and health of building occupants.

HVAC systems are the most common building systems to automate but are also the most complicated. They include several parameters that must be controlled simultaneously and that are closely interrelated. Changes in one conditioned air parameter, such as temperature, can affect another parameter, such as humidity. Therefore, these systems use the most sophisticated control logic and must be carefully designed and tuned. A well-automated HVAC system, however, operates efficiently and with little manual input from occupants or maintenance personnel.

Plumbing Systems

A plumbing system is a system of pipes, fittings, and fixtures within a building that conveys a water supply and removes wastewater and waterborne waste. Plumbing systems are designed to have few active components.

That is, most plumbing fixtures operate on water pressure and manual activation alone. However, there are a few applications in which certain parts of a commercial building's plumbing system can be actively controlled, such as water temperature, water pressure, and the supply of water for certain uses. These systems aim to use water the most effectively and provide a water supply with the optimal characteristics.

Fire Protection Systems

A *fire protection system* is a building system for protecting the safety of building occupants during a fire. Fire protection systems include both fire alarm systems and fire suppression systems. Fire protection systems automatically sense fire hazards, such as smoke and heat, and alert occupants to the dangers via strobes, sirens, and other devices. They also monitor their own devices for any wiring or device problems that may impair the system's proper operation during an emergency.

Fire protection systems are highly regulated due to their role as a life safety system. Fire-sensing devices and output devices can only be connected to the fire alarm control panel (FACP). Fire protection systems can be integrated with other building systems, but also only through the FACP. For example, some FACPs include special output connections that can be used to share fire alarm signals with other systems that have special functions during fire alarms.

Security Systems

A *security system* is a building system that protects against intruders, theft, and vandalism. Security systems are similar to fire protection systems in that they are typically designed as separate systems and allow connection with other building systems only at the security control panel. The control panel provides special output connections that can be used to initiate special control sequences in other systems, such as unoccupied modes, access, and security monitoring with surveillance systems.

Access Control Systems

An *access control system* is a system used to deny those without proper credentials access to a specific building, area, or room. Authorized personnel use keycards, access codes, or other means to verify their right to enter the restricted area.

Access control systems can be integrated through their control panels with other systems in much the same way as fire protection systems and security systems. In fact, due to their similarity, access control and security systems are often highly integrated. When a person's credentials have been verified by the access control system, the control panel can initiate the subsequent response of other systems, such as directing the surveillance system to monitor the person's entry.

Voice-Data-Video Systems

A *voice-data-video (VDV) system* is a building system used for the transmission of information. Voice systems include telephone and intercom systems. Data systems include computer and control device networks, but can also carry voice or video information that has been encoded into compatible data streams. Video systems include closed-circuit television (CCTV) systems.

Building automation integration typically involves the VDV systems being controlled by other building systems. For example, a telephone line may be seized by the fire alarm system in order to alert the local authorities to a fire in the building. Alternatively, the CCTV system that is used for surveillance may be controlled by the security or access control systems to monitor certain areas where people have been detected.

Facts

The NEC® provides requirements for safe electrical practices. It uses Informational Notes throughout the Code to explain information that is contained within the Code. Informational Notes do not contain any mandatory provisions but do provide supplemental material.

Elevator Systems

An *elevator system* is a conveying system used for transporting people and/or materials vertically between floors in a building. Elevator systems operate largely on their own and with their own control devices but can accept inputs from other systems to modify their operating modes. A prime example of this is the integration of the fire alarm system with the elevator system. Since elevators can be dangerous to use during a fire, the elevator cars must be removed from service if the fire alarm is activated. An output on the FACP is connected to the elevator controller. In the event of a fire alarm signal, the elevator controller switches over to a fire service mode and disables the elevator car.

REVIEW QUESTIONS

1. Briefly describe the three primary benefits of a building automation system.

2. Describe the power supply of a self-contained control system.

3. Why are line-voltage control systems rarely used in commercial building spaces?

4. Describe a bimetallic element.

5. List the four main groups of components that pneumatic control systems can be grouped into based on their functions.

6. Describe the power supply of a system-powered control system.

7. Explain the use of a resistive bridge circuit in an electronic control system.

8. What is a hybrid control system?

9. How do the roles of the consulting-specifying engineer and the controls contractor differ with respect to designing the building automation system?

10. What is the *CSI MasterFormat®?*

Chapter 1 Review and Resources

Control Concepts 2

Building automation systems use controllers to adapt dynamically to changing conditions and needs of the occupants. Controllers are designed to provide comfort to the occupants of building spaces. Each individual controller includes the logic to make decisions within its own system to change the operation of a building system. Together, all of these controllers compose a system that operates smoothly and seamlessly.

CONTROL DEVICES

A *control device* is a building automation device that monitors or changes system variables, makes control decisions, or interfaces with other types of systems. A *variable* is some changing characteristic in a system. Control devices include sensors, controllers, actuators, and human-machine interfaces.

Sensors

A *sensor* is a device that measures the value of a controlled variable, such as temperature, pressure, or humidity, and transmits a signal that conveys this information to a controller. The sensor output signal is commonly electrical, such as voltage or current, but may also be conveyed by air pressure, optics, or radio frequency. Even though they output signals, sensors are considered input devices, which is relative to the point of view of the controllers that manage the systems.

As individual devices, sensors allow variables to be measured in any location or type of environment, while other control devices are installed in convenient and centralized places. For example, a controller may be connected to a number of individual temperature sensors that are installed throughout an HVAC unit. **See Figure 2-1.** Sensors may also be integral with a controller, as in a room thermostat.

Sensors must be carefully selected to provide the appropriate information of the variables to be sensed in a way that is understandable by the receiving control devices. Also, they must be located where they can properly sense the variables, such as in a duct, pipe, or room remote from the controller. Factors to consider when selecting sensors are sensor type, range, resolution, accuracy, calibration, signal type, power requirements, operating environment, mounting configuration, size, weight, construction materials, and agency listings.

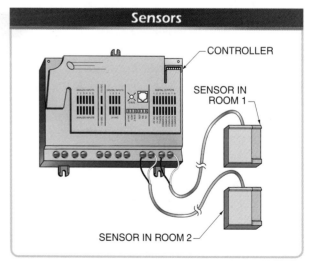

Figure 2-1. Sensors provide the inputs into building automation controllers.

Sensors that are compatible with modern automation systems are typically either electromechanical or electronic sensors. They are distinguished by the way in which the sensors monitor their environment. Electromechanical devices use mechanical means to measure a variable but include electrical components to convert motion into electrical signals and process and transmit the signals. Some variables that may require mechanical capabilities to be measured include pressure, position, and fluid level.

Purely electronic sensors are typically smaller and more reliable than electromechanical devices, since they have no moving parts. **See Figure 2-2.** The variable is measured by detecting the direct changes the variable has on the electrical properties of a solid-state component. Then other electronics process and transmit the information as a signal. Sensors are connected to terminals on a controller.

Modern controllers are all electronic devices with processors, memory, and software that can execute control sequences very quickly and accurately. The electronics also allow relatively simple modification of the programming and settings to optimize building system operation. However, the electronics may be less rugged than some sensors.

Some controllers are designed for specific control applications, such as controlling an air-handling unit. These provide many features and programs developed by the manufacturer to simplify installation and commissioning of the system, but they may not allow much flexibility to change inputs, outputs, or decision making. Generic controllers, on the other hand, include only the electronics to read from standard input connections, run software, and write to standard output connections. Any type of device can be connected to the terminals and any type of decision can be made. These devices allow the greatest flexibility in applications, but the program must be custom-written and tested by the controls contractor to ensure correct operation and reliability.

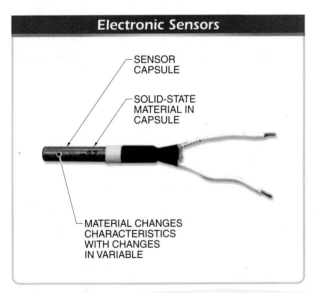

Electronic Sensors

SENSOR CAPSULE

SOLID-STATE MATERIAL IN CAPSULE

MATERIAL CHANGES CHARACTERISTICS WITH CHANGES IN VARIABLE

Figure 2-2. Electronic sensors detect changes in a variable from corresponding changes in the electrical properties of the sensor.

Controllers

A *controller* is a device that makes decisions to change some aspect of a system based on sensor information and internal programming. The controller receives a signal from the sensor, compares it to a desired value to be maintained, and sends an appropriate output signal to an actuator.

Facts

A high limit or low limit may be used on a sensor signal when using multiple measurements of the same object. This strategy eliminates errors due to failed sensors.

Actuators

An *actuator* is a device that accepts a control signal and causes a mechanical motion. This motion may actuate a switch, rotate a valve stem, change a position, or cause some other change that affects one or more characteristics in a system. For example, an actuator may open a valve to allow more steam into a heating coil, causing the temperature of the airflow across the coil to rise. From the point of view of the controller, actuators are the output devices.

A large number of controlled devices may be used to change system characteristics, such as dampers for regulating airflow, valves for regulating water or steam flow, refrigeration compressors for delivering cooling, and gas valves and electric heating elements for regulating heating. Many of these are actually controlled by just a few different types of actuators. Common types of actuators are relays, solenoids, and electric motors. **See Figure 2-3.**

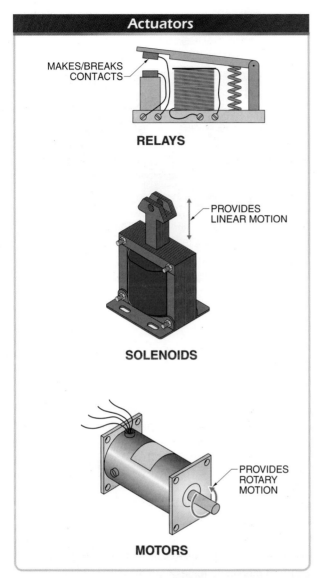

Figure 2-3. Common types of actuators found in controlled devices are relays, solenoids, and motors.

A *relay* is an electrical switch that is actuated by a separate electrical circuit. It allows a low-voltage control circuit to open or close contacts in a higher-voltage load circuit. Many relays are electromechanical types, but some are completely solid-state, using no moving parts. Relays can be used with any device that is controlled by turning it on or off. They are common for switching lighting circuits and HVAC package units, as well as sharing on/off information between controllers.

A *solenoid* is a device that converts electrical energy into a linear mechanical force. Solenoids are typically used in quick-acting valves and for locking and unlocking doors.

An *electric motor* is a device that converts electrical energy into rotating mechanical energy. Electric actuator motors can be used to turn something completely open or completely closed, but are particularly useful for actuating devices with multiple positions, such as dampers and valves.

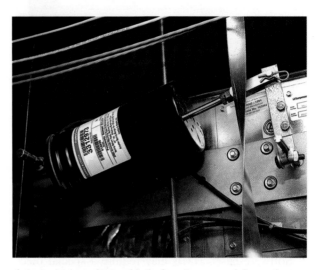

Actuators are used to provide the force to open and close a damper.

Human-Machine Interfaces

A human-machine interface is connected into a building automation system to allow personnel to view and modify the information being shared between control devices. A *human-machine interface (HMI)* is an interface terminal that allows an individual to access and respond to building automation system information. Many HMIs show information graphically and include data over time so that current values and longer-term trends are easily visible. HMIs may also allow a user to interact with the system by manually inputting changes to values or device behavior.

HMIs are either hardware-based or software-based. **See Figure 2-4.** Hardware-based HMIs are basically small computers that are specially designed to gather, process, and display system data. They are typically self-contained units with integral monitors, memory, connection terminals, and communication ports. If they are intended to be installed in extreme environments, they may be housed in special dust-resistant, moisture-resistant, rugged enclosures. If the HMI accepts inputs into the system from the user, it typically includes a keypad, touchscreen display, or I/O port for connecting a separate keyboard or mouse.

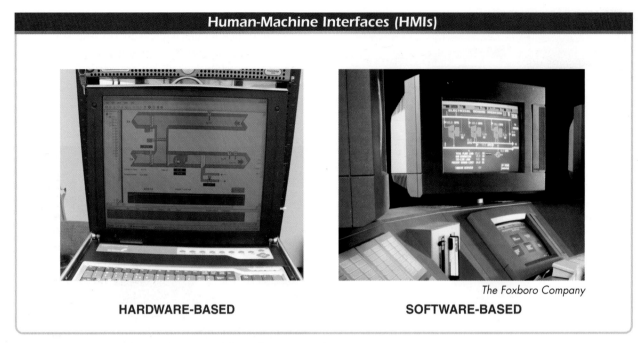

Human-Machine Interfaces (HMIs)

HARDWARE-BASED

The Foxboro Company

SOFTWARE-BASED

Figure 2-4. HMIs may be hardware-based or software-based.

Software-based HMIs may require a special piece of hardware to physically connect with the building automation system and retrieve data, but the interface portion relies on software running on separate personal computers. Some packages require special software (a "thick client" application) on the computer to communicate with the hardware. However, the current trend is to design the interface to be viewed with a standard web browser, which is common software on all computers and requires no additional installation (a "thin client" application).

The hardware unit acts as a web server, delivering the building system information in a way that is similar to any other Internet site. These HMIs can be configured to provide this information to only computers within the building or to any computer anywhere in the world. Either type of software-based HMI is capable of being programmed to accept system inputs from the user. Security features ensure that only authorized personnel are able to make these changes.

CONTROL SIGNALS

A *control signal* is a changing characteristic used to communicate building automation information between control devices. Control signals are typically electronic signals but can also be transmitted by other media such as air pressure, light, and radio frequency. There are different types of control signals, depending on the type of information to be shared. The three common types of control signals are digital signals, analog signals, and protocol messages.

Digital Signals

A *digital signal* is a signal that has only two possible states. For this reason, digital signals may also be known as binary signals. Digital signals convey information as either one of two extremes, such as completely ON or completely OFF, or completely open or completely closed.

A digital signal is typically conveyed with a change in voltage. For example, 0 VDC can represent an OFF state and 5 VDC an ON state. **See Figure 2-5.** Alternatively, +12 VDC can represent an ON state and –12 VDC an OFF state. Any pair of two different voltages can be used to send digital signals, though they are commonly DC voltages of 24 VDC or less. Also, different digital signal voltage levels can be used within a system, as long as each device is compatible with the type of signal it is sending or receiving.

The devices also must agree on the voltage levels that define the two states. For example, due to a slight voltage drop in the signaling conductor, a 5 VDC (ON) signal may appear to the receiver as slightly less, such as 4.6 VDC. In order for this voltage to still be registered as an ON state, the two states may be more precisely defined as 0 V to 0.8 VDC for an OFF state and 2.0 VDC to 5.25 VDC for an ON state. Voltages between these two levels may be read as an erratically fluctuating ON and OFF state, producing unpredictable results.

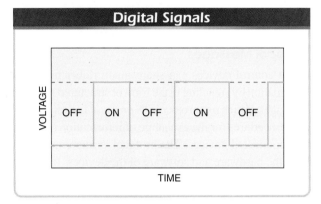

Figure 2-5. Digital signals are produced by a pair of voltage levels that represents either ON or OFF.

Digital signals can be generated by a device with a power supply by applying the necessary voltage to a signaling conductor. **See Figure 2-6.** Alternatively, digital signals can be generated by a set of contacts in an input device, which does not require an external power supply. The contact terminals are wired to a set of controller terminals with a pair of conductors. The controller senses the continuity or discontinuity through the terminals as closure or opening of the contacts in the input device.

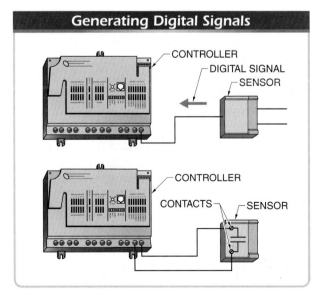

Figure 2-6. Digital signals can be generated by devices with their own power supply or by devices with switch contacts.

Digital signals are commonly used as an output to turn devices on and off, typically through relays that switch the power needed to operate the device. For example, digital signals can turn electric motors, electric heating stages, and valves on and off. Digital signals can also be used to initiate different functions in package units. These units are energized by a separate, manual means, but a controller's digital signals may enable/disable the unit's primary function.

Analog Signals

An *analog signal* is a signal that has a continuous range of possible values between two points. **See Figure 2-7.** Analog signals can convey information that has units of measurement, such as degrees Fahrenheit, cubic feet per minute, meters per second, and inches of water column. They can also provide any value between 0% and 100% of some controllable characteristic. For example, analog signals can be used to control fan speed from 0% (stopped) to 100% of its full rated maximum speed.

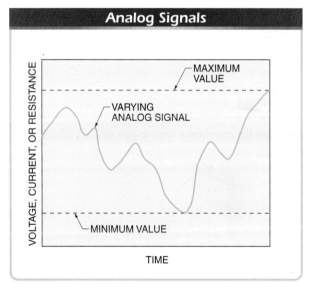

Figure 2-7. Analog signals can vary continuously between two points.

The most common electrical properties used to convey analog signals are voltage, current, and resistance. The most common analog signal ranges include 0 VDC to 10 VDC, 2 VDC to 10 VDC, 4 mA to 20 mA, and 0 kΩ to 10 kΩ. Devices may use other analog signal ranges, as long as they are compatible between the sender and receiver and represent the same values. For example, 0 VDC to 5 VDC, 1 mA to 10 mA, and 0 Ω to 1000 Ω ranges are common.

Analog signals differ from digital signals in that small fluctuations in the signal are meaningful. The range of possible signals is mapped to a range of possible values of the measured variable or the range of actuator positions. For example, an analog signal of 4 mA to 20 mA may be used to indicate a temperature between 32°F and 212°F. **See Figure 2-8.** Therefore, the range of 16 mA (20 mA – 4 mA = 16 mA) represents the range of 180°F (212°F – 32°F = 180°F). Each 1°F of change in temperature is indicated by a change of 0.089 mA. Units must be carefully noted in analog signal mapping, as many quantities can be mapped to other scales. For example, the same temperature analog signal also has the characteristic of 0.16 mA/°C (16 mA ÷ [100°C – 0°C] = 0.16 mA/°C).

The same scaling calculations are done for 0 VDC to 10 VDC and for 2 VDC to 10 VDC signals. Analog signals that have a starting value higher than zero provide a reference feedback signal to the controller as an input verifying the position of the controlled device.

A disadvantage to analog signals is that electrical noise, which is normal to some degree in any electrical system, can affect the signals by changing their intended value slightly. While digital signals address this problem by providing ranges of values that correspond to the intended information, this cannot be done with analog signals, where every small change in value is potentially important.

Depending on the system, noise can cause significant operational problems. Noise can be reduced to some degree by shielded conductors. Proper connections and careful design of the automated system, such as the length of conductor runs and proximity to noise-inducing electrical components, can help mitigate noise problems.

Protocol Messages

Some control devices can share much richer pieces of information by signaling in the form of structured network messages, which use protocols. A *protocol* is a set of rules and procedures for the exchange of information between two connected devices. The protocol is implemented in both the hardware and software of the devices. It defines the format, timing, and signals for reliable and repeatable data transmission between two devices.

Each structured network message includes information on the sending and receiving devices, the identification of the shared variable, its value, and any other necessary parameters. The information is encoded and transmitted via a series of digital signals, composing one complete message. **See Figure 2-9.** Almost any type of information can be shared in this way, as long as all the devices involved can work with the same protocol.

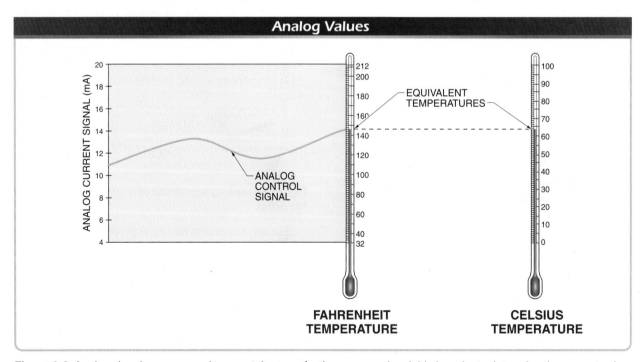

Figure 2-8. Analog signals are mapped to a certain range for the measured variable in order to determine the current value.

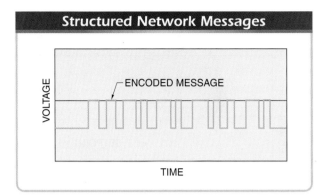

Figure 2-9. Structured network messages are several pieces of control information encoded into a series of digital signals.

Systems using digital and analog signals are typically connected point-to-point. That is, each device is connected with only the devices that need to receive its signal. However, systems using structured network messages are connected together in a similar way to a computer network. In this configuration, any device can communicate with any other device on the network. In fact, some systems can communicate over the same wiring and routing infrastructure as the building's computer local area network (LAN). There are also advantages to keeping the building automation system on a completely separate network, though this increases installation and maintenance costs.

CONTROL INFORMATION

Control devices are used in every type of control application to provide the required information for proper system operation. Hardware and software used to provide control information include control points, virtual points, setpoints, offsets, and deadband.

Control Points

The input information from sensors and output information to actuators creates control points in the building automation system. A *control point* is the actual value that a control system experiences at any given time. Control points change over time and may be different from the setpoints, or desired values. **See Figure 2-10.** Controllers use input control points to make decisions about changes needed in the system and then determine the best value for the output control points.

For example, a temperature sensor in a building space measures a temperature of 74°F, which is an input control point for the HVAC controllers. The information that

the controllers generate and use to cause changes in the system is the output control point. For example, a damper actuator accepts a control point signal that corresponds to the position it must maintain.

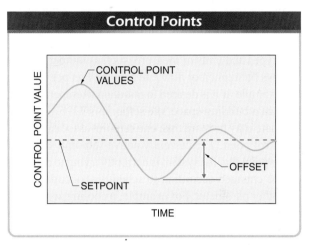

Figure 2-10. Controllers modify the building system until the control point variables equal the setpoint.

Virtual Points

A *virtual point* is a control point that exists only in software. It is not a hardwired point corresponding to a physical sensor or actuator. For example, programmed schedules produce virtual points that correspond to the status of the planned occupancy of a room. The resulting virtual point may have three states: OCCUPIED, UNOCCUPIED, and PREOCCUPANCY (meaning that the room systems are getting ready for OCCUPIED mode). This virtual point may be used by controllers to change the behavior of the HVAC and lighting systems, depending on whether people are expected to be using this room. Since they exist only in software, virtual points can only be monitored with some type of HMI.

Virtual points can also be used to hold snapshots of the values of other control points for different circumstances, such as a timestamp of an event, an extreme (high/low) value of a control point, or the extreme of multiple control points. For example, a series of three control points represent the current temperatures of three different rooms. A separate virtual point may represent the highest temperature any of the rooms reached during the last 24 hr period. Another virtual point may hold the timestamp for that event. Lists of system inputs and outputs in the contract documents contain both hardwired control points and virtual points, though they will be distinguished from each other in some way.

Setpoints

The purpose of the control system is to achieve and maintain certain conditions, which are quantified as setpoints. A *setpoint* is the desired value to be maintained by a system. Setpoints can be stated in different variables including temperature, pressure, humidity, light level, dew point, enthalpy, etc. Setpoints are matched to a corresponding control point variable, sharing the same variable types and units of measure, such as temperature in degrees Fahrenheit or flow rate in cubic feet per minute. For example, if it is desired to maintain a temperature of 72°F in a building space, the setpoint is 72°F.

There may be more than one setpoint associated with a control point. The currently active setpoint, which is the setpoint that the system is currently trying to achieve, can be changed according to schedules, occupancy, or any other parameter. For example, a cooling setpoint is raised from 74°F during the day to 85°F at night or when a building is unoccupied. This saves energy by reducing the system loads when a building is unoccupied, but still prevents the building conditions from getting excessively far from the occupied setpoint. Alternatively, setpoints can be reset based on other conditions. For example, a hydronic water heating setpoint is raised as the outside air temperature falls.

Offsets

Ideally, a control point is equal to its corresponding setpoint. However, there is often some difference, though it should be minimized as much as possible. The *offset* is the difference between the value of a control point and its corresponding setpoint. Offset values measure the accuracy of the control system and may also be called the error. **See Figure 2-11.**

The offset can be either positive or negative, depending on whether the process variable is greater than or less than the setpoint. No matter the sign of the offset, the controller still performs the same basic function. The controller adjusts something in the process until the process variable equals the setpoint and the offset is zero.

Expectations of control system accuracy have increased as the available technology has become more sophisticated. In the past, an accuracy of ±2°F was considered acceptable. With the widespread use of building automation systems, much tighter accuracies (such as ±0.5°F) are possible. Control system accuracy is a function of the accuracy of the controllers, the precision of the sensors and actuators, and the quality of the system design and tuning.

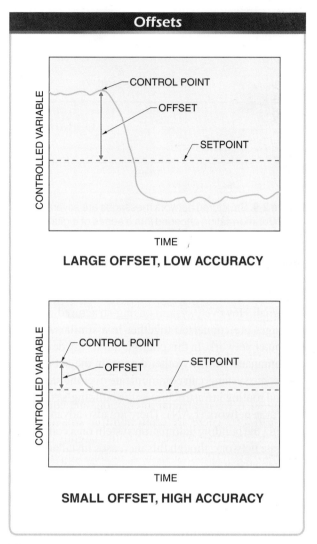

Figure 2-11. A control system with large offsets has low accuracy, and a system with small offsets has high accuracy.

Deadband

Sometimes it is not necessary to maintain a condition at exactly one setpoint. Rather, the variable must only remain within a certain range. The range is defined by a pair of setpoints. For example, the temperature within a space must be maintained between 70°F and 74°F. At 70°F, the heating system operates, and at 74°F, the cooling system operates, with a deadband in between. A *deadband* is the range between two setpoints in which no control action takes place. **See Figure 2-12.**

As long as it does not adversely affect occupant comfort, a deadband arrangement reduces energy use by the control system. It also reduces oscillation and frequent cycling of the system.

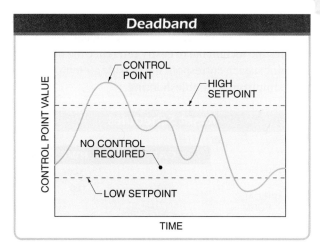

Figure 2-12. Deadband is defined by a pair of setpoints between which no control action is required.

CONTROL STRATEGIES

A *control strategy* is a method for optimizing the control of building system equipment. The optimum outcome is one that fulfills the requirements of the sequence of operations, such as maintaining a comfortable indoor environment or providing adequate lighting, while minimizing energy use, manual interaction, and equipment wear and tear. There are often multiple strategies that can be used to achieve this, each with a slightly different method for controlling the energy-using equipment in a building. Strategies are chosen based on the particular layout of the building; the type, number, and location of the control devices; the use of the building; and the building owner's priorities.

Setpoint Control

Setpoint control is a control strategy that maintains a setpoint in a system. **See Figure 2-13.** Setpoint control is the most common building automation control strategy. The setpoint, which is the desired value to be maintained by a system, can be one of many controlled variables, such as temperature, humidity, pressure, light level, dew point, and enthalpy. The setpoint and the desired stability are programmed into a building automation controller. For example, if a building automation system is required to maintain a temperature of 72°F in a building, the 72°F temperature is the setpoint of the building automation system.

Setback Control

Most setpoints are meant for when the building space is occupied. However, it is sometimes necessary to

maintain certain conditions when the building is unoccupied. Setback control uses setpoint values that are active during the unoccupied mode of a building automation system. **See Figure 2-14.**

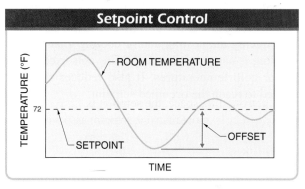

Figure 2-13. Control systems make adjustments in order to minimize the offset between a control point and its setpoint.

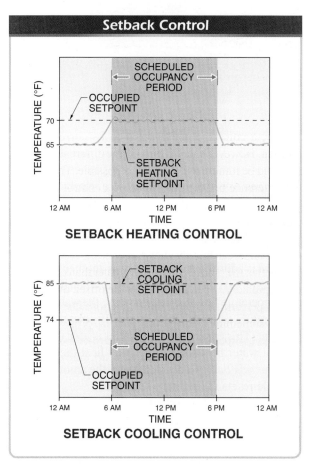

Figure 2-14. Setbacks are used as setpoints during unoccupied periods.

A *setback* is the unoccupied heating or cooling setpoint. For example, if a heating setpoint is lowered from 70°F during the day to 65°F at night, then the setback heating setpoint is 65°F. If a cooling setpoint is raised from 74°F during the day to 85°F at night, then the setback cooling setpoint is 85°F.

Setback control is commonly used with building space temperature setpoints. This strategy saves energy by reducing the heating and/or cooling load when a building is unoccupied, but still prevents excessively hot or cold temperatures. It also reduces the time needed to reach the occupied setpoint.

The simplest type of thermostat uses only a setpoint without any other strategies.

Reset Control

Reset control is a control strategy in which a primary setpoint is adjusted automatically as another value (the reset variable) changes. For example, when the outside air temperature falls to a certain point, the water heating temperature setpoint is automatically reset to a higher setpoint. This ensures that the boiler is able to provide the hot water needed to effectively heat the space. A *setpoint schedule* is a description of the amount a reset variable resets the primary setpoint.

A *reset schedule* is a chart that describes the setpoint changes in a pneumatic control system. A reset schedule is usually listed on the pneumatic system print. A reset schedule consists of the extremes of the setpoint change listed for each variable and can be shown in tabular or a graphical form. **See Figure 2-15.** These

forms can be copied and taped inside the temperature control panel for reference by the technician. The graph enables the technician to easily reference the hot water setpoint at each corresponding outside air temperature, thus simplifying troubleshooting.

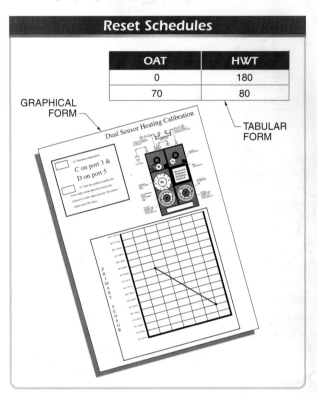

Figure 2-15. A reset schedule describes the setpoint in a pneumatic control system.

Low-Limit/High-Limit Control

Low-limit and high-limit controls ensure that a control point remains within a certain range. **See Figure 2-16.** *Low-limit control* is a control strategy that makes system adjustments necessary to maintain a control point above a certain value. Low-limit control is commonly used with mixed-air damper controls. For example, low outside air temperatures cause low mixed-air temperatures. If the mixed-air temperature drops below 45°F, the controller overrides the normal control logic and forces the outside air dampers closed.

High-limit control is a control strategy that makes system adjustments necessary to maintain a control point below a certain value. This is similar to low-limit control and is commonly used with temperature, pressure, or humidity control points. For example, a high-limit control can be used to prevent a water temperature from exceeding 210°F.

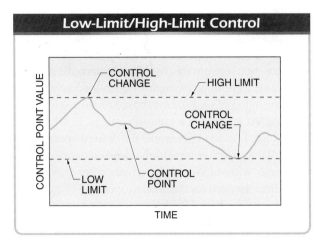

Figure 2-16. Low-limit and high-limit controls are used to keep a control point above or below certain values, respectively.

Lead/Lag Control

Lead/lag control is a control strategy that alternates the operation of two or more similar pieces of equipment in the same system. The most common application of lead/lag control is the control of multiple hot water or chilled water pumps. For example, a primary (lead) pump is energized by the building automation system when it is required by a sequence. A backup (lag) pump is energized if the lead pump fails to start. **See Figure 2-17.** Often, lead/lag pump operation is rotated on a time schedule to have equal run time on the two pumps. Refrigeration and air compressors are also common applications of lead/lag control.

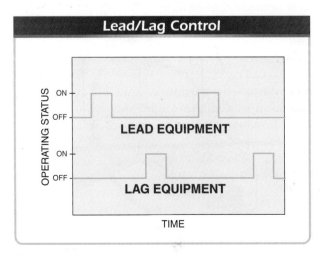

Figure 2-17. Lead and lag loads are alternated in order to equalize the wear and tear on each unit.

High/Low Signal Select

High/low signal select is a control strategy in which the building automation system selects the highest or lowest values from among multiple inputs for use in the control decisions. The most common application of high/low signal select is the control of a building space temperature using multiple temperature sensors at different locations within the zone. The highest signal represents the warmest area and the lowest signal represents the coolest area. The building automation system uses these signals and reset control to determine the setpoint. The highest signal may be used to reset a cooling function to satisfy the warmest space. The lowest signal may be used to reset a heating function to satisfy the coolest space.

Averaging Control

Averaging control is a control strategy that calculates an average value from multiple inputs, which is then used in control decisions. Averaging control is used with a group of sensors that may include high and low values that do not accurately represent the overall conditions in a building. For example, a temperature sensor located in a foyer senses outside air temperatures whenever a door is open and returns the coldest or warmest space temperatures from among a group of sensors in the area. **See Figure 2-18.** It would not be desirable for this sensor to control the heating or cooling reset setpoint most of the time. Instead, an average value from several temperature sensors is more representative of zone conditions. Correct placement of the averaging control sensors is required for the best results.

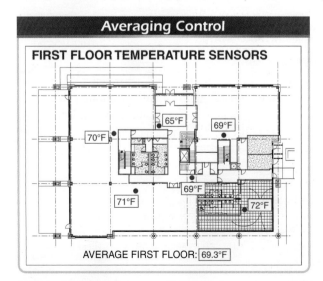

Figure 2-18. Averaging is an effective way to manage areas with multiple sensors because it moderates the effect of very high or low readings from one sensor.

CONTROL LOGIC

The decisions that controllers make to change the operation of a building system involve control logic. *Control logic* is the portion of controller software that produces the calculated outputs based on the inputs. There are many different ways in which these decisions can be made, depending on the inputs used to make the decisions and the algorithms used to produce the results. An *algorithm* is a sequence of instructions for producing the optimal result to a problem. The decision-making process is described with a control loop. A *control loop* is the continuous repetition of control logic decisions. Control systems are categorized as either open-loop control or closed-loop control.

Open-Loop Control

Open-loop control is control in which no feedback occurs between the controller, sensor, and controlled device. *Feedback* is the measurement of the results of a control action by a sensor. An *open-loop control system* is a control system in which decisions are made based only on the current state of the system and a model of how it should work. An example of an open-loop control system is a controller that turns a chilled water pump on when the outside air temperature is above 65°F. The controller has no feedback to verify that the pump is actually on. **See Figure 2-19.**

Open-loop control requires perfect knowledge of the system and assumes there are no disturbances to the system that would otherwise change the outcome. There is no connection between the controller's output and its input.

The most common example of open-loop control in a building automation system is based on time schedules. *Time-based control* is a control strategy in which the time of day is used to determine the desired operation of a load. Time-based control turns a load on or off at a specific time, without knowledge of any other factors that may affect the need for that load to operate. For example, open-loop time-based landscape irrigation control may activate the sprinklers based on a schedule. However, if it had recently rained, the system is overwatering the landscape and wasting water. Without a moisture sensor to provide any input based on the system output, the system has only open-loop control.

Closed-Loop Control

To address the limitation of open-loop control, most essential control loops include feedback. This makes them closed-loop. *Closed-loop control* is control in which feedback occurs between the controller, sensor, and controlled device. A *closed-loop control system* is a control system in which the result of an output is fed back into a controller as an input. For example, a thermostat controls the position of a valve in a hot water terminal device to maintain an air temperature setpoint. The thermostat in the building space provides the feedback of the air temperature that is used to continually adjust the hot water valve. **See Figure 2-20.**

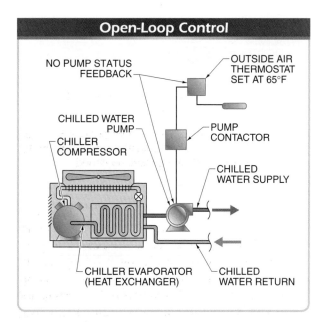

Figure 2-19. Open-loop control makes changes to a system without receiving feedback on the actual state of the system.

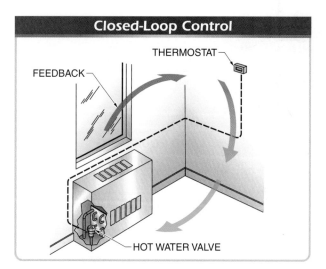

Figure 2-20. Closed-loop control makes decisions based on information fed back into the controller from the system.

However, a malfunction of one component in a closed-loop control system results in other components within the system having an incorrect value or position. This can cause such problems as uncomfortable indoor environments, wasted energy, or even equipment damage.

Control Algorithms

Algorithms are used to determine the necessary output value based on the inputs, enabling a building automation system to achieve a high level of accuracy. Control algorithms are selected when a control device is initially installed and configured. To achieve proper control, the correct algorithm must be selected and accurate setpoints and other parameters must be input into the device. Common algorithms used in building automation systems include proportional, integral, derivative, and adaptive control algorithms. Each algorithm has different characteristics of accuracy, stability, and response time.

Proportional Control Algorithms. A *proportional control algorithm* is a control algorithm in which the output is in direct response to the amount of offset in the system. **See Figure 2-21.** For example, a 10% increase in room temperature results in a cooling control valve opening by 10%. Proportional controllers output an analog signal, which requires compatible actuators.

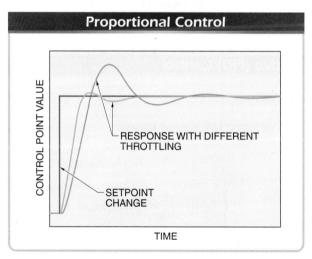

Figure 2-21. In order to bring a control point to a new setpoint, a proportional algorithm adjusts the output in direct response to the current offset.

The algorithm is based on the setpoint offset and a desired proportion (throttling) parameter. Proportional control systems have a lower tendency to undershoot or overshoot than other algorithms, but may not offer precise control. They are used successfully in most applications, but may be inaccurate if not set up properly. When the system reaches the setpoint, the controller outputs a default actuator position, typically a 50% setting. However, a load may require a different position when at the setpoint, resulting in increased offset and energy use.

Integral Control Algorithms. An *integral control algorithm* is a control algorithm in which the output is determined by the sum of the offset over time. Integration is a function that calculates the amount of offset over time as the area underneath a time-variable curve. **See Figure 2-22.** This offset area is then used to determine the output needed to eliminate the offset. The time period used for the calculation changes the results. Integral control algorithms tend to move the system toward the setpoint faster than proportional algorithms.

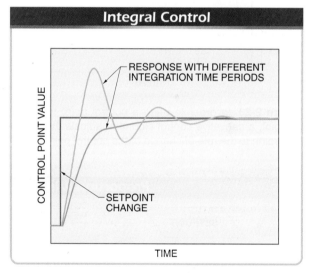

Figure 2-22. In order to bring a control point to a new setpoint, an integral algorithm adjusts the output according to the sum of the offsets in a preceding period.

However, since the integral is responding to accumulated errors from the past, it can cause the present value to overshoot the setpoint, crossing over the setpoint and creating an offset in the other direction. *Proportional/integral (PI) control* is the combination of proportional and integral control algorithms. This combination is generally more stable and accurate than the integral-only algorithm.

Derivative Control Algorithms. A *derivative control algorithm* is a control algorithm in which the output is determined by the instantaneous rate of change of a variable. **See Figure 2-23.** The rate of change is then used to determine the output needed to eliminate the offset. As the input approaches the setpoint, then the output change is reduced early to allow the input to coast to the setpoint. If the input moves rapidly away from the setpoint, extra change is applied to the output to maintain the setpoint. The amount of derivative control in an algorithm affects the overall response. However, this algorithm amplifies noise in the signal, which can cause the system to become unstable.

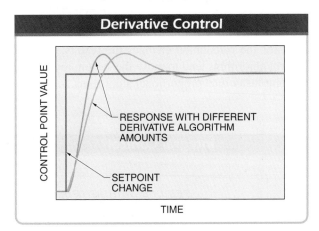

Figure 2-23. In order to bring a control point to a new setpoint, a derivative control algorithm adjusts the output according to the rate of change of the control point.

A control valve can be used to modulate the amount of steam that flows through a heating coil.

Proportional/integral/derivative (PID) control is the combination of proportional, integral, and derivative algorithms. **See Figure 2-24.** The offset is calculated from feedback from the building system that is used as an input. The offset is then used in separate proportional, integral, and derivative calculations. The results of the three separate calculations are added together to determine the output value. The relative proportions of each algorithm are controlled by gain multipliers.

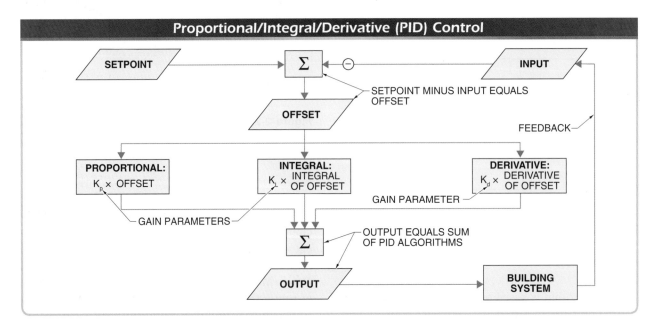

Figure 2-24. PID control systems include contributions from the proportional, integral, and derivative algorithms.

The PID combination improves stability and precise control. The derivative control algorithm moderates the effects of the integral control algorithm, which is most noticeable close to the controller setpoint. It reduces the magnitude of the overshoot produced by the integral component. Only extremely sensitive control applications require PID control. PI control is normally sufficient to achieve a setpoint.

Adaptive Control Algorithms. An *adaptive control algorithm* is a control algorithm that automatically adjusts its response time based on environmental conditions. This results in increased accuracy and stability, and it is simpler to implement. Adaptive control algorithms are the most sophisticated control algorithms because they are a self-calibrating form of PID control. **See Figure 2-25.** Adaptive control algorithms require less calibration because they can adjust their parameters to load changes or incorrect programming. Not all control devices provide adaptive control algorithms.

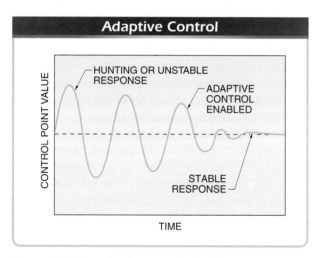

Figure 2-25. An adaptive control algorithm can automatically determine the best calibration parameters to bring a hunting or unstable response back to the setpoint.

Adaptive control is often used with air-handling unit dampers. If the outside air temperature is very close to the return air temperature, the outside air damper must move a lot to change the mixed-air temperature. However, during winter, small adjustments of the outside air damper can impact the mixed-air temperature significantly. Adaptive control algorithms adjust the tuning parameters as needed to account for these types of seasonal operational changes.

Calibration

Control logic, especially PI and PID control must be carefully calibrated by a building automation technician to ensure accuracy and stability. *Calibration*, also known as tuning, is the adjustment of control parameters to the optimal values for the desired control response. The parameters are the relative contributions (gains) of each algorithm to the final output decision, plus any parameters used within each calculation. **See Figure 2-26.** The control logic must be calibrated during commissioning of the system and routinely checked and adjusted if the response is incorrect.

Each system responds differently and must be calibrated individually. The following are several guidelines by which experienced control professionals can determine the best parameter values:

1. Calibrate the control loop during a heavy load demand, which satisfies the setpoint under the most adverse conditions.

2. Calibrate the control loop using proportional control only, disabling the integral and derivative algorithms. The proportional control algorithm sets the control system in the desired range. Therefore, a control loop cannot be calibrated at all if it cannot be calibrated approximately using only proportional control. Continue the calibration process only after achieving good control with the proportional-only algorithm.

3. Once the proportional control algorithm is stable, double the throttling parameter.

4. Adjust the gain of the integral control algorithm. Start with long integration times, which provide stable conditions. Short integration times result in quick responses but increase the chance of cycling.

5. Shorten the integration time slowly, checking the system response. As the times are shortened, the system becomes unstable. Lengthen the time until the system retains stability.

6. When the integral algorithm has been calibrated in a stable manner, the derivative control algorithm can be adjusted to ensure a quick response in the event of a rapid load change. A small gain of the derivative algorithm makes the system stable but react slowly. A larger gain makes the system unstable but react quickly. When increasing the derivative algorithm gain, also increase the integration time.

Control loops can be calibrated manually by introducing disturbances into the system and adjusting variables accordingly. Software is also available for automating the process.

Effects of Increasing PID Gain Parameters				
Algorithm	Rise Time	Overshoot	Settling Time	Steady-State Error
Proportional	Decreases	Increases	Small change	Decreases
Integral	Decreases	Increases	Increases	Eliminates
Derivative	Small decreases	Decreases	Decreases	None

Figure 2-26. Calibration involves adjusting the relative contributions of the proportional, integral, and derivative algorithms to quickly stabilize a control point at a setpoint.

When properly calibrated, control loops provide optimal control of building systems. The optimal control behavior varies depending on the application. Some logic must not allow an overshoot of the output beyond the setpoint if, for example, it would create an unsafe situation. Other processes must minimize the energy expended in reaching a new setpoint. Generally, long-term stability is required and the response must not oscillate for any combination of conditions and setpoints.

Hunting

If the control-loop parameters are not well calibrated, the control logic can be unstable. The output may diverge from the setpoint or hunt excessively. *Hunting* is an oscillation of output resulting from feedback that changes from positive to negative. **See Figure 2-27.** Positive feedback increases the offset, while negative feedback decreases the offset. The alternation between the two causes the output to oscillate above and below the setpoint. In some cases, the oscillation can worsen over time. Then, the only limits to the extremes in oscillation are saturation or mechanical limits.

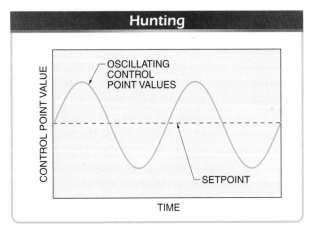

Figure 2-27. Hunting is an oscillating response that does not quickly settle at the setpoint.

SUPERVISORY CONTROL

Modern electronic control devices include the control logic and programming to efficiently operate individual building equipment. These networked control systems do not rely on centralized controllers for normal building system operation. However, some control functions affect the overall operation of the entire building automation system and may use centralized (supervisory) controllers or interface software for their configuration.

A *supervisory control strategy* is a method for controlling certain overall functions of a building automation system. Supervisory control strategies typically override the control logic decisions of individual control devices. For example, a lighting controller's inputs may indicate the need to increase the lighting level in a building area. However, if a supervisory control strategy indicates that the lighting in the area should be off, the local control is overridden and the lighting is turned off.

Multiple supervisory control strategies can be integrated together into an overall approach for efficiently controlling the same loads during the same period. In fact, this is very common. Each control strategy has a priority relative to the others, so that the highest priority strategy overrides all others. Supervisory control strategies include life safety, scheduled, optimum start/stop, duty cycling, and electrical demand control.

Life Safety Control

Life safety control is a supervisory control strategy for life safety issues such as fire detection and suppression. Life safety control strategies have the highest priority of all control strategies. A building automation technician must be familiar with life safety system wiring, software, and codes.

Scheduled Control

Scheduled control is a supervisory control strategy in which the date and time are used to determine the desired operation of a load or system. Based on programmed schedules, this strategy turns loads on or off, adjusts setpoints, or changes the occupancy states of building zones, which are then used by the control system to control building systems. Scheduled control strategies are among the most basic and common supervisory control strategies for building automation systems.

Many different types of scheduled control strategies were created to manage a variety of operating situations. Building automation technicians set up these schedules during system commissioning, though most systems provide a user-friendly way to make future schedule changes. Schedule control strategies include seven-day, daily multiple time period, holiday and vacation, timed overrides, temporary scheduling, alternate scheduling, and schedule linking.

Seven-Day Scheduling. *Seven-day scheduling* is the programming of time-based control functions that are unique for each day of the week. **See Figure 2-28.** Seven-day programming is common, but some building automation systems use a 5+2 schedule. A 5+2 schedule system recognizes Monday through Friday (5 days) as normal workdays with the same daily schedule, with Saturday and Sunday (2 days) treated separately. For maximum efficiency and flexibility, building automation systems should provide the capability to program each day independently, which can accommodate the specific needs of any building.

Seven-Day Scheduling	
Day	**Normal Occupancy Schedule**
Sunday	None
Monday	07:00 – 18:00; 20:00 – 22:00
Tuesday	07:00 – 20:00
Wednesday	07:00 – 18:00
Thursday	07:00 – 20:00
Friday	07:00 – 16:00
Saturday	10:00 – 18:00

Figure 2-28. Seven-day scheduling allows occupancy periods to be programmed individually for each day of the week.

Daily Multiple Time Period Scheduling. *Daily multiple time period scheduling* is the programming of time-based control functions for atypical periods of building occupancy. With this function, building systems or individual loads can be scheduled to operate during multiple independent time periods. For example, the normal scheduled operating hours of a rooftop unit in a commercial building are 8 AM to 5 PM. However, the building is also used for a continuing education class from 8 PM to 10 PM three times a week, so this time period is added to the schedule. Daily multiple time periods can be programmed for certain hours and

for certain days. Building automation system software typically provides for several separately programmable time periods per day.

Holiday and Vacation Scheduling. *Holiday and vacation scheduling* is the programming of time-based control functions during holidays and vacations. This is typically used to reduce loads or turn them off completely since the building is expected to be unoccupied. This overrides the normal occupancy schedules that would have operated the equipment.

A comprehensive yearly operation calendar is required when programming holiday and vacation scheduling. A *permanent holiday* is a holiday that remains on the same date each year. A *transient holiday* is a holiday that changes its date each year. For example, New Year's Day, which is on January 1 each year, is a permanent holiday. Memorial Day, which falls on the last Monday in May of each year, is a transient holiday.

Timed Overrides. A *timed override* is a control function in which occupants temporarily change a zone from an UNOCCUPIED to OCCUPIED state. During this period, the controller uses the setpoints for the OCCUPIED state. The state reverts back to UNOCCUPIED after a programmed time period elapses. Timed overrides provide a quick response to unanticipated changes in building occupancy. A timed override can be activated by a pushbutton or other type of user input. **See Figure 2-29.** Activating the input a second time within the override period may cancel the timed override. Some building automation systems record the amount of time spent in the override mode each month. This information can be used to investigate ways to improve the normal scheduled control functions so that overrides are needed less frequently.

Timed Overrides

Figure 2-29. Timed override inputs can include pushbuttons for occupants to temporarily add a new occupancy period to the control system.

Temporary Scheduling. *Temporary scheduling* is the programming of time-based control functions for a one-time temporary schedule. Temporary schedules are commonly associated with a specific calendar date with unique occupancy needs, accommodating a specific event in a building without using a timed override. Temporary schedules take priority over normal time schedules. At the end of a temporary schedule, it is erased and the normal time schedules resume. Temporary scheduling is commonly used for regularly scheduled weekly or monthly events.

Alternate Scheduling. *Alternate scheduling* is the programming of more than one unique time schedule per year. Alternate scheduling is commonly used during seasonal changes in building operations. For example, a retail business may use alternate scheduling during the holiday shopping season in December. One time schedule may extend from January through November. Beginning on the day after Thanksgiving, the alternate time schedule would take effect, overriding the yearly schedule until the end of the holiday shopping season.

Schedule Linking. *Schedule linking* is the association of loads within the building automation system that are always used during the same time. For example, when a rooftop unit is energized for a particular zone, the lighting load for that zone is also energized. Schedule linking enables both loads to be energized simultaneously from the same schedule.

Facts

Communication cables are often subject to electromagnetic interference (EMI). Possible solutions to EMI problems include fiber-optic cables, balanced lines, shielded cables, and proper routing of cables.

Optimum Start/Stop Control

Schedules do not always represent the actual operation of a load, but instead the period of occupancy. For HVAC systems in particular, there is a lag between the start of load operation and the reaching of a setpoint. Therefore, it is necessary to determine when the loads need to operate in order for the setpoint conditions to be fully achieved during the occupied period. **See Figure 2-30.** Optimizing these start and stop times fulfills this requirement without operating the loads any more than necessary. This maximizes energy savings while maintaining comfort levels.

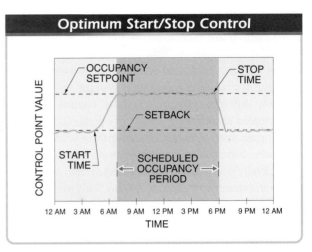

Figure 2-30. Optimum start/stop control determines the best actual start and stop times for HVAC equipment to meet the occupancy requirements.

Optimum Start Control. *Optimum start control* is a supervisory control strategy in which the HVAC load is turned on as late as possible to achieve the indoor environment setpoints by the beginning of building occupancy. The actual start times of the HVAC equipment are calculated based on building and other conditions and may change daily. The current outside air temperature influences the heating or cooling load, and the current indoor air temperature and setpoint determine the temperature change required within a building space. Two methods used in building automation systems to determine the actual start time are adaptive start time control and estimation start time control.

Adaptive start time control is a process that adjusts the actual start time for HVAC equipment based on the building temperature responses from previous days. This method tries to achieve the optimum start time for each day. A 7-day, 10-day, or 14-day history is commonly used to determine the success rate of previous start times and as a guide for adjusting new calculations. **See Figure 2-31.** Adaptive start time control is the most common optimum start method used in building automation systems.

Estimation start time control is a process that calculates the actual start time for HVAC equipment based on building temperature data and a thermal recovery coefficient. A disadvantage of this control method is that the estimated coefficient can be calculated or input incorrectly, causing the HVAC system to start late so that the building space temperature is not at the setpoint when occupancy begins.

Adaptive Start Time Control			
Previous Attempts	Start Time	Time When Setpoint Reached	Start of Occupancy
1	05:18	07:01	07:00
2	05:17	07:02	07:00
3	05:16	06:59	07:00
4	05:17	07:03	07:00
5	05:15	06:57	07:00
6	05:17	07:05	07:00
7	05:15	06:53	07:00
8	05:20	07:03	07:00
9	05:30	07:10	07:00
10	05:10	06:50	07:00

Figure 2-31. A record of the start-up responses from several previous days is used to help calculate the best start times.

Thermal Recovery Coefficients. A *thermal recovery coefficient* is the ratio of a temperature change to the length of time it takes to obtain that change. A thermal recovery coefficient is expressed in temperature degrees per unit of time, such as degrees Fahrenheit per minute (°F/min). Thermal recovery coefficients are used to calculate the actual start time of HVAC systems in commercial buildings. For example, a rooftop unit may start operation at 7 AM for building occupancy at 9 AM. During this time, the indoor temperature increases from 60°F to 72°F. Therefore, it takes 120 min to increase the temperature 12°F. The thermal recovery coefficient is 0.1°F/min (12°F/120 min).

Thermal recovery coefficient values can also be used as indicators of HVAC equipment efficiency and/or mechanical problems. For example, an HVAC system having a thermal recovery coefficient of 0.2°F/min in one month and 0.1°F/min the next month may require a filter replacement or preventive maintenance.

Optimum Stop Control. *Optimum stop control* is a supervisory control strategy in which the HVAC load is turned off as early as possible to maintain the proper building space temperature until the end of building occupancy. For example, in winter, optimum stop control allows the building temperature to gradually decline until the end of occupancy. Optimum stop control is commonly limited to a specific length of time, such as 15 min or 30 min, and may be implemented independently of optimum start control.

Duty Cycling Control

Duty cycling control is a supervisory control strategy that reduces electrical demand by turning off certain HVAC

loads temporarily. When HVAC units are oversized, they can be cycled in this way without adversely affecting building space temperature. The on cycles of the loads are staggered so that only one load is operating at a time. For example, if two HVAC units that operate simultaneously have an electrical demand of 15 kW each, the electrical demand is 30 kW. By implementing duty cycling control, the operation of each unit is alternated so that the electrical demand is 15 kW. **See Figure 2-32.**

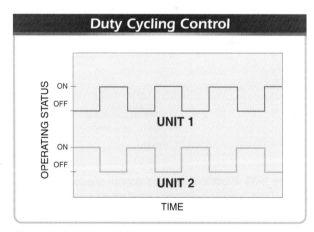

Figure 2-32. Duty cycling control reduces peak electrical demand by alternating the operation of multiple units.

Duty cycling control is effective for reducing electrical demand and the associated expenses. However, duty cycling increases motor wear due to the frequent motor starts. Duty cycling also results in the loss of temperature control and ventilation during the duty cycle's off time. In addition, the cycling of HVAC equipment may cause excessive noise.

Duty cycling control is typically used in commercial buildings that have a large number of electric baseboard heaters or small exhaust fans. The duty cycling sequence for the baseboard heaters can be programmed to alternate around the building. For example, a heater in one office is duty cycled for a short period, followed by another on the opposite side of the building.

Electrical Demand Control

Commercial building owners are typically charged for the highest period of electrical power demand for the month. Therefore, controlling the operation of electrical loads to lower the electrical demand can significantly lower utility bills. *Electrical demand control* is a supervisory control strategy designed to reduce a building's overall electrical demand. This strategy shuts off certain electric loads in a way that reduces building electrical demand without adversely affecting occupant comfort or productivity. **See Figure 2-33.**

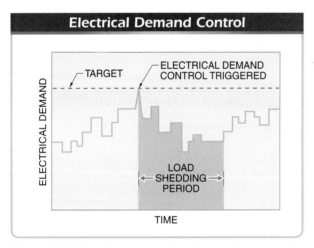

Figure 2-33. Electrical demand control sheds low-priority loads if the building's demand reaches a certain target.

A *shed load* is an electric load that has been turned off for electrical demand control. A *restored load* is a shed load that has been turned on after electrical demand control. Loads are shed (turned off) as the building electrical demand increases to a specific limit (target). The electrical demand decreases below the target when loads are shed. The building automation system must have a method of measuring the electrical power at the building service.

Shed Tables. A *shed table* is a table that prioritizes the order in which electrical loads are turned off. Building automation systems commonly provide low-priority and high-priority shed tables. A *low-priority load* is a load that is shed first for electrical demand control. A *high-priority load* is a load that is important to the operation of a building and is shed last when electrical demand goes up. High-priority loads are restored (reenergized) first when the electrical demand drops. A thorough analysis of the electrical loads in a building must be performed to document the priority of each load. Not all building loads have to be included in shed tables. Some loads are essential to the efficient operation of a commercial building and cannot be shed.

When building electrical demand is above the target, the loads in the low-priority shed table are shed in order. If the building's electrical demand stabilizes or drops, the loads are restored in reverse order. However, if the electrical demand continues to increase, all loads in the low-priority shed table are shed and the building automation system begins to shed loads in the high-priority shed table. The low-priority shed table is often referred to as first off/last on. The high-priority shed table is often referred to as last off/first on.

Rotating Priority Load Shedding. Load shedding reduces electrical demand in a commercial building but can also create problems. For example, if the first load in a low-priority shed table is an electric water heater, it is always the first load to be shed. If the building has frequent high electrical demand periods, this load is often off and not available for use. The solution is to rotate the loads within the shed table. *Rotating priority load shedding* is an electrical demand control strategy in which the order of loads to be shed is changed with each high electrical demand condition.

In rotating priority load shedding, if load 1 is the first load shed for one high electrical demand period, load 2 is the first load shed for the next high electrical demand period. **See Figure 2-34.** Some systems rotate both low- and high-priority shed tables while other systems rotate one shed table and leave the other shed table fixed.

Rotating Priority Load Shedding					
Shed Order	Shed Loads				
	Cycle 1	Cycle 2	Cycle 3	Cycle 4	Cycle 5
1	Lighting circuit 5	Lighting circuit 7	VAV unit 12	Fan 2	Pump 7
2	Lighting circuit 7	VAV unit 12	Fan 2	Pump 7	Lighting circuit 5
3	VAV unit 12	Fan 2	Pump 7	Lighting circuit 5	Lighting circuit 7
4	Fan 2	Pump 7	Lighting circuit 5	Lighting circuit 7	VAV unit 12
5	Pump 7	Lighting circuit 5	Lighting circuit 7	VAV unit 12	Fan 2

Figure 2-34. The order in which equal-priority loads are shed can be rotated so that no one load is always the first to be turned off.

Shedding Strategies. Electrical demand control features often include additional parameters that can be programmed for efficient control of shed loads. The building automation software typically includes timers for the maximum and minimum shed times. The maximum shed time timer causes a load to be restored after it has been shed for a certain length of time. This load is restored regardless of the electrical demand in a commercial building at the time. The maximum shed time timer is commonly used with loads that are essential to the building operation. These loads would not be involved in electrical demand control without the maximum shed time timer.

The minimum shed time timer ensures that a shed load cannot be restored until a specific time period has elapsed. The minimum shed time timer reduces the possibility that a load is cycled on and off repeatedly by the building automation system.

Electrical demand control can also be programmed to shed certain loads only while other loads are operating. For example, the cooling compressor of a package unit can be turned off during supply fan operation. This allows air to be circulated, providing some relief while the compressor is off. Demand is still reduced because the compressor uses more electrical power than the supply fan.

Electrical demand control can be programmed to temporarily change the setpoint when electrical demand is high. Instead of shedding a load, the normal setpoints are changed by a certain number of degrees. For example, a rooftop unit is programmed for a normal cooling setpoint of 72°F. However, 4°F is added to the cooling setpoint when the building is in a high electrical demand period. The new, temporary setpoint is 76°F. This setpoint change causes the unit compressor to shut off. If the temperature in the building space changes outside of the new setpoint, the compressor turns on regardless of the building's electrical demand. This feature can also be programmed for heating applications.

Electrical Demand Targets. An effective electrical demand control strategy requires accurate monthly electrical demand targets. Load shedding is reduced if electrical demand targets are high and increased if electrical demand targets are low. **See Figure 2-35.** The development of accurate electrical demand targets requires experience in evaluating prior electric bills. After the targets are developed and the actual building electrical demand is evaluated, the targets may require adjustment. The electrical demand targets can be lowered incrementally over a period of time until the optimum level is reached.

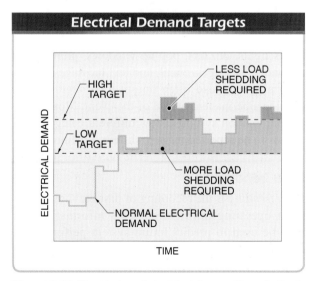

Figure 2-35. The choice of electrical demand target affects how much load shedding is required.

BUILDING SYSTEM MANAGEMENT

The operational data gathered by a building automation system can also be used by maintenance technicians to help manage building system equipment. The collection of information by a single system helps operators and technicians efficiently monitor abnormal conditions, document system performance, and perform preventive maintenance.

Alarm Monitoring

Alarm monitoring is the detection and notification of abnormal building conditions. The most common alarms are associated with temperature sensors. Alarms can also indicate abnormal levels of humidity and pressure or failure of a fan or pump. Alarms quickly alert maintenance personnel to equipment failures and/or problems.

Alarm Classification. Most modern building automation systems allow the operator to classify alarms into different categories. The most common alarm-classification categories are critical alarms and noncritical alarms. Critical alarms concern devices vital to proper operation of the building. Critical alarms are reported immediately to multiple maintenance personnel for a quick response. Noncritical alarms concern elements of building operation that are not vital. Noncritical alarms may or may not be reported immediately because a quick response is not necessary to the proper operation of the equipment or building.

Alarm Notification. Alarm conditions can also be configured with a type of notification. Building automation systems typically provide a number of ways to notify personnel when an alarm is triggered, such as status lights, buzzers, pop-up windows, prerecorded telephone messages, email, and text messages. Alarm information is also often output to printers, which provide a permanent record of the event. Most building automation systems allow different categories of alarms to be reported through different methods.

Data Trending

Data trending is the recording of past building equipment operating information. This information can then be used to predict future system performance, calibrate a control loop during commissioning, certify or publicize energy-efficient buildings, troubleshoot equipment problems, and other uses. Data trending functions can record temperature, humidity, power,

and any other equipment or building condition variable. **See Figure 2-36.** Any control point used by the building automation system can be recorded. Multiple inputs and outputs can be recorded simultaneously, which can show relationships between control points, such as the outside air temperature and heating/cooling equipment operation.

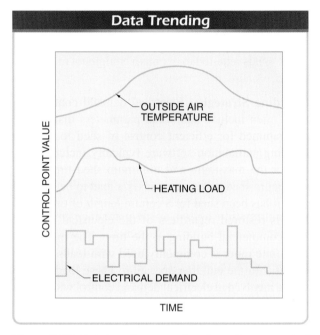

Figure 2-36. Data trending records values of many control points over a long period. This information is saved for later analysis or troubleshooting.

The control point values are recorded at a specific time interval or minimum change. The time interval used to record these values is programmable. A time interval of 20 min is commonly used for long-term data trending. A minimum change setting indicates how much the value must change before a new entry is made into the data log. Long-term data trending is used by a technician to view temperatures and other values that occur overnight or during weekends. For troubleshooting equipment problems, the time interval is often shortened to 1 min or 2 min in order to diagnose abnormal equipment operation.

Data trending can be started and stopped at specific times and dates. For example, there may be complaints that a building space is excessively cold when occupants arrive in the morning. A data trend is created to record the building space temperature, outside air temperature, and equipment on/off status beginning before occupancy and ending at midmorning. The data trend interval time is 5 min. The results of this data trend are used to change the actual start time of a unit or correct mechanical equipment problems. Data trends can also be used to make a decision regarding the purchase or replacement of equipment. Data trend records can be imported into spreadsheet software to create graphs and charts.

Preventive Maintenance

Preventive maintenance is scheduled inspection and work that is required to maintain equipment in peak operating condition. Corrective (breakdown) maintenance is performed after equipment has failed. A preventive maintenance program is typically less expensive and causes less downtime than corrective maintenance.

Many building automation systems can be integrated with computerized maintenance management systems (CMMSs). When integrated, the separate CMMS software uses operational data from the building automation system to help manage preventive maintenance tasks, such as generating work orders based on certain control point information. **See Figure 2-37.** For example, a work order may be automatically generated for lubricating a piece of equipment

based on the number of hours it has operated as logged by the building automation system. Other values that are commonly used to trigger preventive maintenance work orders are the number of motor starts or the pressure drop across filters.

Predictive maintenance is the monitoring of wear conditions and equipment characteristics in comparison to a predetermined tolerance to predict possible malfunctions or failures. Predictive maintenance attempts to detect equipment problems with vibration analysis and other methods. If the building automation system includes the necessary sensors, it may be possible to also integrate that data with predictive maintenance software.

Graphical Interfaces

Many building automation systems use graphical interfaces to visually communicate building and equipment conditions. Building systems and major equipment are represented in illustrations and icons, which are often easier to understand. **See Figure 2-38.** Illustrations can be created by the building automation technician or other vendor. Some systems use photographs of equipment. The actual temperature, humidity, pressure, or status values are superimposed onto the graphics.

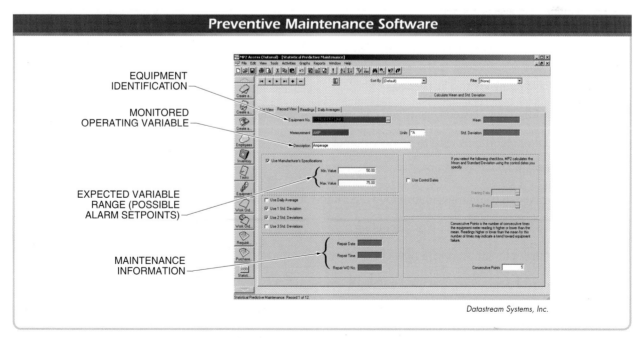

Preventive Maintenance Software

EQUIPMENT IDENTIFICATION

MONITORED OPERATING VARIABLE

EXPECTED VARIABLE RANGE (POSSIBLE ALARM SETPOINTS)

MAINTENANCE INFORMATION

Datastream Systems, Inc.

Figure 2-37. Some preventive maintenance software can be integrated with building automation systems to use control points for equipment information, such as status, run time, and current draw.

Graphical Interfaces

10%

END SWITCH

EXHAUST AIR DAMPER

73°F

SINGLE ZONE

72°F

RETURN AIR DAMPER

DIFFERENTIAL PRESSURE SENSOR

END SWITCH

SUPPLY AIR FAN

HEATING COIL

85°F

75°F

61°F

10%

OUTSIDE AIR DAMPER

80%

CLOSED   OPEN

COOLING COIL

Figure 2-38. Graphical interfaces help convey important system information at a glance.

Graphical interface software falls into two categories: thin client or thick client. Thin client interfaces do not require the use of any software other than a standard web browser. These interfaces are available on nearly any computer. Thick client interfaces are software programs specifically designed to display information about a building's systems. Thick client software is required on a computer before using the interface.

REVIEW QUESTIONS

1. What is the difference between the types of information shared between digital signals and analog signals?

2. How does setback control help optimize building operations?

3. When may averaging control be needed?

4. What is the difference between open-loop control and closed-loop control?

5. How are control algorithms used to produce output values?

6. What is calibration?

7. What is involved in calibrating control loop algorithms for optimal control response?

8. What is the advantage of seven-day scheduling?

9. Why might the optimum start/stop times be different from the occupancy schedule?

10. How is electrical demand control used to reduce utility costs?

Chapter 2 Review and Resources

Modern Control Systems **3**

Electronic control systems use solid-state components that operate at low power levels to control HVAC equipment. Solid-state components include diodes, transistors, and thyristors. Integrated circuits consist of thousands of solid-state devices all contained in a single package or chip.

Precise and reliable communication between electronic control systems is accomplished through networked building automation. It is even possible for control systems from different manufacturers to share information on the same network. In addition, open protocols offer variety and flexibility to allow building owners to optimize every aspect of the building's control systems. With ever-greater emphasis on energy efficiency and occupant comfort, the use of control systems for building automation is becoming a significant factor in both new and existing commercial building operations.

ELECTRONIC CONTROL SYSTEMS

An *electronic control system* is a control system in which the power supply is 24 VDC or less. Electronic control systems are similar to electrical, pneumatic, and automated control systems. For example, electronic control systems use a power supply similar to electrical control systems. In addition, many pneumatic control system components such as thermostats, switching relays, and damper actuators have electronic control system counterparts. Also, the solid-state components used in electronic control systems are also used to construct automated control system controllers.

Electronic control systems use semiconductors (solid-state devices) to control HVAC equipment. Sensors are used to measure the temperature, humidity, or pressure in an HVAC system. These variables are represented by resistance values of 0 Ω to over 1000 Ω, voltage values of 0 VDC to 10 VDC, or current values of 0 mA to 20 mA. The electronic controller compares the input signal from the sensor to an internal setpoint. The controller also has an adjustment to determine the amount that the controller output changes when the measured variable changes. The output of the electronic controller is sent to a motor-driven damper or valve actuator. Electronic control systems may also include switching, averaging, or other electronic controls that complete the sequence of operation for the HVAC equipment.

Electronic control systems are often referred to as analog electronic systems because they provide variable (analog) switching. With the advent of modern automated systems, analog electronic systems have lost some of their popularity. Early electronic devices used vacuum tubes to switch or amplify a signal. Modern electronic devices are made of semiconductor materials.

Facts

Many modern electronic devices, such as personal computers, DVD players, and cell phones, were made possible by the development of semiconductor devices containing transistors that replaced vacuum tubes in older devices.

Vacuum Tubes

In the mid-1940s, the most common electronic device was the AM radio. Immediately after World War II, the military began building the first electronic computers. However, the development of computers was hindered because the main component used for electrical switching was the vacuum tube. A *vacuum tube* is a device that switches or amplifies electronic signals. Vacuum tubes perform these functions by allowing electrons to flow to plates that are located inside a glass tube. **See Figure 3-1.** The glass tube contains plates that are in a vacuum. A vacuum is required because air is an insulator and interferes with the function of the plates. The anode and cathode (main plates) become electrically charged and attract or repel charges at the other plate. The control grid placed between the two main plates modifies the signal between the two main plates. The control grid enables a small control signal to control a large signal between the main plates.

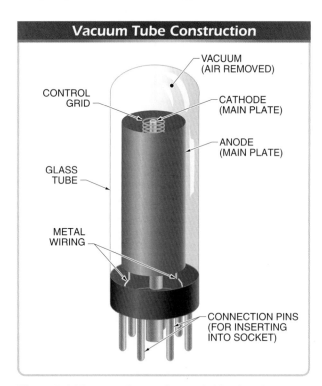

Figure 3-1. Vacuum tubes perform switching functions by allowing electrons to flow to plates located inside a glass tube.

Vacuum tubes consume large amounts of power, produce excessive amounts of heat, and have high failure rates. These characteristics limit the usefulness of vacuum tubes. The search for smaller, more efficient, and more reliable electrical switching and amplifying components led to the development of semiconductor devices.

Semiconductors

A *semiconductor* is a material in which electrical conductivity is between that of a conductor (high conductivity) and that of an insulator (low conductivity). The application of a voltage to the semiconductor material causes the material to change electrical characteristics. Devices constructed from semiconductor material eliminate or reduce many of the problems associated with vacuum tubes. For example, semiconductors use small amounts of power and produce little heat. The small amount of heat produced increases the reliability of the devices. The systems used for component cooling can also be reduced or eliminated. Semiconductors are continually being manufactured in smaller sizes, allowing increased capability in a smaller package.

The most common material used for semiconductor devices is silicon. Silicon is one of the most common naturally occurring elements, which makes devices made from silicon inexpensive. The main costs involved are in the engineering and design of a semiconductor device and not in the raw materials. Other elements and/or materials are added to silicon to create semiconductor materials with various characteristics. The various materials cause the silicon to act differently under various voltage and current flow conditions, such as alternating current (AC) and direct current (DC).

Doping is the addition of material to a base element to alter the crystal structure of the element. The addition of material to the crystal structure of silicon creates N-type material or P-type material. *N-type material* is material created by doping a region of a crystal with atoms from an element that has more electrons in its outer shell than the crystal. N-type material has extra electrons in its crystal structure available for current flow. *P-type material* is material created by doping a region of a crystal with atoms from an element that has fewer electrons in its outer shell than the crystal. P-type material has empty spaces (holes) in its crystal structure into which electrons can be placed. **See Figure 3-2.** Most semiconductor devices are constructed from different arrangements of P-type and N-type materials. Semiconductor devices include diodes, transistors, thyristors, and integrated circuits.

Diodes. A *diode* is a semiconductor device that allows current to flow in one direction only. Diodes are the most common semiconductor devices. A diode is constructed by placing one layer of P-type material and one layer of N-type material back to back. The area where the materials contact is referred to as the PN junction (PN region). **See Figure 3-3.**

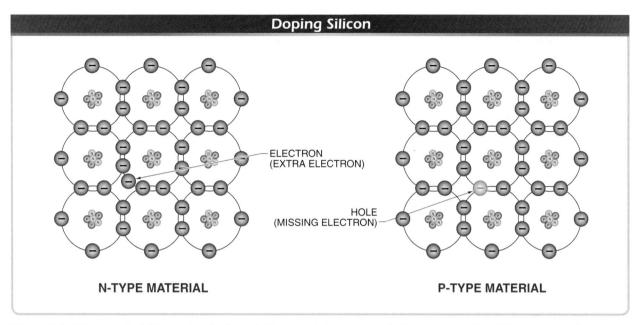

Figure 3-2. N-type material has extra electrons in its crystal structure available for current flow. P-type material has empty spaces (holes) in its crystal structure into which electrons can be placed.

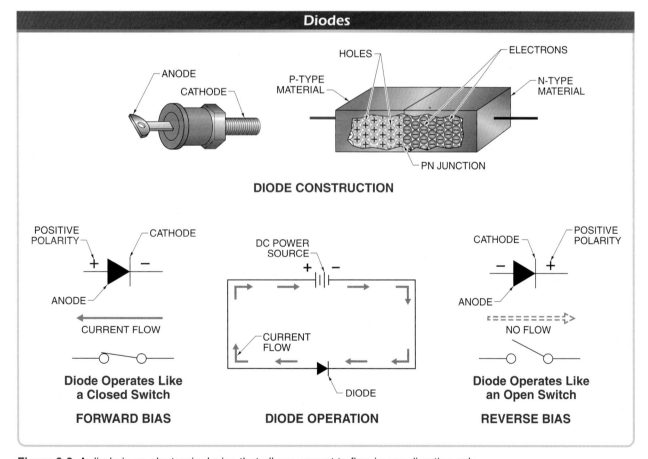

Figure 3-3. A diode is an electronic device that allows current to flow in one direction only.

The P-type material of a diode has holes that are available to receive electrons, and the N-type material has extra electrons available. When a voltage source of 0.6 VDC or more is connected to the N-type material, the extra electrons of the N-type material jump across the PN junction (current flow) and fill the holes of the P-type material. The 0.6 VDC is used to forward bias the diode (allow current flow). The 0.6 VDC required is a function of the doping process performed to the silicon.

A voltage supply connected to the P-type material is used to reverse bias a diode. When a diode is reverse biased, electrons are drawn away from the PN junction, preventing current flow. A high voltage connected to the P-type material of a diode makes the diode conduct current temporarily. After a couple of milliseconds, the diode melts or explodes. The low voltage (0.6 VDC) normally used by semiconductor devices allows low current flow and creates low amounts of heat. Small power supplies can also be used because of the low current flow.

Diodes are used as electronic check valves and as rectifiers. A *rectifier* is a device that changes AC voltage into DC voltage. Diodes are used in power supplies to change 120 VAC supply voltage into rectified low-voltage DC for an electronic circuit. Rectifiers include half-wave, full-wave, and bridge rectifiers. **See Figure 3-4.**

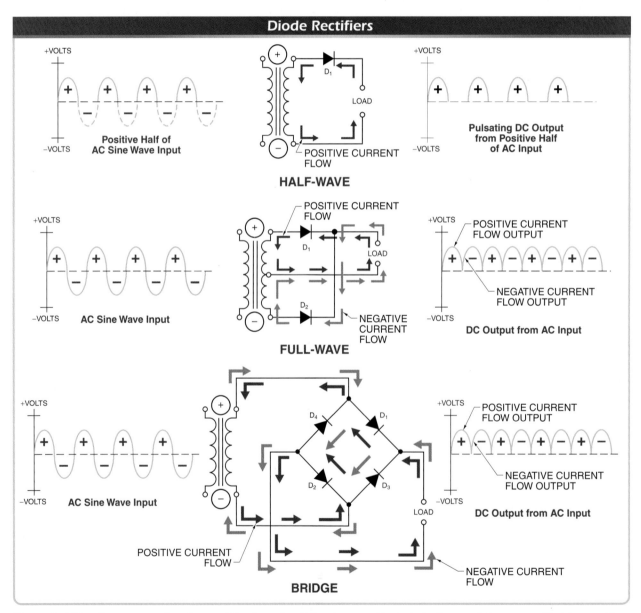

Figure 3-4. Rectifiers are used to change AC voltage into DC voltage and include half-wave, full-wave, and bridge rectifiers.

A *half-wave rectifier* is a circuit containing one diode that allows only half of the input AC sine wave to pass. Half-wave rectification is accomplished because current is allowed to flow only when the anode terminal has a positive polarity with respect to the cathode. Current is not allowed to flow through the rectifier when the cathode has a positive polarity with respect to the anode. Half-wave rectification is inefficient for most applications because one-half of the input sine wave is not used.

A *full-wave rectifier* is a circuit containing two diodes and a center-tapped transformer that permits both halves of the input AC sine wave to pass. Full-wave rectification is accomplished by one diode passing the positive half of the AC sine wave and the second diode passing the negative half of the AC sine wave. A full-wave rectifier is more efficient than a half-wave rectifier because both halves of the input sine wave are used.

A *bridge rectifier* is a circuit containing four diodes that permits both halves of the input AC sine wave to pass. A bridge rectifier is more efficient than a half-wave or full-wave rectifier and is the most common rectifier circuit used in rectification circuits. The output of a bridge rectifier is a pulsating DC voltage that must be filtered (smoothed) before it can be used in most electronic equipment.

Diodes are used for rectification and to block current flow. Some diodes are designed with certain characteristics and are used to perform specific tasks. These diodes include zener diodes, light-emitting diodes, and photodiodes.

A *zener diode* is a diode designed to operate in a reverse-biased mode without being damaged. **See Figure 3-5.** Different types of zener diodes are available that use various voltages to trigger (turn on) the diode. Zener diodes are commonly used as a voltage shunt or electronic safety valve. When a power supply voltage to an electronic circuit increases due to a malfunction, a zener diode passes the unwanted voltage to ground, protecting the sensitive electronic circuit from overvoltage.

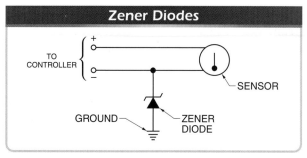

Figure 3-5. Zener diodes are designed to operate in the reverse-biased mode at a certain voltage level.

Fluke Corporation
Digital multimeters can be used to check solid-state components in modern HVAC systems.

A *light-emitting diode (LED)* is a diode designed to produce light when forward biased. The light emitted from most diodes is very small and not visible, but LEDs are specifically designed to create the maximum amount of light. An LED consists of a small diode chip enclosed in an epoxy housing that contains light diffusing particles. When an LED is forward biased, light is released and diffused in different directions by the epoxy housing. The amount of light produced is directly proportional to the current flow through the LED. **See Figure 3-6.** To prevent LEDs from being damaged by overcurrent, a resistor is normally placed in series with the LED to limit current flow. The light wavelength determines the color of the LED, with the most common colors being red and green. The wavelength is determined by the forward-bias voltage of differently doped LEDs.

LEDs are commonly used to visually indicate conditions of equipment. For example, LEDs are used to indicate equipment conditions such as power being supplied, a device operating properly, or communication with other devices taking place. Many devices use LEDs that flash in various patterns to indicate to an operator or technician a particular problem with a machine or piece of equipment. The pattern is explained in the technical manual of the machine.

A *photodiode* is an electronic device that changes resistance or switches on when exposed to light. In many ways, photodiodes are the opposite of LEDs. **See Figure 3-7.** An adequate light source striking the surface of the PN junction of a photodiode chip causes electrons to jump the PN junction and current to flow. Photodiodes can be designed to switch only when exposed to specific amounts of light or wavelengths of light, allowing photodiodes to be used as light detectors.

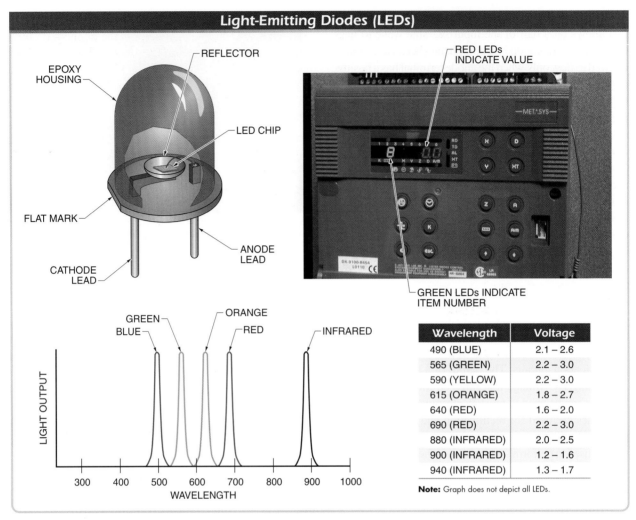

Light-Emitting Diodes (LEDs)

Wavelength	Voltage
490 (BLUE)	2.1 – 2.6
565 (GREEN)	2.2 – 3.0
590 (YELLOW)	2.2 – 3.0
615 (ORANGE)	1.8 – 2.7
640 (RED)	1.6 – 2.0
690 (RED)	2.2 – 3.0
880 (INFRARED)	2.0 – 2.5
900 (INFRARED)	1.2 – 1.6
940 (INFRARED)	1.3 – 1.7

Note: Graph does not depict all LEDs.

Figure 3-6. LEDs are designed to produce light when forward biased.

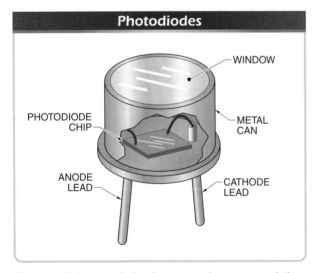

Photodiodes

Figure 3-7. A photodiode changes resistance or switches on when exposed to light.

Specific models of photodiodes are designed to switch when exposed to infrared (heat) energy. Infrared photodiodes are used as motion detectors for alarm systems. In many cases, photodiodes have the P-type material and N-type material formed into flat segments to expose the maximum amount of material to the light source.

Transistors. A *transistor* is a three-terminal semiconductor device that controls current according to the amount of voltage applied to the base. Transistors are used as amplifiers or DC switches, with some transistors able to be used as either one. Transistors include bipolar junction, field-effect, photo, and photodarlington transistors.

Bipolar junction transistors were the first transistors developed. A *bipolar junction transistor (BJT)* is a transistor that controls the flow of current through the emitter (E) and collector (C) with a properly biased base (B). BJTs have three separate layers of P- and N-type material and can be layered in a PNP configuration or an NPN configuration. **See Figure 3-8.** The layers of a BJT are in physical contact with each other and have terminals connected to the layers. The three terminals are the emitter, collector, and base.

The emitter is the common contact for the base and collector circuits. The collector is the connection that is connected to the load. The base is a low-powered connection that functions as a valve (switch). The base is very thin with a low amount of doping material. The base is used to control the flow of electrons between the emitter and collector. A very small flow of electrons between the base and emitter can cause a much larger flow of electrons between the emitter and collector, thus functioning as an amplifier. When a variable resistor is placed in the base circuit, the collector output signal can be changed. **See Figure 3-9.** BJTs are commonly used to amplify small control signals into large signals that are used to drive large-capacity devices, such as electronically controlled valve and damper actuators.

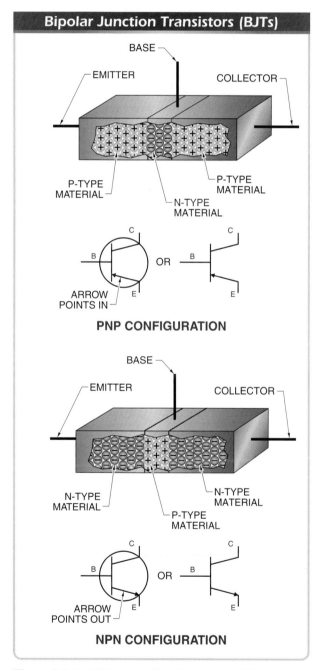

Figure 3-8. A BJT controls the flow of current through the emitter and collector with a properly biased base.

Modern building automation system controllers use different transistors for circuit operation.

A *field-effect transistor (FET)* is a transistor that controls the flow of current through a drain (D) and source (S) with a properly biased gate (G). These transistors include junction field-effect transistors (JFETs) and metal-oxide semiconductor field-effect transistors (MOSFETs). The JFET was developed as a better switching device than the BJT. It is formed from a single piece of N-type or P-type material. A layer (channel) of opposite material is formed around the center section of the P-type or N-type material. **See Figure 3-10.**

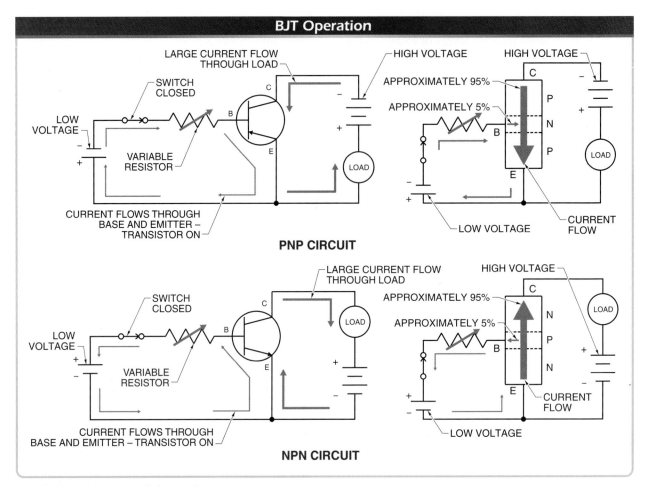

Figure 3-9. BJTs are commonly used to amplify small control signals into large signals that are used to drive large-capacity devices.

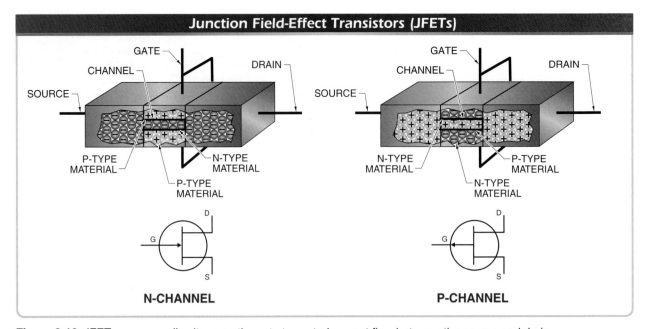

Figure 3-10. JFETs use a small voltage on the gate to control current flow between the source and drain.

JFETs may be N-channel or P-channel JFETs. When a voltage is applied to the material that surrounds the channel, the effect of the electrons being present causes the channel to have various states of conductivity. The channel acts like a valve or switch. The action takes place with a very low current and voltage at the gate. An advantage of JFETs over BJTs is that current and voltage requirements are much lower in JFETs. The low power consumption of JFETs makes them ideal for use in a wide variety of circuits.

The most common transistors used today are metal-oxide semiconductor field-effect transistors (MOSFETs). MOSFETs are constructed from two separate sections of either P-type or N-type material at opposite ends of the transistor. Between the two sections is a gate lead that is connected to the transistor by a metal-oxide material and insulator. The insulator allows no physical contact between the gate and the rest of the transistor. A voltage present at the gate of a MOSFET causes conductivity between the source and drain leads. **See Figure 3-11.**

A MOSFET has almost infinite resistance between the gate and either the source or drain leads because of the insulator. Because of the high resistance, a tiny current flow at the gate can cause the device to switch on. MOSFETs are used as switches and cause on/off signals in large numbers of computer chips. Static electricity or overcurrent can cause the insulator of a MOSFET to be destroyed.

A *phototransistor* is an NPN transistor that has a large, thin base region that is switched on when exposed to a light source. When light strikes the base region, current flow is induced that causes the emitter-collector current to switch on. **See Figure 3-12.** The drawback of a phototransistor is that the large, thin base region causes the current-carrying ability of the transistor to be reduced. In many cases, a device must have a large current-carrying ability to function properly in a circuit.

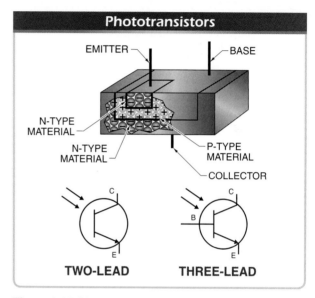

Figure 3-12. Phototransistors are designed to allow current flow when exposed to a light source.

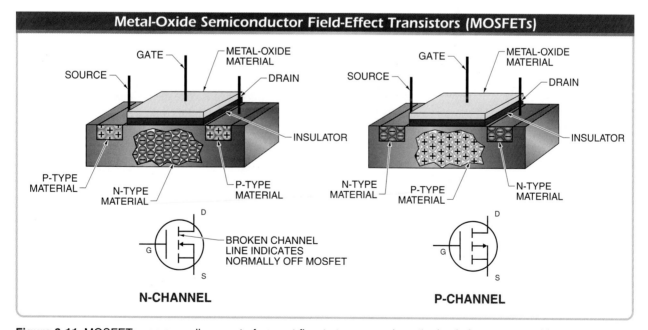

Figure 3-11. MOSFETs use a small amount of current flow to turn on and are the basis for computer chips.

A *photodarlington transistor* is a transistor that consists of a phototransistor and a standard NPN transistor in a single package. In a photodarlington transistor, the sensitive phototransistor is used to switch the standard NPN transistor. **See Figure 3-13.** The phototransistor provides high sensitivity, while the standard NPN transistor provides the high current flow necessary for specific applications, such as energizing relays.

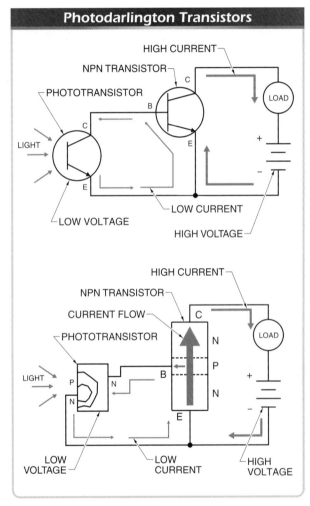

Figure 3-13. Photodarlington transistors are used in applications in which high sensitivity to light is required with high current flow.

Thyristors. Another major category of semiconductor devices is thyristors. A *thyristor* is a solid-state switching device that switches current on by a quick pulse of control current. Thyristors act as solid-state relays that can switch devices, such as heaters, compressors, motors, and relays, on and off. The two major categories of thyristors are silicon-controlled rectifiers (SCRs) and triacs.

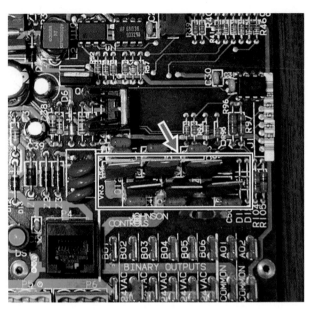

Building automation system controllers use thyristors as switches to turn equipment on and off.

A silicon-controlled rectifier (SCR) is a thyristor that is capable of switching DC. SCRs have four layers of P-type and N-type material. **See Figure 3-14.** One of the layers is referred to as the gate. When a small voltage is introduced at the gate, the middle layer switches on, and the SCR allows current flow in one direction only. The SCR continues to conduct current even when the gate voltage is removed. To stop an SCR from conducting, system power must be disconnected.

SCRs are used in HVAC applications, such as electric heating element control, and in variable-frequency drives (VFDs) used for motor speed control. They have a high current-carrying capability that allows them to create high amounts of heat. The heat must be removed by heat sinks or fans. The disadvantage of SCRs is that they can be used with DC only. However, two SCRs in parallel may be used for modulating AC devices.

A *triac* is a thyristor used to switch AC. A triac is triggered into conduction in either direction by a small voltage to its gate. A triac consists of two SCRs connected in a reverse-parallel configuration. **See Figure 3-15.** One SCR conducts current in one direction, while the other SCR conducts current in the opposite direction. Triacs are used as solid-state relays. Unlike electromechanical relays, triacs have no moving parts and are capable of millions of on/off cycles without failure. Triacs are used in automated systems to turn electrical devices on and off and are limited only by the amount of current they can carry.

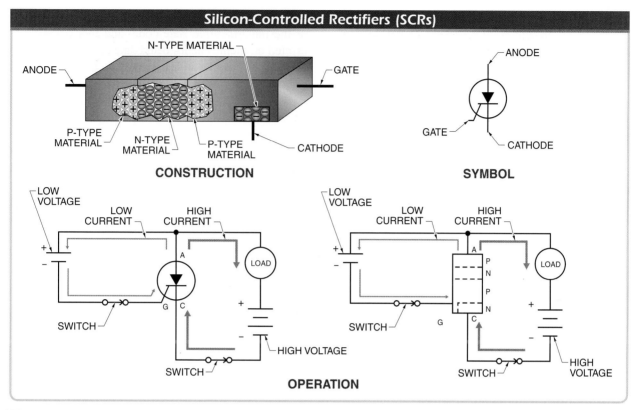

Figure 3-14. SCRs are semiconductor devices used to switch DC.

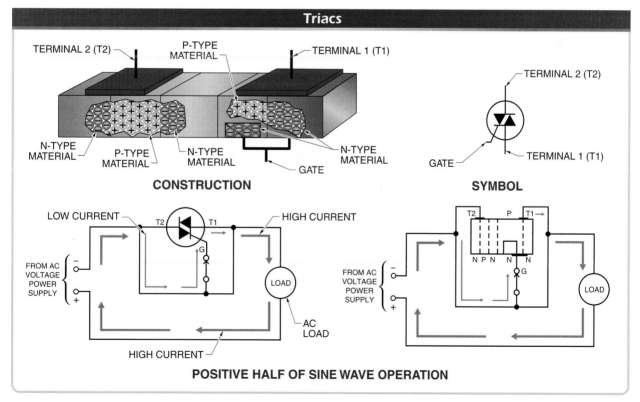

Figure 3-15. Triacs are semiconductor devices used to switch AC.

Integrated Circuits. An *integrated circuit* is an electronic device in which all components (transistors, diodes, and resistors) are contained in a single package or chip. Integrated circuits were developed because as electronic devices were being used to perform more and more functions, it was discovered that many transistors, diodes, and other devices could be deposited microscopically on a small chip of silicon. **See Figure 3-16.**

Microscopically depositing many semiconductor devices on a small chip of silicon enables many functions to be performed by a single semiconductor chip. One function is to represent on/off (digital) decisions. The on/off decision devices are referred to as logic gates and follow a set of logic rules known as Boolean logic. Logic gates include AND gates, OR gates, NOT gates, NOR gates, etc. Combining logic gates onto a single chip in the minimum amount of space so that they use the minimum amount of power enables a ¼″ square chip of silicon to contain hundreds of thousands or millions of individual semiconductor devices.

These devices are deposited by different processes such as photolithography (etching) or by burning with a laser. A chip is attached to a chip holder that protects the chip from damage and attaches the chip to wire leads that carry the electrical signals to and from the other devices in the circuit. Chip holders have the logo of the manufacturer, date code, and index mark imprinted on them. The logo and date code help identify the chip, while the index mark indicates the correct orientation of the chip when it requires replacement.

Some chip holders have windows which enable the chip inside to be erased and reprogrammed. This can occur by exposure to light, for example. Many manufacturers also use stickers on the chips with version numbers for easy identification. Care must be taken when replacing chips in control circuits. Chips must never be forced or have excessive pressure placed on the fragile pins. A chip removal tool is used to aid in chip removal. Integrated circuits are used in many HVAC applications such as chiller control, boiler control, and building automation systems.

Troubleshooting Semiconductor Devices

HVAC technicians require an overview of electronic component troubleshooting procedures and methods. Before troubleshooting any semiconductor device, it is important to always verify that the power is disconnected. If power is connected, serious injury, death, or equipment damage can result. Even low-voltage power can damage electronic circuit components.

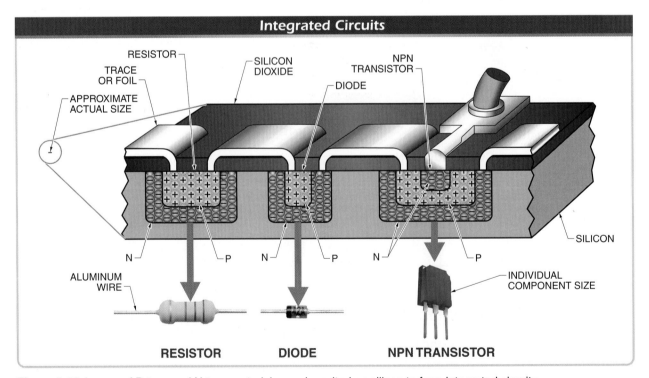

Figure 3-16. Layers of P-type and N-type materials are deposited on silicon to form integrated circuits.

Adequate personal grounding should always be verified. An inexpensive wrist strap can be connected to ground to protect against static electricity discharge to electronic components. Electronic components are damaged or destroyed by even small static electricity discharges. When the cover is removed from an electronic device, the device is unprotected and vulnerable to static electricity discharges.

Quality test meters should always be used. The use of a poor test meter can cause faulty or inaccurate readings. Digital multimeters measure small voltages and currents properly and display the readings in an easy-to-read format. Many digital multimeters have an autoranging function that automatically displays the correct value. The digital readout helps to eliminate reading guesswork. Analog meters are used on certain circuits where circuit loading is required to obtain accurate readings.

The most common electronic circuit troubleshooting situation is to find the correct printed circuit (PC) board, determine if the board has a problem, and replace the board if necessary. A failed PC board is normally returned to the manufacturer for replacement or repair. PC board troubleshooting requires that a technician find the defective semiconductor device on the board. A close visual inspection often reveals burn marks on components or foils (traces), indicating a failed component. It is possible that a semiconductor device has been completely destroyed and is missing. In many cases, an overheated component causes a burning smell. The sense of touch (a short time after power is removed) can often be used to detect a component that is operating excessively hot.

Electronic Control System Applications

Electronic control systems use various semiconductor devices to control HVAC units. Electronic control systems include the sensors, controllers, switching components, and output devices such as dampers and valve actuators. These components then provide temperature, pressure, and humidity control in the building. Electronic control system components include sensors and electronic thermostats that are used in applications such as multizone unit control, boiler control, and chiller control.

Sensors. Many temperature sensors use electronic components. **See Figure 3-17.** Temperature sensors use semiconductor materials that change resistance characteristics as the temperature around the sensor changes. Humidity sensors change resistance as humidity is absorbed or adsorbed by electronic elements. Pressure sensors may use a piezoelectric crystal, which changes output voltage as pressure is exerted on the crystal.

Figure 3-17. A thermistor is an electronic temperature sensor that uses semiconductor materials to measure temperature.

Electronic Thermostats. Many commercial and residential thermostats are electronic. **See Figure 3-18.** Such applications as rooftop heat pumps and other packaged equipment commonly use electronic thermostats. Electronic thermostats use electronic sensors to sample the building space temperature and use triacs to energize and deenergize heating, cooling, and fan functions. Some electronic thermostats include LEDs to indicate system operation or trouble conditions.

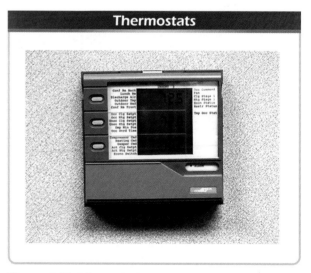

Figure 3-18. Electronic thermostats are used to control HVAC systems.

Multizone Unit Control. In multizone unit control, each zone damper is controlled by a zone thermostat, while the air-handling unit is controlled by several electronic controllers. In this application, the supply fan runs continuously. The hot deck controller controls the hot deck electronic valve to maintain the hot deck setpoint. The setpoint of the hot deck controller is reset from outside air. As the outside air temperature varies from –10°F to 65°F, the hot deck setpoint varies from 120°F to 55°F. The cold deck controller maintains a constant temperature of approximately 55°F. **See Figure 3-19.**

Boiler Control. Electronic controls are used to control boilers and their associated equipment. In this application, the electronic controller changes the hot water setpoint from 100°F to 190°F as the outside air temperature changes from 65°F to –10°F. The electronic controller also modulates a three-way mixing valve to obtain the correct water temperature. The electronic controller also turns on the boiler below a specific outside air temperature to begin to supply heat. **See Figure 3-20.**

Chiller Control. One or more water chillers can be controlled by electronic control systems. In this application, the electronic controller starts and stops the chiller when required. The controller also opens or closes the chiller inlet vane electronic actuator to maintain the correct chilled water supply or return temperature. Electronic outside-air, chilled water return, and chilled water supply temperature sensors measure the temperature and send a signal back to the controller. Low-water and low-oil-pressure safety control circuits ensure that the chiller does not operate if damage to the unit may occur. **See Figure 3-21.**

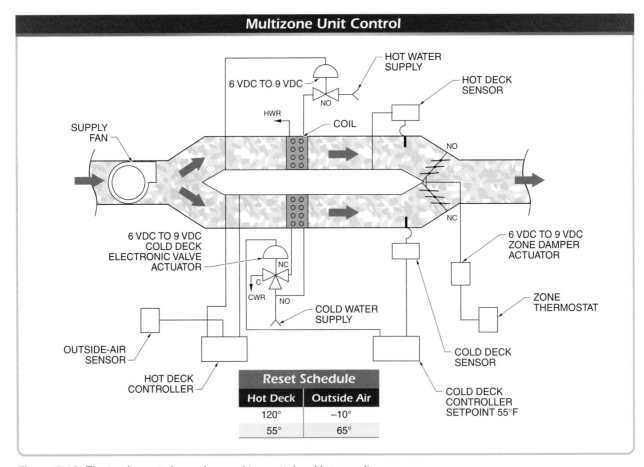

Figure 3-19. Electronic controls can be used to control multizone units.

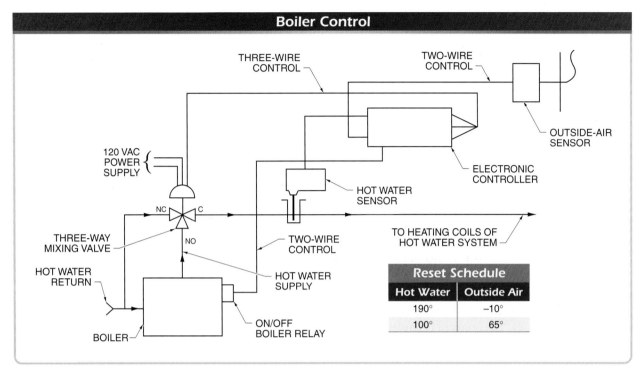

Figure 3-20. Electronic controls can be used to control boilers and their associated equipment.

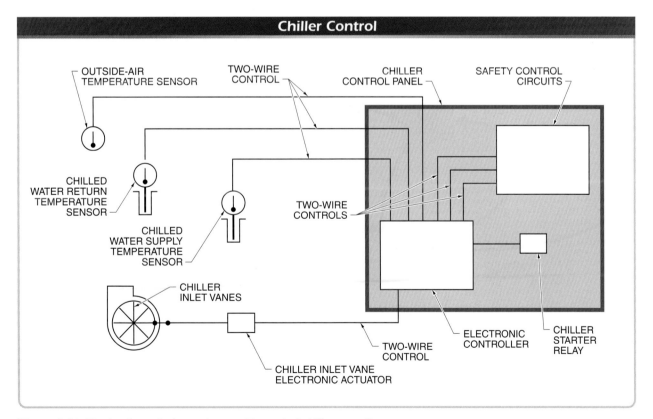

Figure 3-21. Electronic controls can be used to control chiller operation.

BUILDING AUTOMATION COMMUNICATION

Advanced building automation technologies include decision-making ability within the individual control devices, which are linked by a common data communication network. **See Figure 3-22.** These devices are known as "smart" or "intelligent" control devices. These intelligent control devices are radically changing the building automation industry by allowing the integration of multiple building systems that results in lower installation and energy costs and higher levels of comfort, safety, and security.

Building automation systems must structure and share information between control devices in a consistent and reliable way. A communication protocol governs the format, timing, and signals passed between devices to ensure that they are all speaking the same understandable language. A *protocol* is a set of rules and procedures for the exchange of information between two connected devices. The protocol is implemented in both the electronics and software of the control devices

to assemble and interpret structured network messages that contain the detailed control information.

Each structured network message includes the identification of the shared variable and its value, plus other information. **See Figure 3-23.** Much of the content of a message is overhead, such as device-addressing information, message service type, error-checking information, and other parameters. The information is encoded and transmitted via a series of digital signals. Almost any type of information can be shared in this way. The standardization of protocols allows a variety of different control devices from different manufacturers to operate together, as long as they are all compatible with the same protocol.

Building automation systems using this type of communication are connected together in a way similar to a computer network. In this configuration, any device can communicate with any other device on the network. In fact, some systems can communicate over the same wiring and routing infrastructure as the building's local area network (LAN).

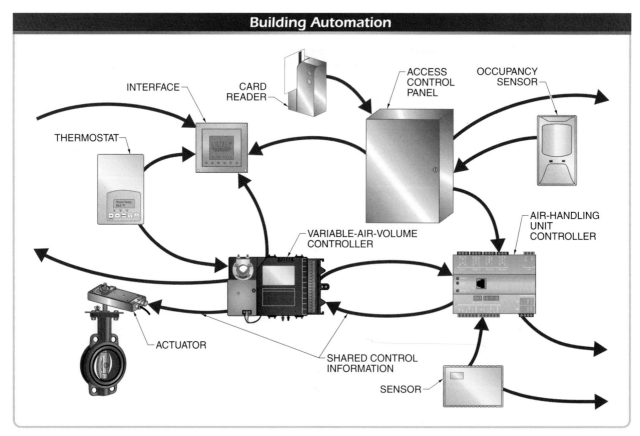

Building Automation

THERMOSTAT

INTERFACE

CARD READER

ACCESS CONTROL PANEL

OCCUPANCY SENSOR

AIR-HANDLING UNIT CONTROLLER

VARIABLE-AIR-VOLUME CONTROLLER

ACTUATOR

SHARED CONTROL INFORMATION

SENSOR

Figure 3-22. Networked building automation systems allow a variety of control devices from many different building systems to communicate and share control information.

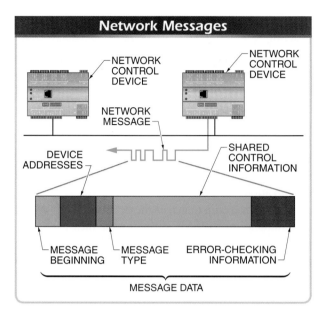

Figure 3-23. Network messages contain addressing, message type, error-checking, and other information, in addition to the control information to be shared.

Proprietary Protocols

Protocols have been used by manufacturers for many years as part of proprietary building automation systems. A *proprietary protocol* is a communication and network protocol that is developed and used by only one device manufacturer. For example, early proprietary HVAC control systems were developed by many manufacturers. However, only the specific manufacturer's equipment and related proprietary software can be used with the system. The details of proprietary protocols are protected so that other manufacturers cannot market compatible devices.

Proprietary protocols also tend to be specific to certain building systems. Buildings utilizing proprietary protocols may include separate control systems for the HVAC, lighting, security, and fire alarm systems. It is not unusual for a building engineer to have separate workstations or interfaces for each of these systems, which require support from different vendors. **See Figure 3-24.**

End users typically are locked in to these vendors and have relatively few choices when it comes to adds, moves, or changes. Plus, these separate systems cannot easily share information without installing expensive and complex gateways or translation devices. Proprietary protocols lack the vendor competition (resulting in lower costs and greater variety) and potentially greater integration opportunities of open protocols.

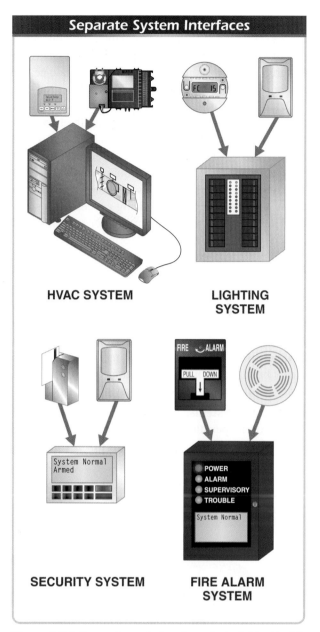

Figure 3-24. Separate control systems for individual building systems may require a separate user interface for each system.

Open Protocols

In the 1980s, a movement in the building automation industry started to create a system that allowed open access to information by devices using universal communication schemes. This introduced the concept of an open protocol. An *open protocol* is a standardized communication and network protocol that is published for use by any device manufacturer. Systems using open protocols must still use protocol-specific devices, but since the protocol standard is publicly available, the device can be made by any manufacturer. Without being tied to a specific manufacturer, open protocols have also fostered the expansion of automation to include nearly any building system. **See Figure 3-25.**

Many open protocols evolved from this effort, though a few have become dominant, including the two most commonly used today: LonWorks® and BACnet®. There is also increased interest in wireless open protocols that are specifically designed for the requirements of wireless communication. Open protocols continue to evolve with technology, but the basic operating concepts are similar. They standardize a common way for very different devices to organize and share information.

Each of these protocols can ultimately be used to achieve similar types of automation, though there are differences in network requirements, device types, communication modes, programming methods, network tools, and other characteristics. Depending on the building and the application, these factors can be either advantages or disadvantages. The protocol chosen for a new building automation system is based on careful analysis of all these considerations.

LonWorks Systems. LonWorks systems are based largely on a specially designed microcontroller chip, called the Neuron® chip, that is part of a LonWorks control device. The Neuron chip, and the associated LonTalk® communication protocol, were developed by Echelon Corporation as a way to embed intelligence into individual control devices as small, inexpensive electronics. The LonTalk protocol can be implemented on other microcontrollers, but the Neuron chip is the most common solution used by control device manufacturers. The ongoing support and development of the LonTalk protocol, as well as device conformance testing, is overseen by the independent LonMark International organization.

The Neuron chip manages the details of the LonTalk communication protocol, including the message transport, addressing, media access, and signaling tasks. **See Figure 3-26.** Depending on the design of the control device, it may also manage the data organization and control logic. All control data is organized into standard variable types, with structures and formats defined in the protocol's standard. Each fundamental decision-making function of a control device application program is represented as a block, which accepts input variables, performs some process or calculation, and provides output variables. Relatively simple devices contain few blocks, while complex controllers may contain many. The operation of each function block can be adjusted by changing its properties.

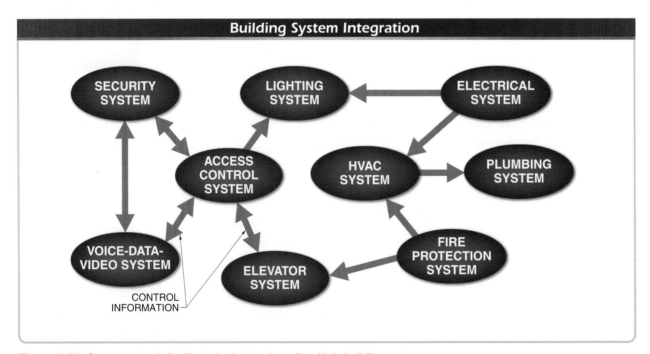

Figure 3-25. Open protocols facilitate the integration of multiple building systems.

Neuron® Chip Responsibilities

Programming
Variables
Network management
Messaging
Addressing
Media access
Signaling

Figure 3-26. The Neuron® chip manages many of the communication and application tasks in a LonWorks® system.

The sharing of control data is represented as bindings (connections) between input and output variables. These variables may be associated with the same function block, between two function blocks within the same control device, or between two function blocks in different control devices. When the connection is one of the former, the processing is internal to a control device. However, if the connection involves two separate control devices, the control information must be formatted and packaged into a network message and transmitted to the second device. **See Figure 3-27.** The LonWorks standard specifies the media, addressing, and other network requirements for efficient and reliable communication between the control devices.

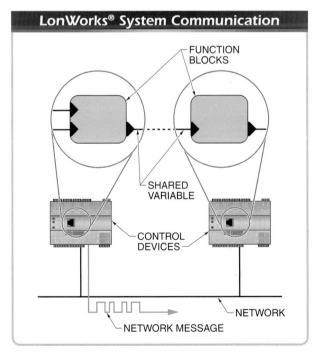

LonWorks® System Communication

FUNCTION BLOCKS

SHARED VARIABLE

CONTROL DEVICES

NETWORK MESSAGE

NETWORK

Figure 3-27. LonWorks® control devices communicate over the network to share variables linking two of their function blocks.

LonWorks systems provide a fully developed platform for the design and implementation of a control system. This includes how the network is designed and how the control devices are commissioned and programmed. Software is available from Echelon for performing all of these tasks, though compatible third-party solutions are also available.

After two decades, the LonWorks platform has proven to be a major player in the building automation industry. Thousands of different products from hundreds of different manufacturers are available for nearly every control application found in commercial buildings. Millions of individual LonWorks-compatible control devices have been deployed worldwide.

BACnet Systems. The BACnet protocol was developed in 1987 by the American Society of Heating, Refrigerating, and Air-Conditioning Engineers (ASHRAE) as a data communication protocol for building automation and control networks. Even though the organization's focus is on HVAC applications, it made a conscious effort to develop a protocol that could potentially be used with any building system. It also maintains a process for continually collecting, evaluating, and implementing proposed changes to the protocol to improve its interoperations and expand its capabilities with new features.

The challenges facing BACnet-based automation systems are similar to LonWorks systems and to any other networked automation system. Control information must be organized into standardized units, which can then be shared over a network using standardized procedures. BACnet uses these two objectives to divide the goal of interoperability into more easily manageable tasks. **See Figure 3-28.**

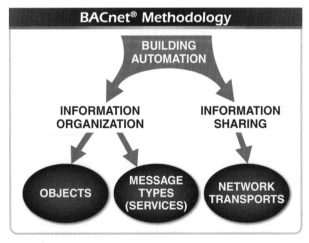

BACnet® Methodology

BUILDING AUTOMATION

INFORMATION ORGANIZATION

INFORMATION SHARING

OBJECTS

MESSAGE TYPES (SERVICES)

NETWORK TRANSPORTS

Figure 3-28. BACnet® divides the challenges of building automation into two primary goals, which are addressed by BACnet-specific solutions.

Like LonWorks systems, BACnet maintains a highly structured information architecture that organizes information as software objects. Each object has a number of configurable properties. **See Figure 3-29.** These objects represent not only control points, but also processes and control functions within the control device application program. Information is identified in a hierarchical scheme by the identity of the device, the identity of the object within the device, the identity of the property of the object, and sometimes the identity of elements within the property.

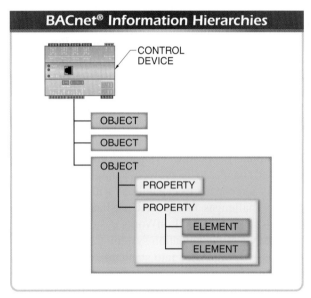

Figure 3-29. The BACnet® information hierarchy divides control information into objects, which are each composed of properties that may be composed of smaller elements.

This addressing is used to identify the sources and/or destinations of information to be shared between control devices. A variety of communication message types are defined by BACnet that standardize the desired actions requested by one control device of another device. In the language of the BACnet protocol, the objects are like nouns and these message types are like the grammar and structure of sentences.

The BACnet standard supports a variety of network types by which these messages are transmitted. These include both types that are unique to BACnet and types that are used by other applications of data networks. Each has a different cost, performance, and other distinct characteristics. The selection allows integrators to choose the most appropriate network for each application.

BACnet is different from LonWorks in that the standard does not specify procedures or outcomes for installing, commissioning, or programming control devices. However, this has by no means prevented BACnet from being implemented in building automation systems worldwide. A number of solutions exist from third-party vendors to accomplish this integration.

Wireless Systems. The categorization of wireless building automation system protocols can be misleading. Many of the traditionally wired protocol systems, including LonWorks and BACnet systems, support a wireless network as one of their message transport options. They can even incorporate both wired and wireless segments on the same network. Therefore, they can technically be wireless systems, although they are not traditionally included in this category.

The protocols that are considered to be in the wireless protocol category, however, rely on the unique characteristics of wireless communication for their operation. Many use a networking technique that involves the relaying of messages between adjacent wireless control devices until the message reaches its ultimate destination. **See Figure 3-30.** The message content of wireless protocols is also reduced to accommodate the low-power and low-data-rate capabilities of wireless transmitters. Since "wireless" can apply to both communication and power wiring, wireless control devices must rely on self-contained power supplies, which limit the performance of the transmitters.

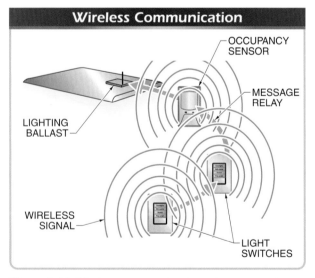

Figure 3-30. Wireless communication relies on the relaying of messages between adjacent control devices in order to reach the final destination.

In addition, wireless mesh networks can be used to reduce the distance a signal has to travel. A *wireless mesh network* is a wireless network where the devices on the network can communicate directly with each other to relay signals. **See Figure 3-31.** Each device can send and receive messages and act as a transmission and receiving router within the network architecture.

In situations that preclude network wiring, wireless systems can be an ideal solution. For example, wireless automation may be the best choice in an existing building if it is not feasible to run new network conductors throughout the building. Also, wireless systems eliminate improper wiring and connector and termination issues as potential sources of communication problems.

However, wireless building automation systems are not as prevalent as wired systems. The reliability, security, and effectiveness of wireless systems is often questioned, though continual system advancements have addressed these concerns in many applications. Some of the more common wireless protocols, such as ZigBee®, are gradually making inroads into mainstream building automation applications.

Interoperability

Unlike proprietary systems, systems based on open protocols provide interoperability. *Interoperability* is the capability of network devices from different manufacturers and systems to interact using a common communication network and language framework. The concept of interoperability is common to many everyday systems. In control systems, interoperable devices from different manufacturers are able to work together in the same system. They speak and understand the same language, store control information in the same way, and use the same group of control actions. These are all defined in the protocol.

There are many different kinds of interoperations. For example, one device can announce the value of a control point to any other interested device, ask another device for information, or ask another device to perform an action. In order for these communications to be interoperable, each device must support the same network type and be fully compliant with the information architecture. There are two primary goals of interoperability: manufacturer/vendor integration and building system integration.

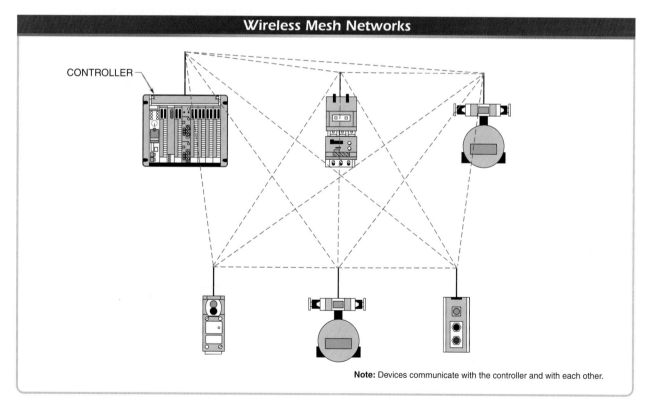

Wireless Mesh Networks

CONTROLLER

Note: Devices communicate with the controller and with each other.

Figure 3-31. Wireless mesh networks relay information from one device to another to reduce the power needed to send messages over long distances.

Manufacturer/Vendor Integration. Since the protocol is open and available for any user, a manufacturer can develop a control device that can operate with any other manufacturer's device, as long as they both follow the specifications in the protocol standard. Therefore, building automation systems using open protocols may consist of devices from a variety of manufacturers.

With multiple manufacturers offering compatible control devices, integrators can select the most appropriate device for each application. For example, a variable-air-volume (VAV) controller from one manufacturer may be the best choice due to size and mounting reasons, while the thermostat with the desired temperature and humidity features are available from another manufacturer. Other factors in choosing a manufacturer's products include interoperability, cost, documentation, support, maintenance requirements, and availability.

The same approach applies to vendors offering integration, programming, maintenance, and troubleshooting support. Any vendor that has the appropriate training on an open protocol can work on the system, which provides additional choices for building owners.

The benefits of interoperability also allow an obsolete or faulty control device to be replaced by another from a different manufacturer. As long as the new device supports the same standard functions and features, the replacement is relatively a one-to-one procedure.

Building System Integration. Open protocols also allow greater opportunities for integrating building systems together. For example, when a person enters a building, the access control system can share information about the entry and the person's identity with the lighting system, which turns on the person's office lights, and the HVAC system, which changes to an occupied mode for the person's work area. Open protocols can provide interoperability without being system-specific. For example, the standardized information infrastructure can be used to structure and share nearly any type of data, regardless of whether it is a temperature reading, damper position, or lighting level.

The combination of interoperability and the resulting variety in compatible vendors has fostered greater integration of building systems and more opportunities for automation. New controls manufacturers find it attractive to enter the control device market for a particular open protocol. Once invested in developing products for open protocols, some manufacturers expand outside of their traditional system markets. For example, a manufacturer of HVAC controls may leverage their

experience in open protocol systems by adding lighting controls to their product line. The result is an increasing availability of control devices for all of the major building systems. Since they are all interoperable, these building systems can then be integrated together on a common control network.

Even if there is no need in a particular building for the systems to share information with each other, the integration of multiple building systems on a single automation system may still be desirable. **See Figure 3-32.** This allows control operations to be monitored from a unified operator interface, which is particularly desirable for a maintenance staff that is responsible for all of these very different building systems. The integrated system can also be used to log control data together.

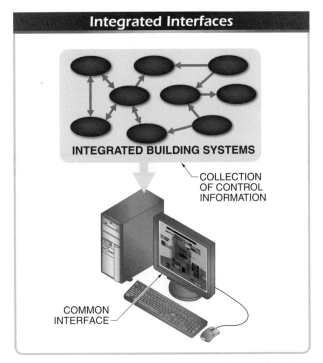

Figure 3-32. Interoperability allows a single user interface to access the information shared among integrated building systems.

Network Architectures

Integration and interoperability requires that all control devices reside on the same network. Messages may need to be able to travel between any two control devices, depending on the level of system integration. This can result in networks of different architectures and even the incorporation of different network types or protocols for some portions.

Hierarchical Network Architectures. *Hierarchical network architecture* is a network configuration in which control devices are arranged in a tiered network and have limited interaction with other control devices. One control device serves as a network manager for the devices attached to it in a lower tier. **See Figure 3-33.** These arrangements are also called supervised or master/slave network architectures, and network managers are also known as central or master controllers. Multiple network manager devices may be used to supervise many groups of subordinate devices.

The network manager typically includes applications for installing, configuring, and programming the network. Additional features often include scheduling, alarm, data logging, programmable control sequences, proprietary protocol drivers, and web-based user interface functions. The presence of the network manager may also be required for the subordinate devices to perform their control functions or communicate. Therefore, this architecture creates single points of failure that may render large portions of a control system, or even the entire system, inoperable if a network manager fails.

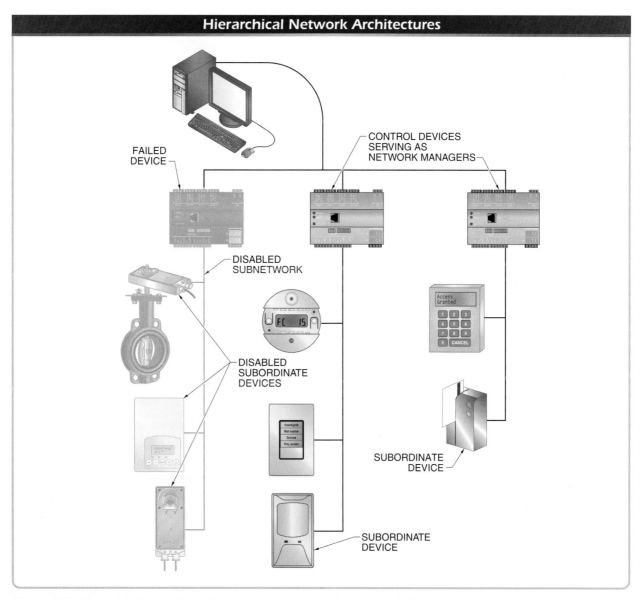

Figure 3-33. Hierarchical network architectures rely on network manager devices to organize the communication of subordinate devices below them.

Several characteristics of hierarchical network architectures are generally considered disadvantages to this type of system. Implementing changes in a hierarchical system often requires proprietary software that can be complex and require specialized training. Network manager devices typically go off-line while program modifications are loaded, resulting in control system downtime. Network manager devices use polling to read sensor values, which is an inefficient use of the network and may miss important control events. Hierarchical architecture systems tend to be limited to specific building systems, making them challenging to integrate with other systems.

Some older control systems employed hierarchical architectures that required hardwired sensors and actuators to have home runs back to a master control panel. Some open-protocol-based systems support hierarchical architectures, but they are typically implemented in flat network architectures when that arrangement is also supported.

Flat Network Architectures. *Flat network architecture* is a network configuration in which control devices are arranged in a peer-to-peer way. This allows any device to communicate directly with any other device on the network without the need for network managers. **See Figure 3-34.** Control devices perform their local control functions independently while sharing data with other control devices on the network media.

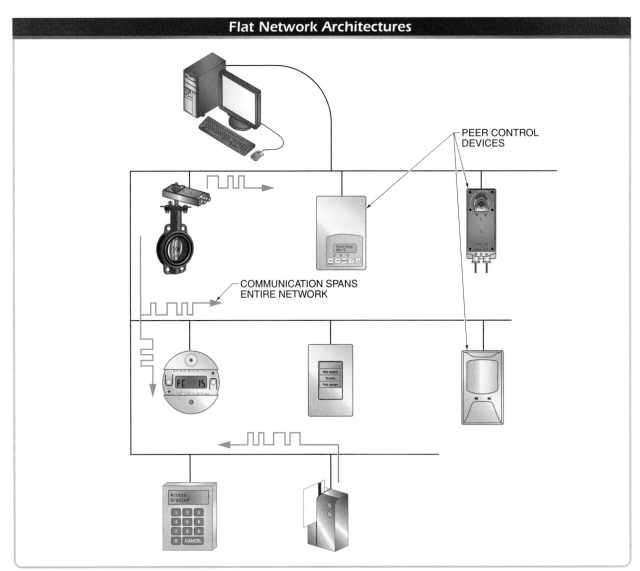

Figure 3-34. In flat network architectures, every control device can communicate directly with every other control device in a peer-to-peer way.

Programming tools may be needed to initially commission and program the control network, but since the individual devices do not require any outside controllers to operate normally, the tools can then be disconnected from the network. They can also be reconnected as needed for moves, adds, or changes without affecting the operation of control devices unassociated with the changes.

Flat network architecture has no single point of failure. Individual control devices may still fail, but they typically do not disable any other devices in the process. A control device relying on information from a failed device is typically programmed to use default fail-safe values until the problem is remedied.

Gateway Network Architectures. A *gateway* is a network device that translates transmitted information between different protocols. *Gateway network architecture* is a network configuration in which a gateway is used to integrate separate control systems based on different protocols. **See Figure 3-35.** Gateway architectures are mergers of these separate control systems, which may be any combination of hierarchical and flat network architectures.

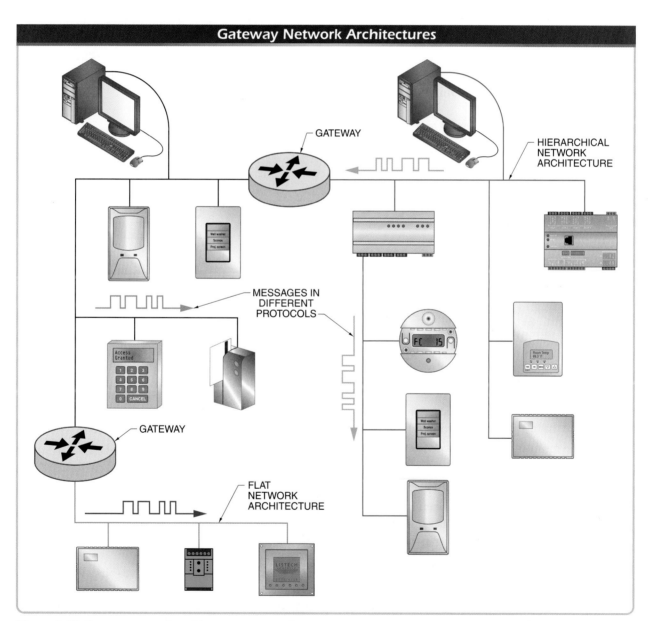

Figure 3-35. Gateway network architectures are used to connect networks using different communication protocols.

The systems may be any mix of open and proprietary protocols. System-level interoperability is achieved by translating the control information between the different control protocols. Corresponding control points must be mapped to each other during programming, typically requiring manufacturer-specific programming tools as well as knowledge of each protocol's format and structure.

APPLICATIONS: Testing Diodes, Transistors, and Thyristors

3-1—DIODE TESTING

Ohmmeter Diode Test

Forward bias is the condition of a diode while it conducts current. The anode in a forward-biased diode has a positive polarity compared to the cathode. *Reverse bias* is the condition of a diode when it acts as an insulator. The cathode of a reverse-biased diode has a positive polarity compared to the anode. A diode acts as a closed switch when it is forward biased and as an open switch when it is reverse biased. An ohmmeter is used to conduct a basic diode check. **See Ohmmeter Diode Test.**

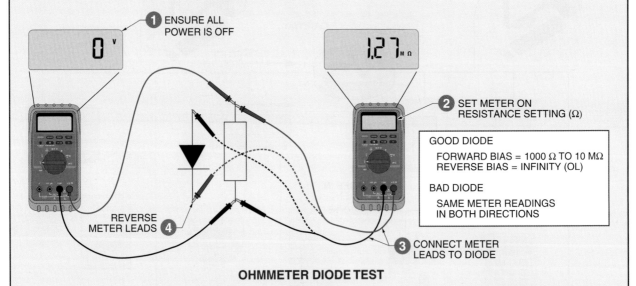

OHMMETER DIODE TEST

To use an ohmmeter to test a diode, apply the following procedure:

1. Ensure that all power in the circuit is OFF. Test for voltage using a voltmeter to ensure power is OFF.

2. Set the meter on the resistance setting.

3. Connect the meter leads to the diode. Record the meter reading.

4. Reverse the meter leads. Record the meter reading.

The diode is forward biased when the positive (red) probe is on the anode and the negative (black) probe is on the cathode. The forward biased resistance of a good diode should be between 1000 Ω and 10 MΩ. The resistance reading is high when the diode is forward biased because the current from the meter's voltage source flows through the diode and results in a resistance measurement.

The diode is reverse biased when the positive (red) lead is on the cathode and the negative (black) lead is on the anode. The reverse biased resistance of a good diode should equal infinity. An infinite resistance is displayed as an overload (OL) on a digital meter. The diode is bad if the meter readings are the same in both directions.

Multimeter Diode Test

Testing a diode using an ohmmeter does not always indicate whether a diode is good or bad. Testing a diode that is connected in a circuit with an ohmmeter may give false readings because other components may be connected in parallel with the diode under test. The best way to test a diode is to measure the voltage drop across the diode when it is forward biased.

A good diode has a voltage drop across it when it is forward biased and conducting current. The voltage drop is between 0.5 V and 0.8 V for the most commonly used silicon diodes. Some diodes are made of germanium and have a voltage drop between 0.2 V and 0.3 V.

A multimeter set on the diode test position is used to test the voltage drop across a diode. In this position, the meter produces a small voltage between the test leads. The meter displays the voltage drop when the leads are connected across a diode. **See Multimeter Diode Test.**

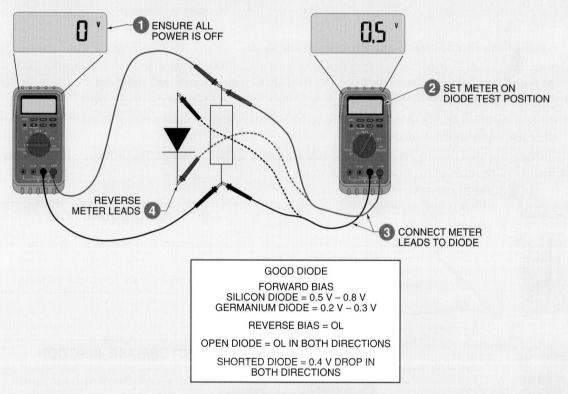

GOOD DIODE

FORWARD BIAS
SILICON DIODE = 0.5 V – 0.8 V
GERMANIUM DIODE = 0.2 V – 0.3 V

REVERSE BIAS = OL

OPEN DIODE = OL IN BOTH DIRECTIONS

SHORTED DIODE = 0.4 V DROP IN
BOTH DIRECTIONS

MULTIMETER DIODE TEST

To test a diode using the diode test position on a multimeter, apply the following procedure:

1. Ensure that all power in the circuit is OFF. Test for voltage using a voltmeter to ensure power is OFF.

2. Set the meter on the diode test position.

3. Connect the meter leads to the diode. Record the meter reading.

4. Reverse the meter leads. Record the meter reading.

The meter displays a voltage drop between 0.5 V and 0.8 V (silicon diode) or 0.2 V and 0.3 V (germanium diode) when a good diode is forward biased. The meter displays an OL when a good diode is reverse biased. The OL reading indicates the diode is acting like an open switch. An open (bad) diode does not allow current to flow through it in either direction. The meter displays an OL reading in both directions when the diode is open. A shorted diode gives the same voltage drop reading in both directions. This reading is normally about 0.4 V.

3-2—TRANSISTOR TESTING

A transistor becomes defective from excessive current or temperature. A transistor normally fails due to an open or shorted junction. The two junctions of a transistor may be tested with an ohmmeter. **See Transistor Testing.**

To test an NPN transistor for an open or shorted junction, apply the following procedure:

1. Connect a multimeter to the emitter and base of the transistor. Measure the resistance.

2. Reverse the meter leads and measure the resistance. The emitter/base junction is good when the resistance is high in one direction and low in the opposite direction.

   Note: The ratio of high to low resistance should be greater than 100:1. Typical resistance values are 1 kΩ (with the positive lead of the meter on the base) and 100 kΩ (with the positive lead of the meter on the emitter). The junction is shorted when both readings are low. The junction is open when both readings are high.

3. Connect the meter to the collector and base of the transistor. Measure the resistance.

4. Reverse the meter leads and measure the resistance. The collector/base junction is good when the resistance is high in one direction and low in the opposite direction.

   Note: The ratio of high to low resistance should be greater than 100:1. Typical resistance values are 1 kΩ (with the positive lead of the meter on the base) and 100 kΩ (with the positive lead of the meter on the collector).

5. Connect the meter to the collector and emitter of the transistor. Measure the resistance.

6. Reverse the meter leads and measure the resistance. The collector/emitter junction is good when the resistance reading is high in both directions.

The same test used for an NPN transistor is used for testing a PNP transistor. The difference is that the meter test leads must be reversed to obtain the same results.

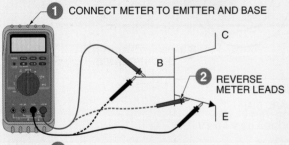

① CONNECT METER TO EMITTER AND BASE

② REVERSE METER LEADS

EMITTER/BASE JUNCTION

GOOD = HIGH RESISTANCE IN ONE DIRECTION, LOW RESISTANCE IN OPPOSITE DIRECTION

SHORTED = LOW RESISTANCE IN BOTH DIRECTIONS

OPEN = HIGH RESISTANCE IN BOTH DIRECTIONS

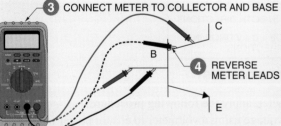

③ CONNECT METER TO COLLECTOR AND BASE

④ REVERSE METER LEADS

COLLECTOR/BASE JUNCTION

GOOD = HIGH RESISTANCE IN ONE DIRECTION, LOW RESISTANCE IN OPPOSITE DIRECTION

SHORTED = LOW RESISTANCE IN BOTH DIRECTIONS

OPEN = HIGH RESISTANCE IN BOTH DIRECTIONS

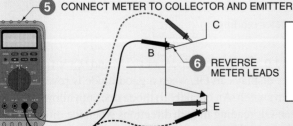

⑤ CONNECT METER TO COLLECTOR AND EMITTER

⑥ REVERSE METER LEADS

COLLECTOR/EMITTER JUNCTION

GOOD = HIGH RESISTANCE IN BOTH DIRECTIONS

SHORTED = LOW RESISTANCE IN BOTH DIRECTIONS

OPEN = CANNOT BE DETERMINED

TRANSISTOR TESTING

APPLICATIONS: Testing Diodes, Transistors, and Thyristors (cont.)

3-3—TRIAC TESTING

Triacs should be tested under operating conditions using an oscilloscope. A multimeter may be used to make a rough test with the triac out of the circuit. **See Triac Testing.**

To test a triac using a multimeter, apply the following procedure:
1. Set the multimeter on the R × 100 scale.
2. Connect the negative meter lead to main terminal 1.
3. Connect the positive meter lead to main terminal 2. The meter should read infinity.
4. Short circuit the gate to main terminal 2 using a jumper wire. The meter should read almost 0 Ω. The zero reading should remain after the lead is removed.
5. Reverse the meter leads so that the positive lead is on main terminal 1 and the negative lead is on main terminal 2. The meter should read infinity.
6. Short circuit the gate to main terminal 2 using a jumper wire. The meter should read almost 0 Ω. The zero reading should remain after the lead is removed.

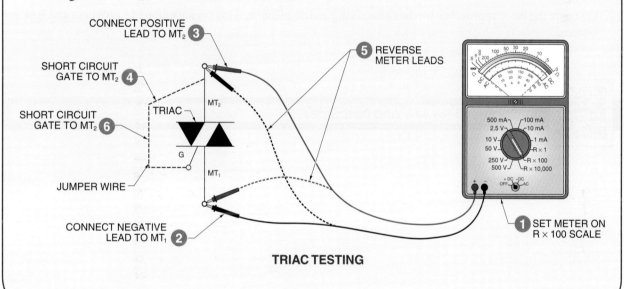

TRIAC TESTING

REVIEW QUESTIONS

1. Define semiconductor and identify the most common material used for semiconductor devices.

2. Differentiate between N-type material and P-type material.

3. List and define the three types of rectifiers.

4. Describe LEDs.

5. List and define the four types of transistors.

6. What is the difference between an SCR and a triac?

7. What are integrated circuits and why were they developed?

8. What is the role of a protocol in building automation?

9. What are the primary differences between proprietary and open protocols?

10. Compare the basic approaches for data structuring and decision-making organization of LonWorks and BACnet systems.

Chapter 3 Review and Resources

Network Data Communications 4

Data communication includes a large number of technologies and concepts for communicating digital data between two or more entities on a network. This information is displayed to users and/or used to control the action of a machine. A familiar example is the computer network used to access the Internet. However, data communication is necessary in a variety of applications, such as the type of machine-to-machine communication used in building automation applications.

DATA COMMUNICATIONS

The role of any data communications technology is to facilitate the exchange of information in predictable and reliable ways. Like the many forms of communication, many methods have been developed for allowing machines to communicate to other machines. Machines that are involved in network communication are known as nodes. **See Figure 4-1.** A *node* is a computer-based device that communicates with other similar devices on a shared network.

Communication Protocols

Before any two nodes can effectively communicate with each other, they must agree on a protocol to govern the manner in which they exchange information. A protocol consists of a set of codes, message structures, signals, and procedures implemented in hardware and software that permits the exchange of information between nodes. In other words, a protocol is a collection of rules that enable the nodes to exchange information in reliable and repeatable ways.

A *signal* is the conveyance of information. Symbols are used to group signals together, often as binary bits representing the concept of zero and one. Such groups of zeroes and ones can also be thought of as a number. **See Appendix.** A code equates symbols or numbers with real ideas that often depend on context. The protocol's

rules must define what kinds of signals and symbols are used to convey information and the structure, encoding, and content of messages that may be exchanged.

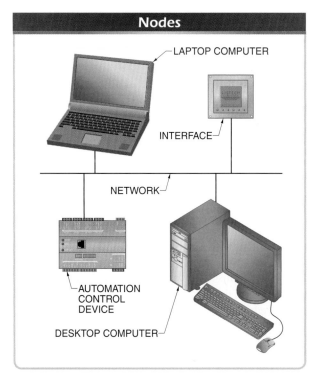

Figure 4-1. Nodes are computer-based devices that are connected together on a data communications network.

Signaling

Signaling is the use of electrical, optical, and radio frequency changes in order to convey data between two or more nodes. Since computer-based devices deal most effectively with binary information (zeroes and ones), most signaling methods involve the sending and receiving of streams of binary bits. Typically, a digital signal uses the presence and absence of voltage or light, or radio frequency manipulation, to represent 1 and 0. **See Figure 4-2.** For example, a common signaling scheme uses a 5 VDC electrical level to represent 1 and a 0 V electrical level to represent 0.

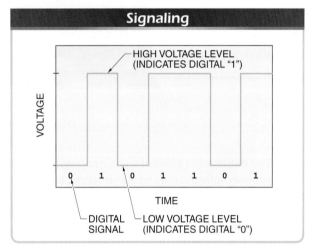

Figure 4-2. Electrical signaling commonly uses changes between two voltage levels to indicate the ones and zeroes of a digital signal.

Signaling between nodes is accomplished with transceivers. A *transceiver* is a hardware component that provides the means for nodes to send and receive messages over a network. The word "transceiver" is a combination of "transmitter" and "receiver" since it handles both sending and receiving messages. Each node on the network must have a transceiver for communication.

Each signaling method has characteristics that make it more appropriate for certain situations. Some methods operate faster but require equipment that is more expensive and harder to install or that has distance limitations. For these reasons, there is no "one size fits all" solution, and many methods have evolved to suit different circumstances.

Bandwidth. *Bandwidth* is the maximum rate at which bits can be conveyed by a signaling method over a certain media type. A *media type* is the specification of the characteristics and/or arrangement of the physical conductors or electromagnetic frequencies used for digital communication.

Although there are many measures that can be used to express bandwidth, for data communications networks, it is common to describe bandwidth in terms of some quantity of bits per second (bit/s). Depending on the signaling technology, the bandwidth may be thousands of bits per second (kbit/s), millions of bits per second (Mbit/s), or even billions of bits per second (Gbit/s).

As a rule, the speed and performance (bandwidth) of a given signaling method is proportional to the cost. **See Figure 4-3.** The higher-speed, higher-performance technologies typically cost more. Also, the maximum distance of total network length tends to be inversely proportional to speed, and consequently cost. Lower-performance technologies generally allow greater distances for wiring without degradation of the information. Higher performance technologies generally have more severe distance limitations. It is sometimes possible to extend the maximum distance with signal repeaters, but these also increase cost.

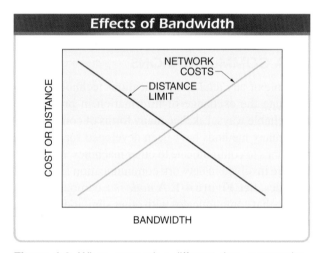

Figure 4-3. When comparing different data communications technologies, the speed of the network tends to be proportional to its cost and inversely proportional to the maximum network distance.

Throughput. *Throughput* is the actual rate at which bits are transmitted over a certain media at a specific time. While bandwidth is the maximum amount of data that can be transmitted over a media type, throughput is the actual measure of data that is transmitted over a specific network route, using certain media, and at a specific time. Factors that affect throughput include the type of data being transmitted, the network topology, the number of network users, and network congestion. Throughput can never exceed the bandwidth of a media type; it is always less than or equal to bandwidth.

Latency. *Latency* is the time delay involved in the transmission of data on a network. In one-way communication, latency is the difference between when a message is transmitted and when it is received. In two-way communication, latency also includes the delay required for the receiving node to respond to the message, such as sending a reply. Increased latency decreases throughput to a fraction of the total bandwidth capacity of the signaling method.

Signaling Directions. Data communication systems are defined partly by the signaling directions enabled by the node and media types. **See Figure 4-4.** *Simplex communication* is a system in which data signals can flow in only one direction. These systems are often employed in broadcast networks, where the receivers do not send any data back to the transmitter. Most computer and node networks use duplex communication, which includes half-duplex and full-duplex communication.

Half-duplex communication is a system in which data signals can flow in both directions but only one direction at a time. Once a node begins receiving a signal, it must wait for the transmitter to stop transmitting before it can reply. An example of a half-duplex system is a set of walkie-talkie-style two-way radios, where one person must indicate the end of transmission before the other can reply.

Full-duplex communication is a system in which data signals can flow in both directions simultaneously. For example, landline telephone networks are full-duplex systems because callers can speak and be heard at the same time. Full-duplex systems often use separate sets of conductors, one for each direction, to accomplish this. Full-duplex communication improves throughput, since there are no collisions that require retransmission. Full bandwidth is available in both directions and nodes do not require media access methods because there is only one transmitter for each twisted pair.

Message Frames

All networks send individual messages in discrete units that may be called packets or frames, though these terms have slightly different meanings. A *packet* is a collection of data message information to be conveyed. A *frame* is a packet surrounded by additional data to facilitate its successful transmission and reception by delineating the start and end (or length) of the packet. Frames can have somewhat different structures depending on the message protocol, but most have similar parts. **See Figure 4-5.**

Each frame has a beginning and an end that mark the frame using a special sequence of bits. The names of the beginning and end markers are not standardized, but there are several common terms. The beginning marker is often referred to as the header, start-of-frame, or preamble, and the end marker is often referred to as the trailer, end-of-frame, or postamble. Between the frame markers, the data is structured to have meaning.

Almost all sections of a frame are transmitted as a sequence of octets. An *octet* is a sequence of eight bits. These are sometimes also referred to as bytes. In this context, the term "byte" is usually accurate but is a less precise term because it has been used for groupings of other than eight bits in older computer systems. Octet, by definition, is always a group of eight bits, so it is always an accurate term for this context.

Frame Fields. Information is conveyed within the boundaries of a frame using logical groupings called fields. Most protocols use fields that are made up of one or more sequential octets. Frames typically include several fields arranged in a certain order.

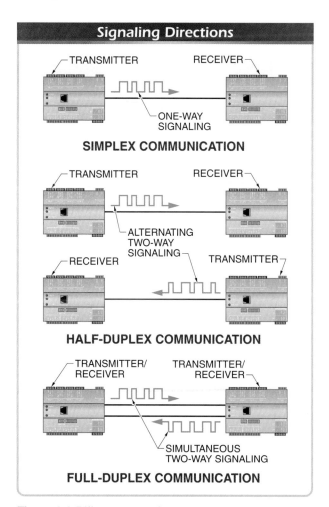

Figure 4-4. Different types of communication arrangements allow signals to be transmitted in one or both directions.

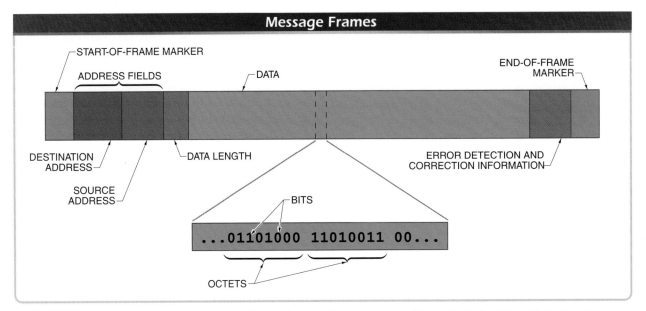

Figure 4-5. Message frames are like envelopes for sharing data between nodes. Frames include defined fields for addresses, data type and length, and error-checking information in addition to the actual data payload.

The two primary fields or groups of fields are address and data information. The address octets encode where a frame is supposed to be going and where it came from. Messages may vary in length, so a portion of this information usually also indicates the length of the data area. In some protocols, though, the length must be implied from the receipt of the end-of-frame marker.

In many cases, frames also contain error-checking information that can be used to detect if the data has been altered or corrupted during transmission. This field is typically called a frame-check sequence (FCS). There are many different schemes for detecting errors using mathematical algorithms, such as a cyclic redundancy check (CRC). Some sophisticated methods can not only detect, but also correct, some errors.

Addresses. The address portion of a frame identifies the sender and the recipient of the message. To prevent confusion, each node on a network segment must have some unique address. Different protocols and frame types use different lengths of the address field. The length of the node address defines the number of unique network addresses, which determines the maximum number of nodes that may communicate on a network. **See Figure 4-6.** For a single-octet address, the maximum number of nodes is only 256 (2^8), which is the number of unique numbers in one octet. For Ethernet, which uses a 6-octet address, the maximum number of nodes is much larger: 281,474,976,710,656 (2^{48}).

Address Length	
Number of Octets	**Number of Unique Network Addresses**
1	256
2	65,536
3	16,777,216
4	4,294,967,296
5	1,099,511,627,776
6	281,474,976,710,656

Figure 4-6. The number of octets reserved for node addresses in a frame determines how many unique addresses (nodes) the network can support.

Some address schemes reserve a portion of the possible addresses for special uses like broadcasting and multicasting. A *broadcast* is the transmission of a message intended for all nodes on the network. Some networks allow nodes to be members of one or more logical groups, which have multicast addresses. A *multicast* is the transmission of a message intended for multiple nodes, which are all assigned to the same multicast group. A multicast is only received by nodes that are configured to be members of that particular multicast group. A given node may receive messages addressed specifically to its unique node address, to a broadcast address, or to a multicast group that it is a member of.

Sometimes a network message must be directed to a certain application program on a node, so a node-only address is not sufficient. In this case, a port number is added to identify the application destination. A *port* is a virtual data connection used by nodes to exchange data directly with certain application programs on other nodes. The most common ports are TCP and UDP ports, which are used to exchange data between computers on the Internet.

Ports are identified by a 16-bit number, which is often specified following the node address. For example, TCP port 80 is used to share hypertext transfer protocol (HTTP) information with web browsers. Some port numbers are officially assigned to certain applications, while others are available for any use. By using port numbers, messages for different applications on the same node can be received and used efficiently within the node.

Segmentation. There is a fixed maximum size for the payload in the data portion of a frame, though this varies between different technologies. Typically, the limit is in the 512 octet to 1500 octet range. However, some data communication applications, including building automation, require relatively large data payloads. Therefore, nodes that need to send more data than the limit allows must break the data stream into segments. At the destination, the segments are reassembled back into one large data unit.

Segmentation is a protocol mechanism that controls the orderly transmission of large data in small pieces. **See Figure 4-7.** Each segment must be marked with a sequence number so that the receiving end knows where each piece goes in the fully assembled data unit. This is important because pieces may arrive out of order.

In most cases, the sender wants assurance that all of the pieces have made it to the destination. Each piece that is successfully received can be individually acknowledged by the receiver. However, this approach is costly in terms of time and network bandwidth. Instead, most protocols allow groups of segments to be acknowledged all at once. This reduces the number of acknowledgments required but retains the efficiency of resending only failed segments.

Media Access

Every protocol must define a method for managing when nodes can transmit to each other. This determines how they access the medium of the network. If more than one node transmits at the same time, the messages can collide. A *collision* is the interaction of two messages on the same network media that can cause data corruption and errors. **See Figure 4-8.** As the traffic of messages on the network increases, the problem of collisions worsens. Several methods have been developed to resolve this problem.

Master/Slave. In master/slave networks, the master node controls the message traffic. Slave nodes transmit only when granted permission by the master. The failure of the master makes all communication impossible. Historically, this has been the most commonly used method of media access control in building automation systems.

Implementing this method can be risky because failure of a single node, the master, blocks communication between the remaining functional nodes. In response to this issue, media access methods have been developed for peer-to-peer networks, where each node has equal rights and responsibilities.

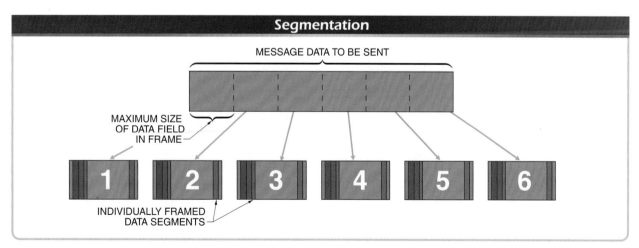

Figure 4-7. Segmentation breaks large messages into smaller pieces that are sent in individual message frames and reassembled at the receiving side.

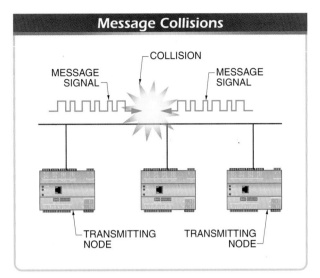

Figure 4-8. Message signals can collide on a network medium if two nodes transmit at the same time.

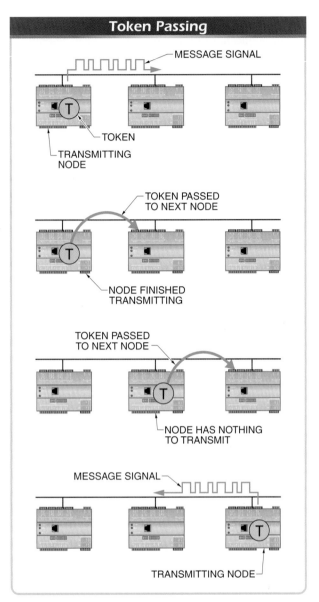

Figure 4-9. Token passing shares the right to transmit between cooperating nodes. The nodes pass the token throughout the network until it is received by a node that needs to transmit.

Contention. Contention is a media access control method that is used with peer-to-peer networks. The contention method allows a node access to the medium at any time, but the node must choose the best time to begin transmitting. In order to prevent collisions, the node listens for activity on the medium and waits for silence before starting to transmit.

However, two nodes may detect silence and then begin transmitting at the same time, causing collisions. There are various contention-based media access control schemes that try to address this possibility. Some can detect the collision and immediately stop transmitting. The node then waits a short time and attempts to transmit again.

Contention schemes degrade in performance as traffic increases. More traffic causes more collisions, retries, and transmission failures. Also, due to the random-length delays, one cannot predict exactly how long it will take a message to reach its destination. Because of this, the contention method is called nondeterministic.

Token Passing. The token-passing method requires nodes to receive a token message before being allowed to access the medium. **See Figure 4-9.** Each node voluntarily passes ownership of the token to the next node so that the medium is shared in a predictable manner. Nodes must wait until they receive the token before they can transmit. Once a node has the token, it may transmit before passing the token again.

The advantage of this orderly technique is that the performance of the network is guaranteed. Since the worst-case transit time for the token can be calculated, this kind of network is called deterministic. However, the disadvantage is that a portion of the network's available bandwidth is wasted on passing the token. Also, a node must wait for the token before transmitting, even when traffic is light. The token-passing method must also be capable of dealing with nodes entering and leaving the network, as well as instances when the token is lost.

OPEN SYSTEMS INTERCONNECTION MODEL

The International Organization for Standardization (ISO) has led an effort to develop a wide assortment of data communications standards. The intent of these standards is to foster the design and implementation of open computer systems that can be interconnected for a variety of applications. The first open-system standard approved by the ISO was the Basic Reference Model (ISO 7498), also known as the Open Systems Interconnection (OSI) Model. The *Open Systems Interconnection (OSI) Model* is a standard description of the various layers of data communication commonly used in computer-based networks. The purpose of this standard was to create a framework that could be used as the basis for defining standard communication protocols.

The concept behind the OSI Model is to divide the very complex problem of computer-to-computer communication into several smaller pieces. Each piece has a carefully defined function to perform and a well-defined interface through which it interacts with the other pieces. This arrangement sometimes even allows replacement of one piece by another, providing the same functions without changing the rest of the communication hardware and software. This is very similar to the way computer programs are divided into subroutines and functions.

The OSI Model arranges these functional pieces in a hierarchy of layers. **See Figure 4-10.** Each layer provides services to higher layers and relies on the services provided by lower layers. A *protocol stack,* also known as a protocol suite, is a combination of OSI layers and the specific protocols that perform the functions in each layer. Protocol stacks may include some or all of the OSI layers.

Each computer has its own protocol stack. Parallel layers in each node are called peer layers. Peer layers communicate with each other through their own matching protocol. In the sending computer, messages flow down one protocol stack, and each layer may add data to the message according to its specific protocol. **See Figure 4-11.** At each layer, the data portion of the message is made up of the original message plus the cumulative overhead from all of the higher layers. Although it is possible for the data itself to be transformed, for example through encryption or translation, most often a layer simply adds a new header and/or trailer before passing the message down to the next layer. The final message is then transmitted across the physical medium to the receiving computer, where it then passes up its protocol stack. Each layer strips out the portion of the message data that is meant for it and passes the remainder on.

Physical Layers

A physical layer is the lowest layer in the OSI Model. The *physical layer* is the OSI Model layer that provides for signaling (the transmission of a stream of bits) over a communication channel. No frame headers or trailers are added to the data by the physical layer protocol. The protocols for the physical layer define the signaling and wiring rules for communications, such as how a binary 0 or 1 is to be represented. For example, a voltage above a certain level represents a 1 and a voltage below another level represents a 0. Physical layer protocols also define the types of physical media to be used to transmit the data (such as twisted-pair conductors, coaxial cable, optical fiber, or infrared transmission), the connectors to be used, and the allowable network arrangements.

Open Systems Interconnection (OSI) Model		
Layer	**Type**	**Function**
Application	End-to-end communication	Interfaces with user's application program
Presentation	End-to-end communication	Converts codes, encrypts/decrypts, and reorganizes data
Session	End-to-end communication	Manages dialog and synchronizes data transfers
Transport		Provides error detection and correction, segmentation, and reliable end-to-end transmission
Network	Point-to-point communication	Manages logical addressing and determines routing between nodes
Data Link	Point-to-point communication	Manages physical addressing and orderly access to physical transmission medium
Physical		Transmits and receives individual bits on physical medium

Figure 4-10. The OSI Model breaks the complex procedure of data communications between network nodes into seven different layers, each with specific responsibilities.

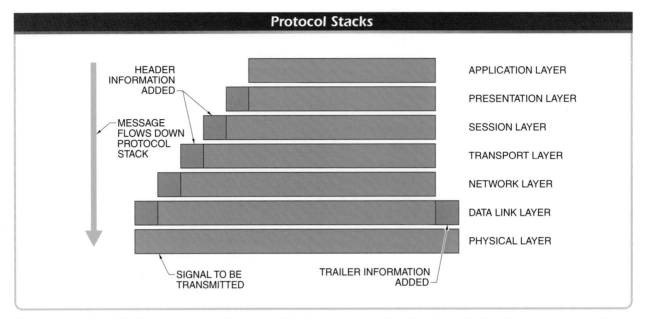

Figure 4-11. Each OSI layer in a protocol stack adds information, usually in the form of a header, to a message before passing it down to the next layer.

Data Link Layers

The *data link layer* is the OSI Model layer that provides the rules for accessing the communication medium, uniquely identifying (addressing) each node, and detecting errors produced by electrical noise or other problems. Some data link protocols also provide a way to correct transmission errors.

MAC Layers. In the OSI Model, the physical and data link layer protocols form a special combination known as the MAC layer. The *MAC layer* is a sublayer of the OSI Model that combines functions of the physical and data link layers to provide a complete interface to the communications medium. This interface presents each message as a collection of bits that is destined for a particular node, which is identified by its MAC address. **See Figure 4-12.** A *MAC address* is a node address that is based on the addressing scheme of the associated data link layer protocol. Incoming messages received by a MAC layer are similarly presented to other layers as a collection of bits that came from a particular MAC address.

The MAC layer structures the data into the appropriate form suitable for the particular underlying framing and signaling technology, gains access to the signaling medium, and then transmits the message (or receives incoming ones). There are many kinds of MAC layer technologies, each using very different schemes for organizing data into frames, detecting and correcting errors, arbitrating media access, and signaling.

Local Area Networks. A *local area network (LAN)* is the infrastructure for data communication within a limited geographic region, such as a building or a portion of a building. **See Figure 4-13.** A LAN uses a particular MAC layer type. A LAN does not specify anything about the content of the messages or the way in which the messages are encoded. Higher-layer protocols provide these specifications. A single LAN can carry messages from several incompatible higher-layer protocols simultaneously. For example, TCP/IP, NetBEUI, and IPX/SPX messages can coexist on the same LAN. In effect, a LAN is a collection of computers using the same MAC layer on a common network.

Network Layers

The *network layer* is the OSI Model layer that provides for the interconnection of multiple LAN types (MAC layers) into a larger network. The LANs may be distinct because of differences in the LAN protocols or because there is a need to segregate address spaces. An *address space* is the logical collection of all possible LAN addresses for a given MAC layer type. For example, an 8-bit MAC address allows for a maximum of 256 possible addresses, and a 48-bit address allows for more than 280 trillion addresses in the address space.

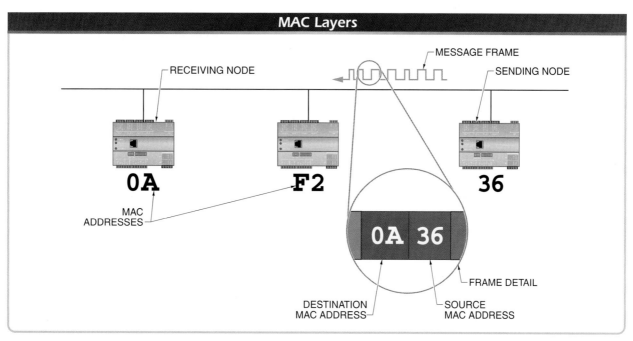

Figure 4-12. The MAC layer is an interface to the communication medium and handles addressing.

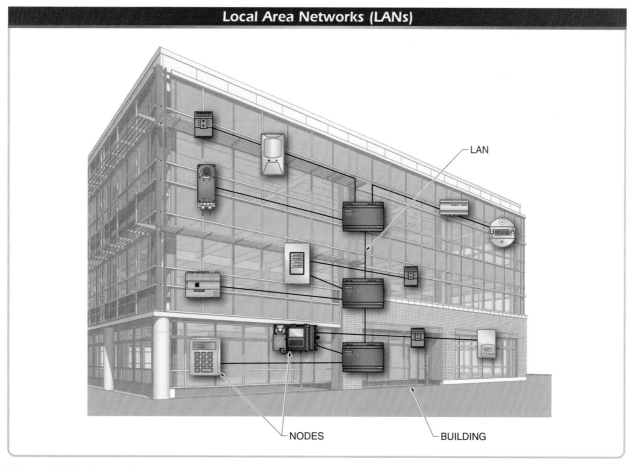

Figure 4-13. LANs use the same MAC layer technologies and are limited to a small geographic area, such as a building.

Some internetworks allow for multiple paths between computers on different LANs. The network layer protocol is primarily responsible for managing the delivery of messages through different possible routes. *Routing* is the process of determining the path between LANs that is required to deliver a message. **See Figure 4-14.** If there are multiple route choices, the network layer protocol may decide on a route based on certain criteria such as the cost, reliability, and speed of the routes. Sometimes network layer protocols provide message segmentation, in which case they are also responsible for message reassembly.

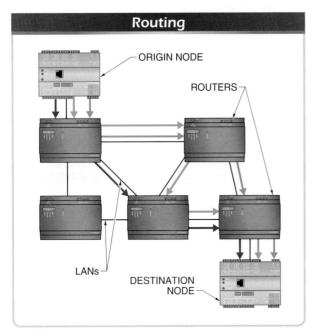

Figure 4-14. If network connections allow more than one route between the sending and receiving nodes, the network layer handles the routing choices.

Because nodes that perform routing functions have limited resources, some routes may become congested if there is a large amount of message traffic. Under these conditions, the network layer protocol may also be responsible for flow control and coordination with other routing nodes to manage and alleviate congestion.

Transport Layers

The transport layer is the first of the upper layers of the protocol stack. The *transport layer* is the OSI Model layer that manages the end-to-end delivery of messages across multiple LAN types. The lower layers (the physical, data link, and network layers) address only point-to-point protocol issues. This distinction is only relevant in internetworks, which transmit messages across multiple LANs.

The transport layer provides end-to-end error detection and correction, which may rely on network layer services. The transport layer protocol may also provide segmentation and reassembly of long messages and a variety of levels of quality-of-service (QOS). QOS may be characterized in terms of throughput, transmit delay, residual error rate, and failure probabilities. The user of the transport service is guaranteed a particular QOS that is independent of any changes in the underlying network service or its quality.

Session Layers

The *session layer* is the OSI Model layer that provides mechanisms to manage a long series of messages that constitute a dialog. **See Figure 4-15.** The transfer of large files is an example of a dialog. The session layer organizes the exchange of data into a series of dialog units. Checkpoints between dialog units provide the ability to resynchronize the communication at intermediate points after a communication failure rather than at the start of the transaction.

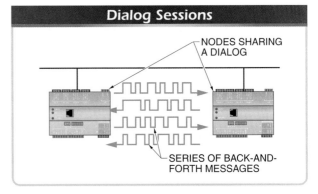

Figure 4-15. The session layer protocol is responsible for back-and-forth communications between two nodes, known as dialogs.

Presentation Layers

The *presentation layer* is the OSI Model layer that provides transformation of the syntax of the data exchanged between application layer entities. The presentation layer protocol permits each local application to represent information using its own syntax but still share information with other applications. The protocol converts this local syntax into one of many possible transfer syntaxes. The appropriate transfer syntax to use is negotiated by the peer-presentation entities. This ensures that the data exchanged can be interpreted appropriately by the two application layer entities. Data compression and encryption are also typically done at the presentation layer.

Application Layers

The application layer is the endpoint of the OSI Model. The *application layer* is the OSI Model layer that provides communication services between application programs. A specific type of application program interfaces with the application layer in order to participate in communications between itself and peer application programs in other nodes. There are many possible applications for computer-to-computer communication, so the details of a particular application are not part of the OSI Model. One can think of the application layer as a choice between different possible applications at a given endpoint, similar to a telephone extension at a company telephone number.

There are many well-known applications as well as unique or proprietary applications that fit the OSI Model. An application protocol is a specific kind of communication, with its own rules, that communicates between peer entities at the application layer. For example, an Internet browser communicates to a website using the application protocol hypertext transfer protocol (HTTP).

The strengths of the OSI Model are its generality and adaptability. It can be applied to almost any data communication application between computers located anywhere in the world. However, it provides for a rich set of functionality that may not be necessary for every application. Since each function and each layer protocol adds size to the message, which reduces the efficiency of the system, it makes sense to consider adapting the model to a particular application by eliminating unnecessary functionality. *Collapsed architecture* is a protocol stack that excludes layers that are not needed for the application. In most collapsed architectures, some of the functionality of the missing layers may be provided by protocols in the other layers.

A good example of a collapsed architecture is the BACnet protocol for building automation and control networks. **See Figure 4-16.** The few aspects of the presentation, session, and transport layers that are needed by BACnet nodes are instead implemented as part of the application layer.

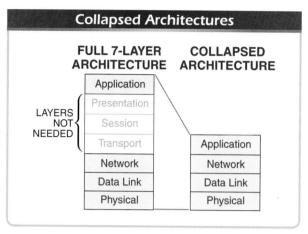

Figure 4-16. If a protocol does not need to implement each layer of the OSI Model, its collapsed architecture eliminates the unneeded layers.

NETWORK ARCHITECTURES

In the simplest network, two or more nodes share a segment. A *segment* is a portion of a network in which all of the nodes share common wiring. **See Figure 4-17.** Segments are usually limited to certain lengths, depending on the characteristics of the particular physical media type. *Network architecture* is the physical design of a communication network, including the network devices and how they connect segments together to form more complex networks.

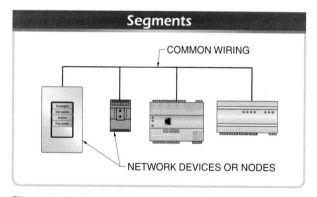

Figure 4-17. A segment is a section of network wiring that is continuous at the physical layer.

Network Devices

Network devices can be used to extend segments, connect segments together, and transition between different types of physical media, such as from twisted-pair to optical fiber. A *logical segment* is a combination of multiple segments that are joined together with network devices that do not change the fundamental behavior of a LAN.

Repeaters. Physical layer protocols often place restrictions on the length of an individual segment. A *repeater* is a network device that amplifies and repeats electrical signals and provides a simple way to extend the length of a segment. **See Figure 4-18.** It does not understand addresses and cannot filter or route traffic. A repeater forwards all traffic, regardless of validity. This means that corrupted packets or noise may also be repeated. There is a slight delay, or latency, from the time the signal arrives to when it is repeated. Since these delays are cumulative, there may be a limit to the number of repeaters that can be used to extend the segment. Repeaters operate at the physical layer of the OSI Model.

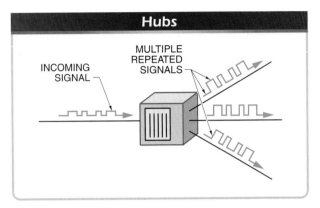

Figure 4-19. Hubs are multiport repeaters that amplify and repeat all signals they receive.

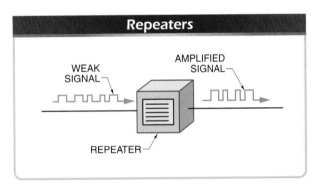

Figure 4-18. A repeater amplifies and repeats message signals at the physical layer.

Hubs. A *hub* is a multiport repeater that operates at the physical layer of an OSI Model and repeats messages from one port onto all of its other ports. When a hub detects a signal on one of its ports, it retransmits the signal through all the other ports of the hub. **See Figure 4-19.** Like repeaters, there may be practical limits on the number of hubs that can be used along any given path because of the cumulative latency. Physical constraints on the electrical characteristics of the signaling technology may also limit the maximum distance of wiring from a node to a hub, or from hub to hub, thus further limiting the total length of a network.

Bridges. A *bridge* is a network device that joins two LANs at the data link and physical layers. **See Figure 4-20.** When two LANs are bridged, the address of each node must remain unique in the extended network. Bridges provide traffic filtering based on destination address. The locations of nodes on the network are learned as the bridge receives messages on its ports.

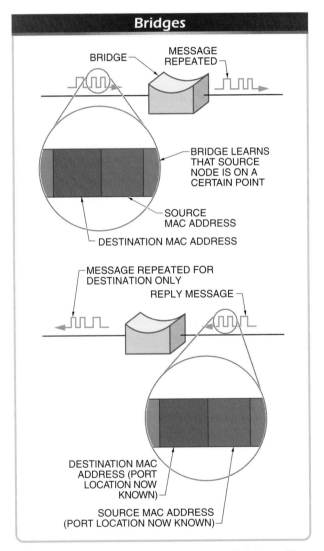

Figure 4-20. Bridges operate at the data link layer. They operate as hubs until they learn the locations of nodes by reading the source addresses. They then forward messages only to the port associated with the known address.

A bridge can be used to extend the length of a LAN as a traffic management device or to connect two different topologies or physical media. For example, a bridge can connect a LAN of standard (thick) Ethernet (10BASE5) with a LAN of fiber-optic Ethernet (10BASE-FL).

The placement of a bridge within a network is very similar to that of a repeater. The key difference is that a bridge understands data link layer addressing and can use that knowledge to provide filtering, while a repeater simply amplifies and retransmits the electrical signals.

Switches. A *switch* is a multiport bridge that can forward messages selectively to one of its other ports based on the destination address. A switch is designed for a particular kind of message frame. It can read addresses from the frames and learn which addresses belong to which ports. Once learned, the switch can immediately repeat the message to the correct port. **See Figure 4-21.** If the destination node's port is unknown, or the destination is a broadcast address, it can repeat the message on all ports.

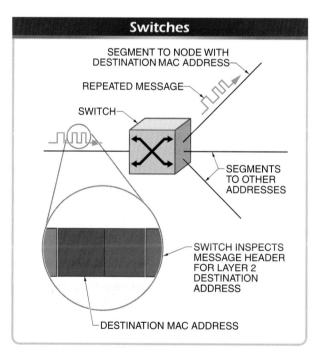

Figure 4-21. Switches forward messages selectively based on addresses in the data link layer.

With contention-based MAC layers, switches allow simultaneous communication between multiple port pairs without collisions. Also, since the switch has a dedicated point-to-point connection to each node, it can simultaneously transmit and receive to that node.

Because signal-timing constraints are generally imposed on signals rather than frames, this can help to mitigate constraints on timing and distance and significantly reduce collisions even in high-traffic situations.

Facts

The concepts and rules of data communication are common to all computer-based networks, including office LANs of personal computers and building automation networks of intelligent controllers.

Routers. Two or more LANs are connected together into a single unified network with a router. A *router* is a network device that joins two or more LANs together at the network layer and manages the transmission of messages between them. The LANs may have fundamentally different physical, electrical, signaling, or behavior characteristics. The LANs may also be similar, requiring the router only to fill a logical requirement to separate the network segments. LANs joined with routers form an internetwork. **See Figure 4-22.** An *internetwork* is a network that involves the interaction between LANs through routers at the network layer. The most famous example of an internetwork is the Internet.

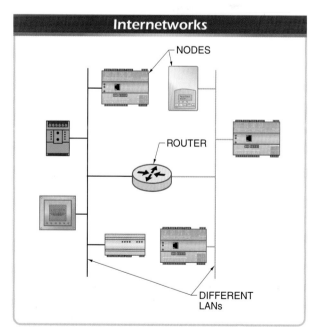

Figure 4-22. Routers connect LANs together into internetworks and provide routing, segmentation, reassembly, and flow control.

A router continuously listens for messages on all of its ports and isolates network traffic by selectively forwarding network messages based on the destination address. When the router receives a message on one LAN segment, it inspects the message for the destination address and completeness. If the message is destined for a node on another LAN segment, it repeats the message on the port for the proper LAN segment. **See Figure 4-23.** If the message is destined for a node within the same LAN segment, the router does not repeat it. Routers also validate messages and do not forward them if they are corrupted. All of these functions help reduce excess communication traffic that can negatively impact network performance.

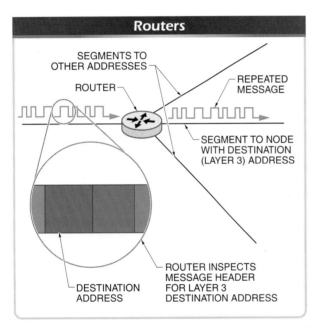

Figure 4-23. Routers forward messages selectively based on addresses in the network layer, provide transitions between LANs, and manage traffic between groups of nodes.

The router does not need to understand, or even look at, the content of a message. It only needs to know the source and destination of each message. This is analogous to the post office delivering a letter while knowing only the destination and return addresses printed on the envelope. The logical segments that represent each LAN each have a separate address, much like the town or city in a mailing address.

Messages between nodes on an internetwork may have many possible paths. The network protocol determines the best route for the message to take. This may involve multiple hops across routers from one LAN to the next, but the network layer usually only needs to know how to direct messages to the nearest intermediate destination. A distinct address for each network provides for the possibility that two nodes have the same MAC address on their respective networks.

Configured routers are the most common type of router. These forward only valid packets to segments defined in the internal routing tables. Routing tables are updated by the network management tool when segments are added to the network design. Alternatively, learning routers dynamically update their internal routing tables based upon the source addresses of incoming message packets. When learning routers are reset, such as during a power cycle, they must relearn the subnet locations, which may cause an initial flood of network traffic.

A *firewall* is a router-type device that allows or blocks the passage of packets depending on a set of rules for restricting access. Modern firewalls include stateful handling, which recognizes the progression of states in a typical TCP conversation and uses the state as part of the rules for determining legitimate conversations between nodes inside and outside of the firewall.

Gateways. The primary goal of peer-to-peer networks is for each node to be able to communicate effectively with its peers, which typically involves nodes that use the same application protocol. However, there are many situations in which more than one protocol is implemented within the same network. This is very common in building automation systems.

A gateway can be used to integrate two or more application protocols that are being used to perform similar functions into one communication system. A *gateway* is a network device that translates transmitted information between different protocols. **See Figure 4-24.**

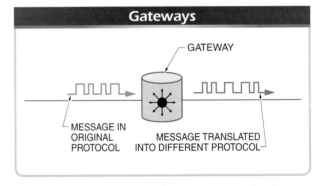

Figure 4-24. Gateways translate between application protocols that have similar functions.

A gateway is similar to a router in that its job is to connect to two or more LANs and manage the necessary communication between them. However, a gateway has an additional and much more complex task. It cannot simply repeat the message on the other LAN because the application language or message content must be translated conceptually from one protocol into the other. The gateway has two independent protocol stacks and possibly even physically separate and different types of physical media. The gateway makes a logical connection between peer entities at the application layer. **See Figure 4-25.**

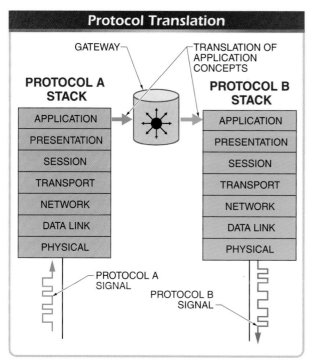

Figure 4-25. Translating between information concepts in two different protocols requires a complete protocol stack for each protocol.

The two application protocols must have common features between them so that it is possible for the gateway to integrate them together. This usually involves the logical association between some concept in one protocol and a similar or equivalent concept in another protocol. *Mapping* is the process of making an association between comparable concepts in a gateway. Some protocols are rich in functionality while others are simpler. When creating gateways between protocols, the protocol designer may sometimes be forced to simulate functionality or to emulate the functions of a more sophisticated device in order to have successful integration between protocols.

Because there are so many possibilities, the OSI Model falls short where gateways are concerned and does not define or model how they might work. As a result, most gateways are very specific to particular application protocols that share common concepts.

Network Topologies

Nodes can be connected together in many different ways. In some cases, the signaling method affects the methods, locations, and shape of the connections. *Topology* is the shape of the wiring structure of a communications network. There are four primary physical topologies used in network technologies: bus, star, ring, and mesh. In some instances, technologies can use a mixture of these topologies by using network devices to connect them together.

Bus Topologies. A *bus topology* is a linear arrangement of networked nodes with two specific endpoints. **See Figure 4-26.** This topology is often depicted as a line connected to nodes with short wiring stubs, or drops. Since most signaling methods use at least two conductors, this topology can also be drawn as a ladder shape.

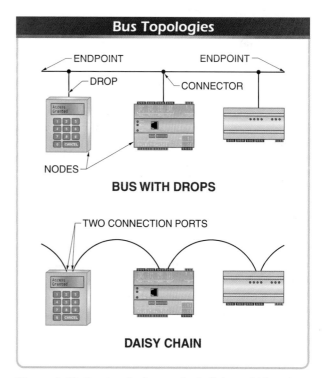

Figure 4-26. Bus topologies can be built by stubbing short connections to nodes off a continuous bus or by daisy chaining the nodes together.

The extra conductor lengths and connectors required for this arrangement can cause unwanted changes in electrical impedance, so this wiring method is rarely practical. A more common configuration of bus topology is a daisy chain. A *daisy chain* is a wiring implementation of bus topology that connects each node to its neighbor on either side. Nodes that can be used in bus topology typically include two connection ports or sets of terminals to facilitate daisy chaining.

Installing a bus topology network is relatively simple and it is easy to troubleshoot because segments can be easily isolated in order to locate failures. However, a disadvantage to bus topology is that it is sensitive to node failures and wiring problems. A single short or open connection along the bus can disable some or all of the other nodes. Also, the expansion of a bus-topology segment can be challenging because the daisy-chain configuration must be opened to add new nodes.

Star Topologies. A *star topology* is a radial arrangement of networked nodes. At the center of a star topology is a hub that includes multiple ports with which to connect to each node. **See Figure 4-27.** Each node has a dedicated connection to a hub port, which overcomes a principal limitation of bus topologies by providing isolation of individual nodes from others if a node or wiring failure occurs. The hubs can also be thought of as an inverted tree structure.

Hubs can be connected to other hubs to form a collection of stars. The hubs may be daisy chained in a bus structure, though this arrangement is vulnerable to the same failure issues as other bus networks. Alternatively, it is quite common to connect hubs in hierarchies. Not only can less expensive twisted-pair wiring be used, but an individual failure is isolated to its segment (and any segments below it).

Ring Topologies. A *ring topology* is a closed-loop arrangement of networked nodes. The network forms a loop, with each node having access to both ends of a segment. **See Figure 4-28.** Some ring networks may also use hubs along the ring, similar to star topology. These hubs facilitate connections to other rings. The principle advantage to the ring topology is that the failure of any one node does not compromise the rest of the nodes on the ring because there are always two paths to each node.

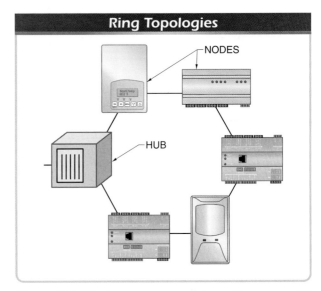

Figure 4-28. Ring topologies are versions of buses in which the ends are both connected to a hub.

Rings are like bus topologies that have both ends connected together or to a hub. Therefore, like buses, the individual nodes can be connected to the ring through either drops or being daisy chained.

Mesh Topologies. Mesh topologies are primarily utilized in wireless networks. Each wireless transmitter has a transmission range defined by a circular limit around the node. The nodes in a wireless network are all part of the same radio frequency grouping and can potentially

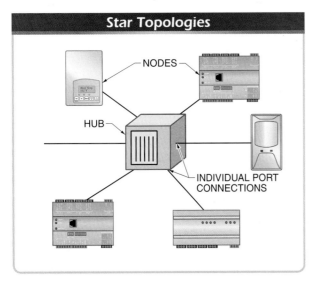

Figure 4-27. Star topologies use hubs to isolate segments and provide each node with a dedicated port.

all receive each other's transmitted messages. In practice, though, this would require relatively high-power transmissions for the farthest nodes to communicate with each other. Instead, the mesh topology takes advantage of the fact that many of the nodes are physically close to each other and the overlapping circles of adjacent nodes' transmission ranges form a mesh. A *mesh topology* is an interconnected arrangement of networked nodes. **See Figure 4-29.** Any radio transmissions of a node must only have the range to reach the nearest other nodes. Each node is connected to multiple other nodes, and messages can travel in a variety of paths to reach destinations.

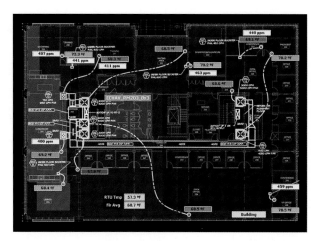

Information from multiple networks can be represented in a single graphical interface.

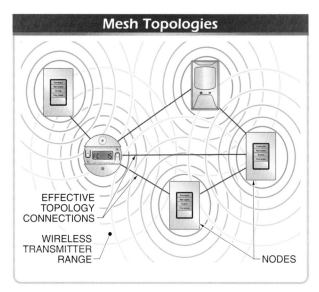

Figure 4-29. The overlapping transmission ranges of wireless nodes allow the nodes to form a mesh topology, which allows multiple connections between most nodes and their neighbors.

The strength of a mesh network is that each node can operate as a simple repeater. If a node receives a message that is not for itself, it passes the message on to its neighbor. Through multiple repeats, the message reaches its destination. Because multiple nodes participate in the meshing, the topology has a built-in redundancy that provides greater reliability and self-healing properties to the network in the event of node failures. Wireless nodes can also easily establish communication links with nearby nodes. In this way, the range of the network is easily extended and the power requirements for each node are reduced.

Free Topologies. A *free topology* is an arrangement of networked nodes that does not require a specific structure and may include any combination of buses, stars, rings, and meshes. **See Figure 4-30.** A common implementation of free topology connects a variety of separate topologies together into a bus configuration. The bus portion is typically designed as a high-speed backbone network that can quickly route messages to individual network segments.

Free topology offers flexibility during installation, as there is no defined structure to the segment wiring. In addition, future expansion of the network is as simple as adding new nodes anywhere along any segment. However, troubleshooting physical wiring issues with a free topology network can be difficult, especially if there is no wiring documentation. Also, since free topology networks have no defined structure, installers can easily exceed a wire-length limitation for the signaling method.

Termination

Electrical signals can reflect from the ends of conductors and travel back along the length of a network segment, which can cause packet collisions and corrupted data. Segment termination is used to avoid this problem. A *terminator* is a resistor-capacitor circuit connected at one or more points on a communication network to absorb signals, avoiding signal reflections. The required values of the resistors and capacitors depend upon segment type and topology.

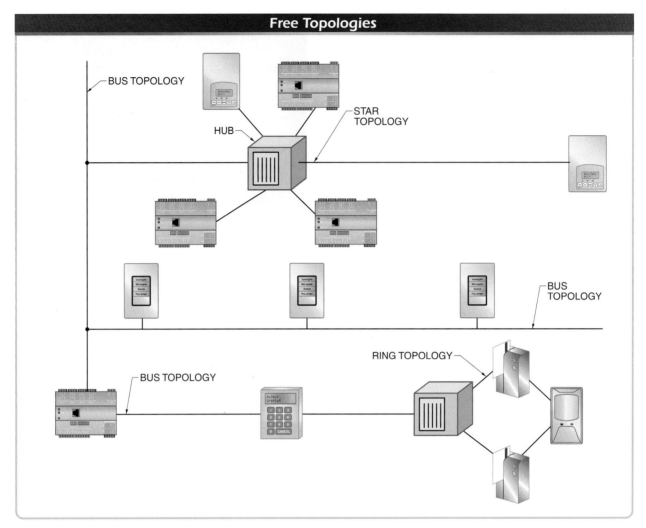

Figure 4-30. Free topology is any mixture of bus, star, ring, and mesh topologies.

Network Tools

The programming and management of a communication network requires software network tools. A *network tool,* also known as a network management tool, is a software application that runs on a computer connected to a network and is used to make changes to the operation of the nodes on a network.

This software is used to define information-sharing relationships, change node communication settings, analyze problems, and perform other management functions. The software saves a representation of the network and its operation on the local computer and loads the necessary settings onto the nodes over the network when changes are made. Once the network is operating, the network tool is not required until changes are needed.

MEDIA TYPES

There are many common types of media used to convey signals for computer-based networks. Media types represent the physical layer characteristics of a particular method of signaling and those physical media (cabling) that may be used with them. These fall primarily into three categories: copper conductors, fiber optics, and radio frequency. Generally, a particular medium is used because it has one or more desirable characteristics such as noise immunity, low power, distance, cost, reliability, or speed.

Copper Conductors

By far, the most common physical media type for computer-based networks, including building automation networks, are copper conductors. The most common types of copper cabling are twisted-pair cable and coaxial cable.

Twisted-pair cable is a multiple-conductor cable in which pairs of individually insulated conductors are twisted together. **See Figure 4-31.** The twisting keeps the conductors a uniform distance apart, which enhances the impedance characteristics and improves noise immunity. Twisted pairs may be shielded with a foil or braided metal wrap around each pair or around the whole group. Twisted-pair cables with different numbers of conductor pairs or types of shielding are readily available. The number of conductors and shielding requirements are determined by the type of signaling to be conveyed with the cable. Category 5 (CAT5) cable, named for its performance specification, is a common twisted-pair cable type that provides four twisted pairs within a common cable jacket.

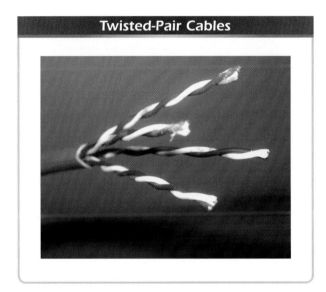

Figure 4-31. Twisted-pair cables are the most common type of copper conductors used for building automation networking.

Coaxial cable is a two-conductor cable in which one conductor runs along the central axis of the cable and the second conductor is formed by a braided wrap. The two conductors are separated by a plastic or foam insulating material. The two conductors share the same central axis, making them "co-axial." This type of cable provides uniform impedance and good capacitance characteristics.

TIA-232. The TIA-232, originally RS-232 and EIA-232, signaling standard defines the electrical characteristics of a type of signaling circuit to be used for computer-to-computer communication based on serially transmitted sequences of data bits. The standard does not define frame structure, only physical characteristics like electrical voltage levels, timing, the slew-rate of signals, and capacitance.

The signals themselves are +15 VDC and –15 VDC, relative to a common ground, although the standard requires receivers to distinguish signals as low as 3 VDC. A negative voltage of –3 VDC to –15 VDC is a logical 1 and is defined as marking or mark level. A positive voltage of 3 VDC to 15 VDC is a logical 0 and is defined as spacing or space level.

The standard defines 20 different signals, but only 9 of them are commonly used. Each uses a different pin in the connector. **See Figure 4-32.** The signals fall into two groups: data and status. The Transmitted Data and Received Data signals use serially transmitted sequences of logical 1/0 bits of data in various framing schemes. The status signals each have logical true/false levels.

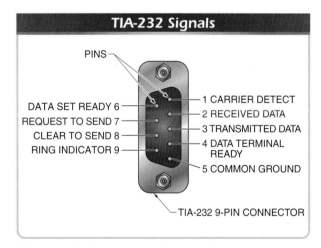

Figure 4-32. TIA-232 signals typically use a 9-pin connector, which carries the standardized data and modem control signals.

In practice, the modem control signals are often not required or used. A valid TIA-232 connection may use only the Transmitted Data, Received Data, and Common Ground connections. Because of the DC voltage level, the distance is also limited by the aggregate voltage drop caused by wire impedance. TIA-232 signaling is typically only used for direct point-to-point connections between two computers.

TIA-485. Like TIA-232, TIA-485 signaling is used to serially transmit sequences of data bits. The standard defines the physical characteristics of the signaling, such as electrical voltage levels, timing, slew-rate of signals, and capacitance. However, TIA-485 signaling is based on a differential voltage across two conductors relative to a common ground connection, which transmits the binary data. One polarity represents a logic 1 level and the opposite polarity represents logic 0. **See Figure 4-33.** Because the signal is differential, it is much more immune to noise than simple voltage-based signaling like TIA-232.

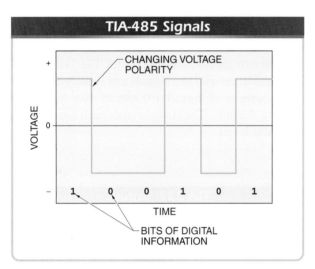

Figure 4-33. TIA-485 signals use two wires (and no ground) to transmit data based on the voltage polarity between the two wires.

The voltage difference must be at least 200 mVDC, but any voltages between +12 VDC and –7 VDC will accurately transmit bits. Although TIA-485 can be used at very high transmission speeds, it is most often applied in automation systems at speeds of 156 kbit/s or lower. Below 100 kbit/s, the signals can be transmitted for a distance of up to 1200 m (about 4000′).

Only one node may transmit at a time. Each node's transceiver contains both a differential driver and a receiver connected to the same two wires. When a node needs to transmit, it enables the appropriate pin, causing the driver to begin transmitting.

In automation systems, TIA-485 signaling is often used as a two-wire, half-duplex, multipoint serial connection. TIA-485 circuits use twisted-pair conductors, with or without a shield conductor. TIA-485 circuits are always arranged in a daisy-chain bus topology. No connector type is specified by the standard.

TP/FT. The *TP/FT* is signaling technology that is only used with LonTalk devices where the type of signaling is a differential Manchester encoded signal for serially transmitted data. **See Figure 4-34.** Differential Manchester encoding combines a synchronizing clock signal and a data signal into a self-clocking bit stream. This media allows free topology, though better performance is available for bus topologies. The data rate is 78 kbit/s for up to a maximum distance of 500 m (about 1640′) under a free topology. The distance is up to 2700 m (about 8858′) if a bus topology with double terminations is strictly enforced.

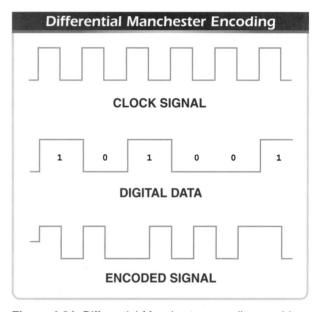

Figure 4-34. Differential Manchester encoding combines a synchronizing clock signal and a data signal into a self-clocking bit stream.

TP/XF. The *TP/XF* is a twisted-pair technology that is used only with LonTalk devices where the signal is a transformer-isolated differential Manchester encoded signal for serially transmitted data. The two forms of this type of technology include TP/XF-78 and TP/XF-1250. Both use twisted-pair wiring. The TP/XF-78 can transmit at 78 kbit/s up to a maximum distance of 1400 m (about 4593′). The TP/XF-1250 can transmit at 1250 kbit/s up to a maximum distance of 130 m (about 426′).

Fiber Optics

Fiber optics is a form of signaling based on light pulses to convey signals. A glass or plastic fiber acts like a pipe to convey light over great distances. **See Figure 4-35.** The ends of each fiber are polished and fitted with a standard connector.

Fiber Optics

Figure 4-35. Glass and plastic fibers can carry optical signals over long distances and are immune to many electrical problems.

Fiber optic cables are used in pairs: one for transmitting and one for receiving. A fiber optic transmitter converts electrical signals into light pulses using a light-emitting diode (LED) or laser diode. The receiver converts the light signals back into electrical signals with a photodiode. Fiber optics are always used point-to-point. Multipoint networks with fiber optics require bridging hubs that convert between the optical signals and electrical signals.

The characteristics of fiber-optic signaling allow for very high speeds, about 14 Tbit/s over 160 km of fiber. In automation systems, though, much slower speeds and shorter distances are typically used. Because the signaling is optical, fiber-optic transmissions are immune to electrical noise, transients, grounding, and lightning. This makes fiber-optic signaling a superior choice for signaling between buildings, especially over long distances.

Radio Frequency

Wireless signaling uses radio waves as the medium for conveying signals. *Radio frequency signaling* is communications technology that encodes data onto carrier waves in the radio frequency range. Signals are conveyed by modulating (changing) the characteristics of a radio wave carrier. A radio wave of a known frequency is used as a baseline, and then digital data is mixed with it to produce a new wave with the data encoded into it. **See Figure 4-36.**

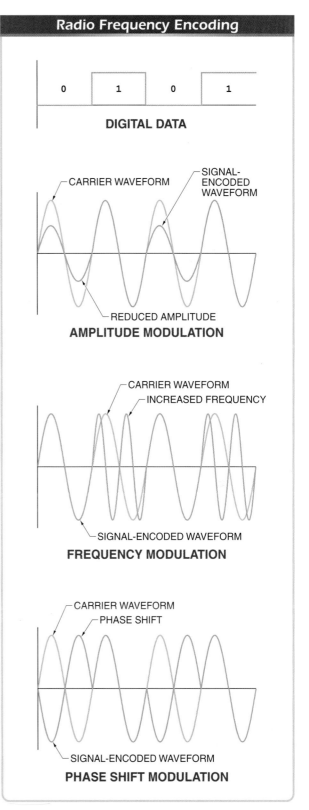

Figure 4-36. Modulating amplitude, frequency, or phase shift are three common methods of encoding digital data onto a radio frequency carrier waveform.

Although it is possible to modulate the amplitude of the carrier signal as a means of encoding data, this is subject to noise and reliability problems. Instead, frequency modulation is a more robust method of encoding data because it alters the carrier signal between two distinct frequencies. There are many other modulation schemes, such as phase shift modulation, that provide even better performance (speed and reliability) but are more complex to implement.

One of the main problems with all of the modulation schemes is that there are many sources of natural and human-made interference and radio noise. Many different radio frequencies are in constant or intermittent use that can interfere with communications. Spread spectrum techniques use multiple carrier frequencies to mitigate these problems. For example, frequency-hopping spread spectrum (FHSS) signaling rapidly switches between several different frequencies in a pseudorandom order known to the sender and receiver.

The bandwidth of wireless signaling is limited by the frequency of the carrier wave. Larger bandwidth requires very high frequencies, which require more power and expensive components. Also, walls and steel structures have an attenuating effect on signals and can restrict the physical range of most wireless technologies to 300′ or less unless repeaters are used to boost the signals.

Powerline

A common drawback to communications wiring is the need for dedicated conductor, either copper or fiber-optic cable. Also, wireless may not be feasible due to interference or bandwidth issues. Powerline technology, however, provides both power and data for each node by using the existing power wiring that is already installed in nearly every part of a building. *Powerline signaling* is a communications technology that encodes data onto the alternating current signals in existing power wiring.

Powerline signaling is similar to the way wireless technologies encode data onto radio frequency carrier waves. In fact, powerline signaling typically uses the same encoding techniques as radio frequency signaling, such as frequency or phase-shift modulation and frequency-hopping spread spectrum. Although there is a lot of development in the area of high bandwidth powerline carrier applications, automation systems mostly use relatively low-speed technology with data rates at or below 38 kbit/s.

Typically, powerline carrier transceivers incorporate both a power supply and data coupling circuit into the same AC connection. **See Figure 4-37.** The coupling isolates the electrical components from noise, transients, and other harmful effects, while passing modulated carrier wave signals through. Although modulation can be achieved through conventional discrete analog circuitry, most powerline carrier hardware uses digital signal processors because of the processors' flexibility and reliability.

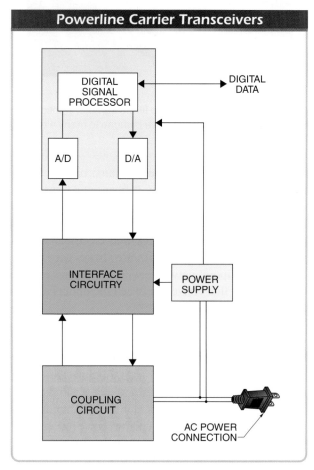

Figure 4-37. Powerline carrier transceivers include a coupling circuit that combines encoded digital data onto AC power lines.

COMMON BUILDING AUTOMATION MAC LAYERS

There are many different types of MAC layer technologies commonly used in building automation applications. Many of these are proprietary, meaning that they were developed by and possibly only used by a single company and its products. However, some technologies are published by the manufacturer as an open technology. There are also MAC layer technologies that are defined by national and international standards.

Standards-based technologies are increasingly popular, especially since they are more likely to interoperate successfully, support system growth and long-term maintainability, and enable the availability of third-party devices and tools. The larger marketplace also drives pricing downward and expands the value of features or services available for the same cost.

The MAC layers commonly used for building automation systems include Ethernet, ARCNET, MS/TP, and wireless technologies. These technologies are families of frame-based protocols that define the physical layer characteristics and signaling, as well as the data link and media access schemes.

Ethernet

Ethernet uses a contention-based scheme for controlling access to the medium that is called carrier-sense multiple access with collision detection (CSMA/CD). Nodes listen to the line for activity and begin transmitting if no traffic is sensed. When a collision is detected, each node stops transmitting and retries again after a short wait. Each node retries after a randomly determined delay, reducing the chance that they will collide again. If a collision occurs again, the node keeps attempting for up to 16 times, which may cause another collision. After the maximum number of tries, the node stops the attempts, so the message never reaches its destination. Therefore, some messages never reach their destination due to repeated collisions.

Ethernet data is transmitted at 10 Mbit/s, 100 Mbit/s, 1 Gbit/s, or 10 Gbit/s on a variety of twisted-pair, coaxial, and fiber-optic cabling. Ethernet is most commonly used in a star topology based on hubs or switches. Many modern hubs and switches can automatically sense the speed and adjust accordingly.

ARCNET

ARCNET uses token passing for media access, which makes it a deterministic network. The ARCNET technology is scalable to arbitrary speeds as high as 10 Mbit/s, though in practice, automation systems only use 2.5 Mbit/s or 156 kbit/s types. ARCNET may be transmitted on a variety of twisted-pair, coaxial, or

fiber-optic cabling and allows bus or star topologies, though this may depend on the media. At 2.5 Mbit/s, ARCNET requires the use of transformer-coupled transceivers, and distances of individual segments are limited. By lowering the speed to 156 kbit/s and using TIA-485 direct-coupled transceivers, it is possible to use longer segments up to 1000 m. Therefore, even though it is slower, the 156 kbit/s version is more commonly implemented.

Master-Slave/Token-Passing

The master-slave/token-passing (MS/TP) MAC layer is defined in the BACnet protocol standard. MS/TP is used when there is only one or a limited number of master nodes that participate in token passing and share the medium with slave nodes. Slave nodes do not participate in token passing and only answer when invited to transmit. The token recovery scheme allows new nodes to join the token passing circle and subcircles to form if the network is broken. Like ARCNET, MS/TP is a deterministic network.

MS/TP is based on TIA-485 signaling at 9600 bit/s, 19.2 kbit/s, 38.4 kbit/s, or 76.8 kbit/s. The standard specifies the use of shielded, twisted-pair cable with specific electrical characteristics. In an MS/TP daisy-chain bus topology, network termination should be installed at the two nodes at the ends of the segment.

Wireless Technologies

Wireless technologies include electromagnetic signals in the infrared, radio, and microwave frequencies, which are becoming practical options for building network infrastructure. Wireless building automation nodes are particularly useful in applications where physical wiring is not practical. They are also easier and faster to install and can be relocated if necessary. However, wireless systems may be less reliable due to range limitations or signal interference.

Wireless technologies often rely on mesh networks to route messages throughout the network. This allows wireless nodes to operate with less power, since each node needs only the signal range to reach its nearest neighbor. Although standards such as IEEE 802.15 and 802.11 cover wireless networking in various forms, there are no standards yet for applying these techniques as MAC layers for automation networking. An emerging consortium called the ZigBee Alliance is beginning to apply their adaptation of the 802.15.4 protocol to automation applications.

REVIEW QUESTIONS

1. Explain the difference between bandwidth and throughput.

2. What are the common fields included in message frames and what is their approximate order?

3. Why does the size of an address field affect the number of nodes allowed in a network?

4. Why is it important to control how nodes access the network media for sending messages?

5. How does the protocol stack affect the composition of the message frames that are transmitted on the network medium?

6. Why do some protocols use collapsed architectures?

7. How does a router filter network traffic?

8. How does a gateway translate between two different protocols?

9. What are the advantages and disadvantages of free topology?

10. What types of information does a signaling standard, such as TIA-232, typically include?

Chapter 4 Review and Resources

Electrical System Control Devices and Applications

Electricity is used for the communication between building automation system devices. Information and instructions are shared through low-voltage analog and digital signals, and control devices are energized or modulated with electrical power. The monitoring and control of the electrical system is required to protect the integrity of automation systems. This ensures the availability of adequate and reliable power, thus making electrical systems central to building automation.

ELECTRICAL SYSTEMS

An *electrical system* is a combination of electrical devices and components connected by conductors that distributes and controls the flow of electricity from its source to a point of use. *Electricity* is the energy resulting from the flow of electrons through a conductor. At the point of use, electricity is converted into some type of useful output, such as motion, light, heat, or sound. **See Figure 5-1.**

Electrical systems encompass the entire electrical infrastructure, from power plants to receptacles. However, a system is commonly divided into the sections that are involved with electricity generation, electricity distribution, and the electrical service and related systems within a facility. Building automation systems deal only with the portion of the system within their facility, though the operations are affected by the outside power sources.

Electricity Generation

Most electrical power is generated by converting energy from one form to another until it ultimately becomes useful energy in the form of electricity. Some processes include several steps. For example, coal-fired power plants change chemical energy (energy in the chemical bonds of coal) into heat energy through burning. **See Figure 5-2.** The heat energy is applied to water, which becomes steam, another form of heat energy. The steam is routed to turbines where it moves the blades, which then rotates a shaft, producing mechanical energy. The shaft drives a generator that uses the mechanical energy together with magnetic energy to produce electricity.

Steam is used to convert heat energy into mechanical energy for the generation of electricity and is used in many types of power plants. Nuclear power plants use nuclear reactions to convert atomic energy into heat energy, which creates electricity by way of steam. Other fossil fuel-powered processes, such as those that use natural gas and oil, also use steam for energy conversion.

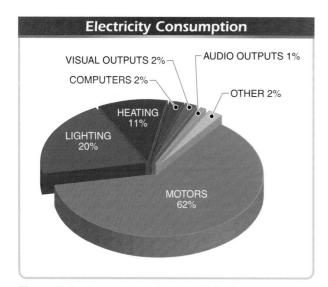

Figure 5-1. The majority of all electricity is consumed by lighting and electric motor loads.

Electricity Consumption

- VISUAL OUTPUTS 2%
- AUDIO OUTPUTS 1%
- COMPUTERS 2%
- OTHER 2%
- HEATING 11%
- LIGHTING 20%
- MOTORS 62%

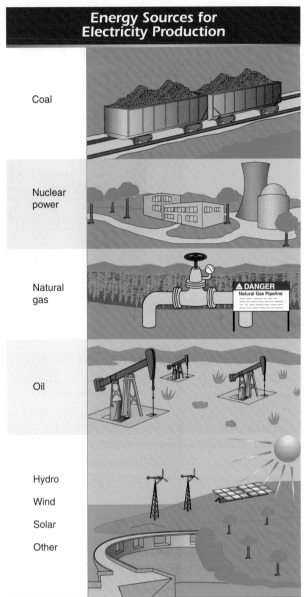

Energy Sources for Electricity Production

Coal

Nuclear power

Natural gas

⚠ **DANGER**
Natural Gas Pipeline

Oil

Hydro

Wind

Solar

Other

U.S. Energy Information Administration

Figure 5-2. Electricity can be produced from a variety of energy sources, although most processes include several intermediate steps.

Energy sources that begin with mechanical energy, however, do not include steam or heat energy in the conversion cycle. In hydroelectric plants, water pressure and flow drive turbine generators. Turbine generators can also be driven by wind, tidal power, or even the movement of fluids in solar thermal systems.

Direct energy conversion systems produce electrical power without steam or mechanical components. For example, fuel cells use electrochemical processes to convert hydrogen into electrical energy, and photovoltaic (PV) systems use semiconductor properties to produce power directly from sunlight.

Primary Power Sources

Most electricity is generated by large power plants that supply millions of consumers. These plants may use a variety of methods to generate electricity, such as coal burning or nuclear reactions, but are all large-scale operations. The power plants are operated by utility companies. A *utility* is a company that generates and/or distributes electricity to consumers in a certain region or state.

Utility-supplied electrical power is very reliable, inexpensive, widely available, and essentially maintenance free. For these reasons, it is the primary, and sometimes only, power source for the vast majority of buildings. The primary power source is the source relied on to provide most or all of a building's power needs. The utility provides a connection to their distribution system, monitors the electricity usage, and bills for the service.

It is possible for a building to use a nonutility power source as its primary power source, usually when the building's remote location prohibits connection to the utility system. This is extremely rare, however, especially for commercial and industrial facilities.

Secondary Power Sources

In addition to the utility-supplied electrical power, many buildings have a secondary source of power. This source is intended to supplement the primary electrical service or supply the building with power in the event of an interruption in power from the electric utility. Secondary power sources are often optional, implemented by the building owner by choice for business or comfort purposes, such as ensuring the continuity of computer systems or production lines. These systems are installed to reduce the potential financial losses caused by a loss of electrical power.

Some secondary power sources are legally required by a governing entity for life-safety purposes. These systems provide illumination and power to critical electrical loads, such as egress lighting and fire pumps, in the event of an outage of the primary power source. Back-up sources of power come from a variety of owner-supplied sources including engine generators, gas-turbine generators, photovoltaic systems, wind turbines, microhydroelectric turbines, and fuel cells.

Engine Generators. An engine generator uses an internal-combustion piston engine and a generator mounted together to produce electricity. The engine burns a compressed air-fuel mixture to provide the mechanical power, and the generator converts mechanical energy into electricity by means of electromagnetic induction.

The engines are typically fueled with gasoline, diesel, propane, or natural gas. Small engine generators typically run on gasoline because of its widespread availability. Larger engine generators can be designed for any of these fuels, though diesel is the most common. The type of fuel used affects the load response of the engine generator. Diesel engines are well suited for constant loads, whereas gasoline engines can respond quickly to changing loads. However, because diesel engines have better partial-load efficiency, run slower, are more robust, last longer, and require less maintenance, they are predominately used for large stationary applications.

Gas-Turbine Generators. A gas-turbine generator compresses and burns an air-fuel mixture, which expands and spins a turbine, converting fluid flow into rotating mechanical energy. The shaft power from the turbine is then used to drive a generator to produce electricity. Gas-turbine generators are essentially aircraft-type jet engines developed for stationary applications.

Gas-turbine generators have a better power-to-weight ratio than internal combustion engines, which means that they can generate more power from smaller, lighter equipment. The main disadvantages of gas-turbine generators are high cost and complexity. They typically run on natural gas, liquefied petroleum gas, diesel, or kerosene fuels. Gas-turbine generators are best suited for constant loads.

Photovoltaic Systems. A photovoltaic (PV) system produces electricity directly from the energy of sunlight. **See Figure 5-3.** Individual PV cells are made from ultrathin layers of semiconductor materials. When exposed to light, one layer releases electrons and the other absorbs electrons. The flow of electrons from one side of the cell to the other can be harnessed as DC electricity. The electrical output of each cell is very small, so they are grouped together to form larger electricity-producing units called modules. Modules are grouped together to form arrays.

PV arrays can be designed for practically any desired electricity output. These systems produce no waste products, require little maintenance, and include no moving parts. However, since sunlight alone is not a reliable energy source, PV systems either require storage batteries or can only be used for supplemental power.

Wind Turbines. A wind turbine harnesses wind power to produce electricity. The wind rotates blades, and the resulting mechanical energy drives a generator to produce electricity. Wind turbines are viable in areas with an average wind speed greater than 10 mph. Small wind turbines range from 1 kW to 20 kW peak output. Larger commercial wind turbines can produce several hundred kilowatts or even a few megawatts with rotor diameters up to a few hundred feet.

Microhydroelectric Turbines. A microhydroelectric turbine produces electricity from the flow and pressure of water derived from streams and rivers. Moving water acts against turbine blades, rotating a shaft that operates an electrical generator. The principles are the same as for large-scale hydroelectric plants, such as the Hoover Dam, but on a smaller scale. Microhydroelectric systems output less than 100 kW.

Facts

Most electric power is produced by alternating current (AC) generators that are driven by rotating prime movers.

Fuel Cells. A fuel cell uses a fuel and oxygen to produce DC electricity, with water and heat as by-products. Although similar to batteries, fuel cells are different in that they require a continual replenishment of the reactants (hydrogen and oxygen). The most common fuel is hydrogen. Other possible fuels include methane and methanol, which can be reformed to remove their hydrogen atoms for use in the fuel cell.

A typical fuel cell element consists of a cathode and anode separated by an electrolytic membrane material. As hydrogen gas flows across the anode, electrons are stripped from the hydrogen and flow through an external circuit, reentering the fuel cell at the cathode. At the same time, positively charged hydrogen ions migrate across the membrane to the cathode, where they combine with oxygen and the returning electrons to form water and heat.

Electricity Distribution

Electricity distribution is the transmission and delivery of electricity outside of consumers' facilities. For utility-supplied electricity, primary power generation is centralized to take advantage of economies of scale. Electricity is then distributed over a large area in a large-scale supply network. Secondary power sources, however, are typically located near consumer facilities and require very little distribution infrastructure.

Photovoltaic (PV) Systems

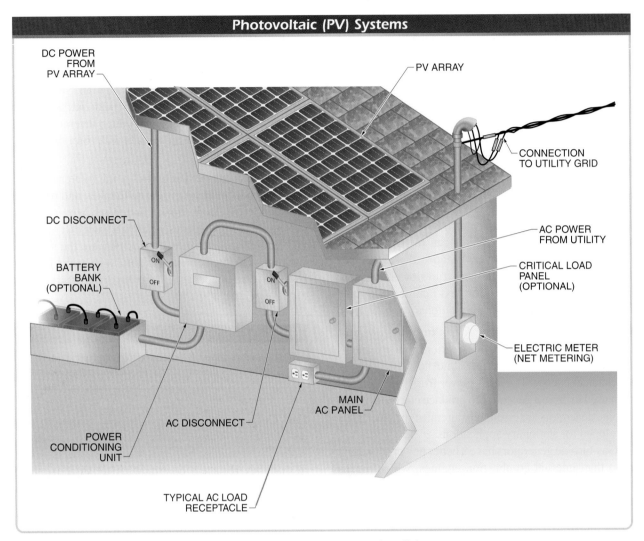

Figure 5-3. A PV system produces electricity directly from the energy of sunlight.

Centralized Generation. *Centralized generation* is an electrical distribution system in which electricity is distributed through a utility grid from a central generating station to millions of customers. By far, most electricity is generated and distributed in this way. A *grid* is the network of conductors, substations, and equipment of a utility that distributes electricity from a central generation point to consumers. **See Figure 5-4.** The grid fans out from the power plants to thousands of homes and businesses within a region. Electricity may travel hundreds of miles before it reaches the end user. Grids may be connected together at certain points so that consumers may still have power if an outage occurs in part of the distribution system. Though rare, outages do still occur, often due to storm damage or an overloaded system.

Distributed Generation. *Distributed generation* is an electrical distribution system in which many small power-generating systems create electrical power near the point of consumption. The electricity may travel only a few feet to the loads. Distributed generation systems include engine generators, PV systems, wind turbines, fuel cells, or other relatively small-scale power systems. **See Figure 5-5.** A distributed generation system may be either a primary or secondary power source, although secondary is by far more common.

> **Facts**
>
> *Voltage sags and swells may indicate a weak power distribution system. In such a system, voltage will change dramatically when a large motor is switched on or off.*

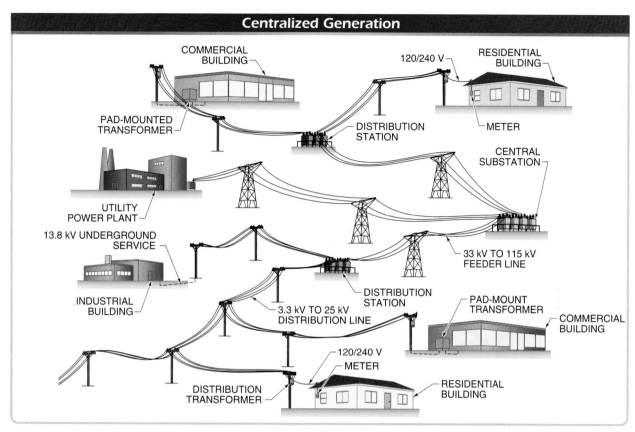

Figure 5-4. Centralized generation systems rely on a large power plant to produce the electricity for many consumers on an interconnected power grid.

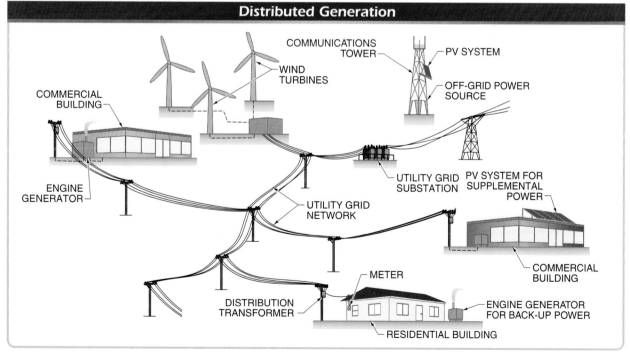

Figure 5-5. Distributed generation systems include many independent power sources to supply electricity close to where it is needed.

If the facility is connected to the utility grid, excess power from the secondary power source can be exported to the grid when it is not needed by the on-site loads. This makes the facility similar to a power plant in a centralized generation system. By metering the amount of exported electricity in addition to the amount of consumed electricity, the facility receives credit from the utility for the excess power added to the grid. This arrangement can reduce costs for the building owner and increase the utility's capacity to serve customers without building new power plants.

Electrical Service

Electrical service is the electrical power supply to a building or structure. **See Figure 5-6.** A smaller-scale power distribution system within a building delivers power from the electrical service to end-use points throughout the building, where it powers individual loads such as motors, lamps, and computers. This system includes switchboards, transformers, panelboards, switches, conductors, and outlets that work together to safely distribute the power throughout a building or to multiple buildings on a site.

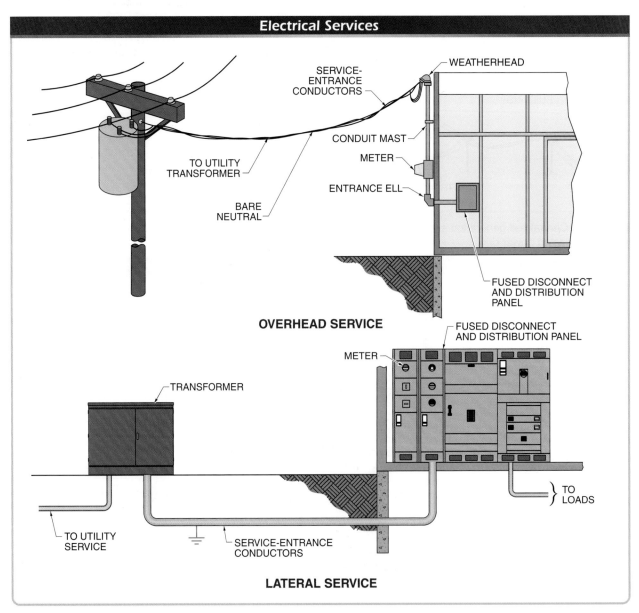

Figure 5-6. The electrical service is the point of connection between the electrical utility company and the electrical system of a building or structure.

Electrical systems must be designed, installed, and maintained in accordance with the National Electrical Code® (NEC®) and/or other applicable codes and regulations adopted by the local authority having jurisdiction (AHJ). All electrical devices must also be rated for both the voltage and current of the application for which they are installed.

An electrical service may be overhead or lateral. An *overhead service* is an electrical service in which service-entrance conductors are run from the utility pole through the air and to the building. A *lateral service* is an electrical service in which service-entrance conductors are run underground from the utility service to the building.

A 120/240 V, single-phase, three-wire service is used to supply power to end users that require 120 V and 240 V single-phase power. This level of service provides 120 V single-phase circuits, 240 V single-phase circuits, and 120/240 V single-phase circuits. **See Figure 5-7.** Because the neutral wire is grounded, it is not fused or switched at any point. A 120/240 V, single-phase, three-wire service is commonly used for interior wiring for lighting and small-appliance use. This service is used to supply most residential buildings and for small commercial applications, although a large power panel or additional panels can be used.

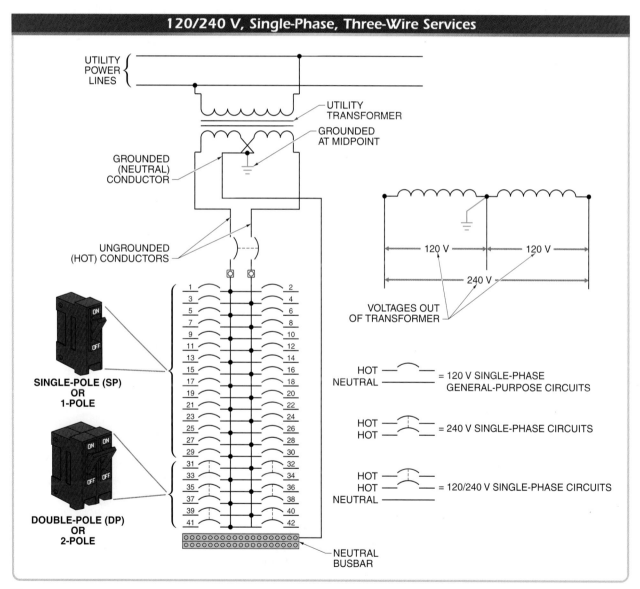

Figure 5-7. A 120/240 V, single-phase, three-wire service is used to supply power to customers that require 120 V and 240 V single-phase power.

A 120/208 V, three-phase, four-wire, wye-connected service is used to supply commercial customers that require large amounts of 120 V single-phase power, 208 V single-phase power, and low-voltage three-phase power. This service level includes three ungrounded (hot) lines and one grounded (neutral) line. Each hot line has 120 V to ground when connected to the neutral line. **See Figure 5-8.**

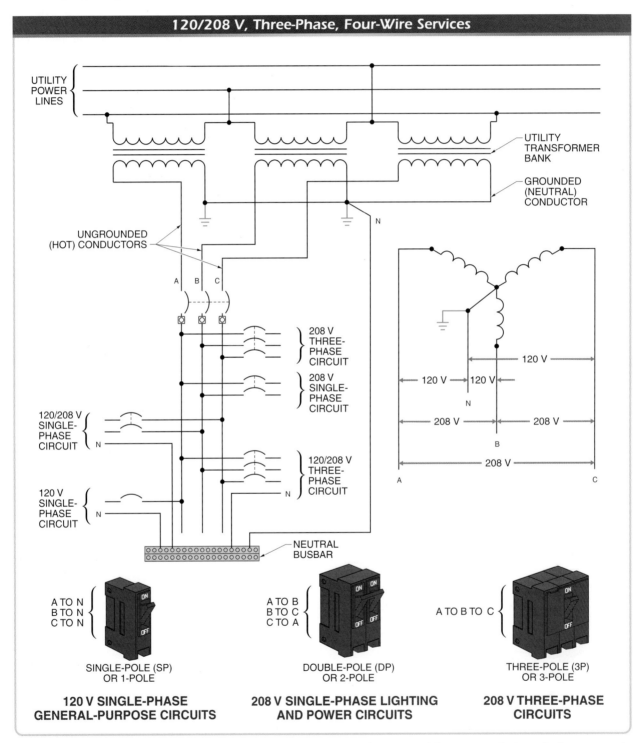

120/208 V, Three-Phase, Four-Wire Services

Figure 5-8. A 120/208 V, three-phase, four-wire service is used to supply commercial customers that require large amounts of 120 V single-phase power, 208 V single-phase power, and low-voltage three-phase power.

Switchboards. A large block of electrical power is delivered from a utility substation to a building at a switchboard, where it is broken down into smaller blocks for distribution throughout a building. **See Figure 5-9.** A *switchboard* is the last point on the power distribution system for the power company and the beginning of the power distribution system for the electrician of the property.

In addition to distributing the incoming power, a switchboard may contain equipment needed for controlling, monitoring, protecting, and recording the electrical use in a building. For example, the addition of motor starters and controls to the switchboard allows motors to be connected directly to the switchboard. This combination allows these high-current loads to be connected to the source of power without further power distribution.

Electricity is distributed out from the switchboard in multiple feeders. A feeder is the circuit conductors between a building's electrical supply source, such as a switchboard, and the final branch-circuit OCPD.

Switchboards

HEAVY-CAPACITY FEEDER

LIGHT-DUTY CONDUIT

SMALL-CAPACITY FEEDER

METERING DEVICES

13519

ON

OFF

MOTOR CONTROLS

INCOMING UNDERGROUND SERVICE

Figure 5-9. As the first component in the electrical system of a building, a switchboard receives electricity from the utility and divides it into smaller feeds to other distribution devices.

Diesel backup generators can be used to provide emergency power to critical systems in the event of a utility power failure.

The switchboard contains overcurrent protective devices (OCPDs) and switches to control the flow of electricity into the building. Then it is up to the building electrical system to further distribute the power to the end-use points within the facility. **See Figure 5-10.**

Transformers. Transformers are electric devices that use electromagnetism to change AC voltage or electrically isolate two circuits. A transformer is composed of two windings of conductors around an iron core. The primary winding draws current from the power source, which induces a magnetic field through the iron core. This causes an electric current to flow in the secondary winding, which delivers the power to the load. **See Figure 5-11.**

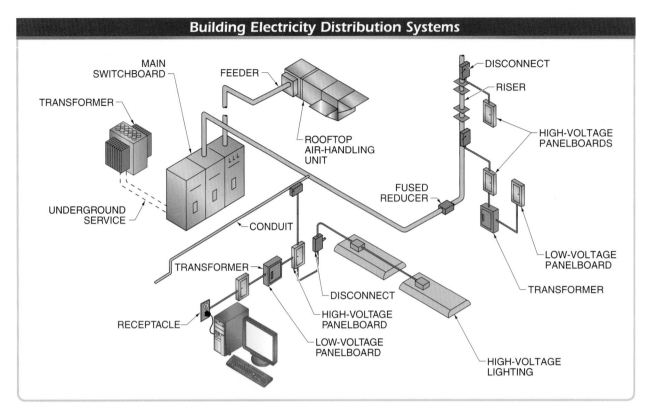

Figure 5-10. The building electricity distribution system includes several parts to control and distribute electrical power among the loads.

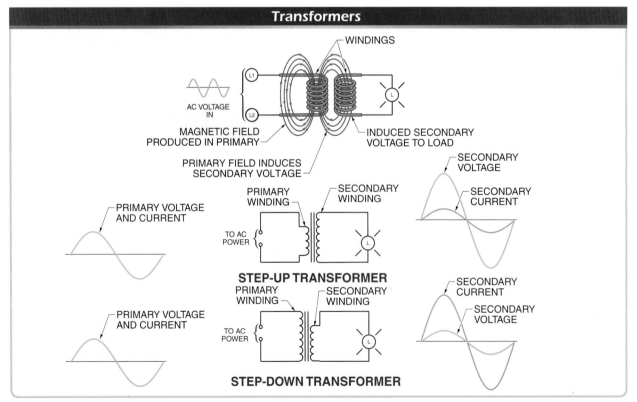

Figure 5-11. Transformers use the magnetic field created by one winding to induce a proportional voltage in another winding.

The ratio of input voltage to output voltage is proportional to the ratio of the number of turns in the two windings. Because the amount of electrical power transferred from one side to the other is nearly constant (there are only small losses from generated heat), the current is inversely proportional. For example, in a 10:1 transformer, if 120 V at 1 A is applied to the primary side, the secondary side will be 12 V at nearly 10 A. The power on both sides of the transformer is approximately 120 W.

Depending on which side of the transformer is treated as the primary or secondary side, a transformer can either step up or step down the input voltage. Utilities use transformers to distribute large amounts of power over long distances efficiently by raising the voltage. Increasing the voltage reduces the current correspondingly, which reduces losses from voltage drop and allows power distribution through smaller gauge wire. Then, additional transformers throughout the grid lower the voltages down to levels usable by loads.

Transformers are often used similarly within the power distribution system of a building. **See Figure 5-12.** High voltage (usually 480 V) is distributed throughout the building, and localized transformers lower the voltage to the 240 V, 208 V, or 120 V levels required by most loads.

Panelboards. A switchboard typically supplies power to multiple panelboards located throughout the building, which divides the power distribution system into smaller units. A *panelboard* is a wall-mounted power distribution cabinet containing overcurrent protective devices for lighting, appliance, or power distribution branch circuits. **See Figure 5-13.** A *branch circuit* is the circuit in a power distribution system between the final overcurrent protective device and the associated end-use points, such as receptacles or loads.

Switches. The loads on branch circuits are controlled primarily by switches that energize or deenergize the entire circuit. A *switch* is a device that isolates an electrical circuit from a power source. Many switches are operated manually, with a hand-operated lever that opens or closes electrical contacts inside the switch and visually indicates the energized state of the circuit. There are also switches that are operated electrically or operated automatically by changes in their environment, such as temperature or pressure.

Facts

For safety reasons, most motor control circuits are powered by step-down transformers, which reduce the voltage to the circuit.

Transformer Applications

UTILITY TRANSFORMER

Fluke Corporation

BUILDING TRANSFORMER

Figure 5-12. Transformers of different sizes and ratios are used extensively in electricity distribution, both inside and outside of buildings.

Panelboards and Branch Circuits

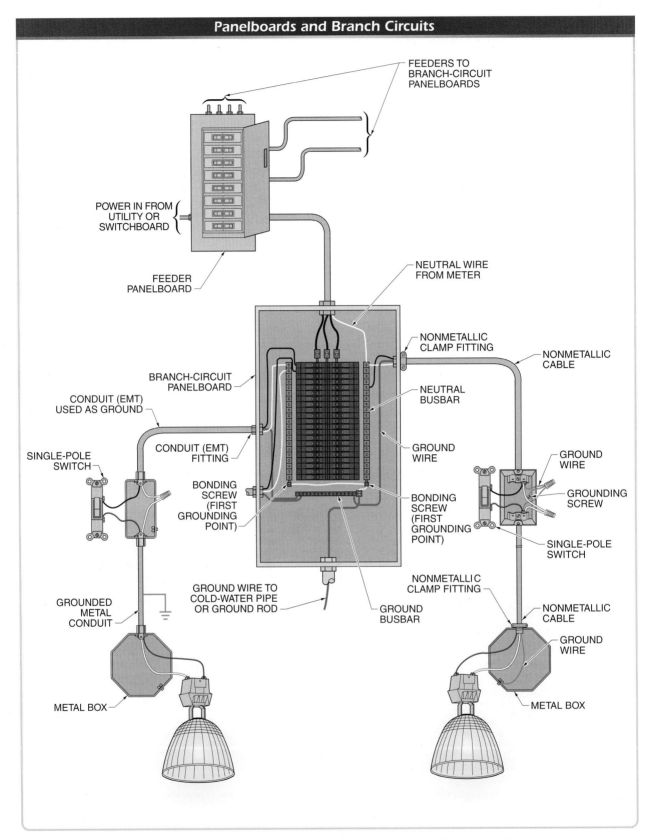

FEEDERS TO BRANCH-CIRCUIT PANELBOARDS

POWER IN FROM UTILITY OR SWITCHBOARD

FEEDER PANELBOARD

NEUTRAL WIRE FROM METER

NONMETALLIC CLAMP FITTING

NONMETALLIC CABLE

BRANCH-CIRCUIT PANELBOARD

NEUTRAL BUSBAR

CONDUIT (EMT) USED AS GROUND

CONDUIT (EMT) FITTING

GROUND WIRE

GROUND WIRE

SINGLE-POLE SWITCH

GROUNDING SCREW

BONDING SCREW (FIRST GROUNDING POINT)

BONDING SCREW (FIRST GROUNDING POINT)

SINGLE-POLE SWITCH

GROUNDED METAL CONDUIT

GROUND WIRE TO COLD-WATER PIPE OR GROUND ROD

GROUND BUSBAR

NONMETALLIC CLAMP FITTING

NONMETALLIC CABLE

METAL BOX

GROUND WIRE

METAL BOX

Figure 5-13. Panelboards include overcurrent protective and switching devices for branch circuits.

Switches are used throughout electrical systems. Common variations of switch types include the method of actuation, voltage and current rating, number and configuration of contacts, and other features. For example, a disconnect is a type of switch commonly used for high-power loads or large circuits. **See Figure 5-14.** Disconnects may also incorporate OCPDs.

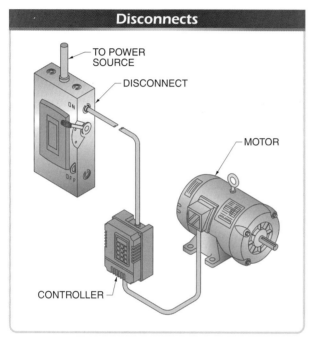

Figure 5-14. A disconnect is used to manually switch electrical power to large loads.

Motor disconnects may be located near the motor, making it easy to remove power during maintenance.

Conductors. Electrical circuits and components are connected using conductors. A *conductor* is a material that has little resistance and permits electrons to move through it easily. Conductors are available as individual wire or in groups, such as cable and cord. **See Figure 5-15.** A *wire* is any individual conductor. A *cable* is two or more conductors grouped together within a common protective cover and used to connect individual components. A *cord* is a group of two or more conductors in one cover that is used to deliver power to a load by means of a plug. Most individual conductors are enclosed in an insulated cover to protect the conductor, increase safety, and meet code requirements. However, some individual conductors, usually ground wires, may be bare.

Conductor material can be copper, aluminum, or copper-clad aluminum. Because of cost, copper (Cu) and aluminum (Al) are the most commonly used materials for conductors. Copper is the most common because it has a lower resistance than aluminum for any given wire size. For this reason, aluminum conductors must be sized one or more sizes larger than copper conductors. *Copper-clad aluminum* is a conductor that has copper bonded to an aluminum core. The total amount of copper used is less than 10% of the conductor. Copper is used to counter the disadvantages of aluminum.

Conductors are sized by using a number, such as No. 12 AWG or No. 14 AWG. The conductor size number is based on the American Wire Gauge (AWG) numbering system. **See Figure 5-16.** The lower the AWG number, the larger the diameter of the conductor and the higher the current-carrying capacity. For example, a No. 12 copper conductor is larger in diameter and can carry 5 A more current than a No. 14 copper conductor. The AWG size used for a circuit depends on the maximum current that the conductor must carry and the conductor material.

Outlets. Devices carry or control electricity but do not use it. Devices include outlets, which are installed at convenient access points on the electrical system. An *outlet* is an end-use point in the power distribution system. Outlets allow for the connection of devices, such as receptacles, switches, and loads, to the circuit. **See Figure 5-17.**

A *receptacle* is an outlet for the temporary connection of corded electrical equipment. Receptacles are available in a variety of current and voltage ratings, connector types, conductor configurations, and colors. Most installed receptacles are of a standard type, though there are many special-use receptacles for certain applications, such as isolated-ground receptacles, hospital-grade receptacles, twist-lock receptacles, and ground-fault circuit interrupter receptacles.

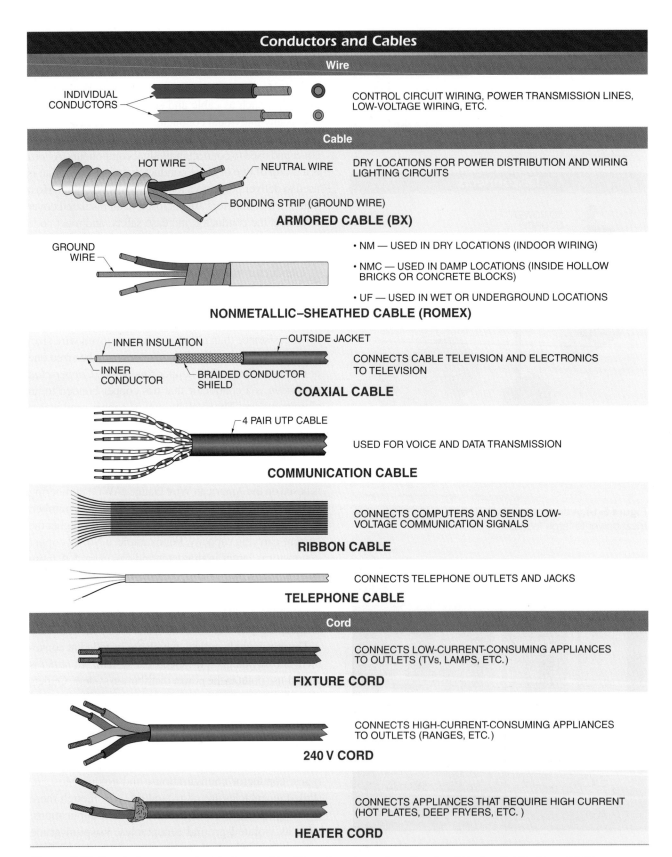

Conductors and Cables

Wire

INDIVIDUAL CONDUCTORS

CONTROL CIRCUIT WIRING, POWER TRANSMISSION LINES, LOW-VOLTAGE WIRING, ETC.

Cable

HOT WIRE — NEUTRAL WIRE

DRY LOCATIONS FOR POWER DISTRIBUTION AND WIRING LIGHTING CIRCUITS

BONDING STRIP (GROUND WIRE)

ARMORED CABLE (BX)

GROUND WIRE

• NM — USED IN DRY LOCATIONS (INDOOR WIRING)

• NMC — USED IN DAMP LOCATIONS (INSIDE HOLLOW BRICKS OR CONCRETE BLOCKS)

• UF — USED IN WET OR UNDERGROUND LOCATIONS

NONMETALLIC–SHEATHED CABLE (ROMEX)

INNER INSULATION — OUTSIDE JACKET

INNER CONDUCTOR — BRAIDED CONDUCTOR SHIELD

CONNECTS CABLE TELEVISION AND ELECTRONICS TO TELEVISION

COAXIAL CABLE

4 PAIR UTP CABLE

USED FOR VOICE AND DATA TRANSMISSION

COMMUNICATION CABLE

CONNECTS COMPUTERS AND SENDS LOW-VOLTAGE COMMUNICATION SIGNALS

RIBBON CABLE

CONNECTS TELEPHONE OUTLETS AND JACKS

TELEPHONE CABLE

Cord

CONNECTS LOW-CURRENT-CONSUMING APPLIANCES TO OUTLETS (TVs, LAMPS, ETC.)

FIXTURE CORD

CONNECTS HIGH-CURRENT-CONSUMING APPLIANCES TO OUTLETS (RANGES, ETC.)

240 V CORD

CONNECTS APPLIANCES THAT REQUIRE HIGH CURRENT (HOT PLATES, DEEP FRYERS, ETC.)

HEATER CORD

Figure 5-15. Conductors carry electrical current to different equipment and loads within the distribution system.

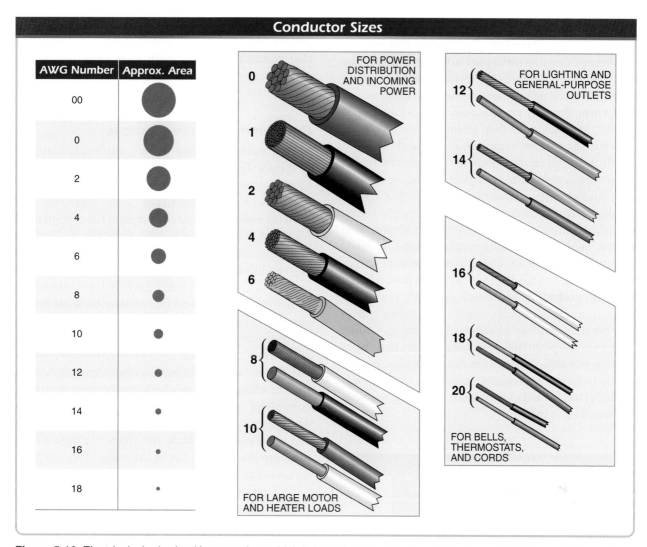

Figure 5-16. Electrical wire is sized by a number, which is based on the American Wire Gauge (AWG) numbering system.

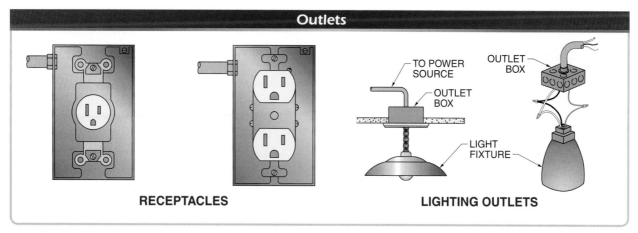

Figure 5-17. Outlets include receptacles and outlet boxes, which allow for the connection of individual loads to the building electrical system.

Grounding Systems

Grounding is the intentional connection of all exposed non-current-carrying metal parts to the earth. Grounding provides a direct path to the earth for unwanted fault current to dissipate without causing harm to persons or equipment. Properly grounded electrical circuits, tools, motors, and enclosures help safeguard equipment and personnel against the hazards of electrical shock.

Most electrical components are bonded to the grounding system with equipment grounding conductors. An *equipment grounding conductor (EGC)* is a conductor that provides a low-impedance path from electrical equipment and enclosures to a grounding system. **See Figure 5-18.** Grounding is then established by connecting the grounding system at the main service equipment to a metal electrode that is in good contact with the ground. A *grounding electrode conductor (GEC)* is a conductor that connects a grounding system to a buried grounding electrode. The total grounding path from any point in the electrical system must be as short as possible, sufficient in conductor size, permanently installed, and uninterrupted from the electrical circuit to the ground. Appropriate grounding methods are specified in the NEC®.

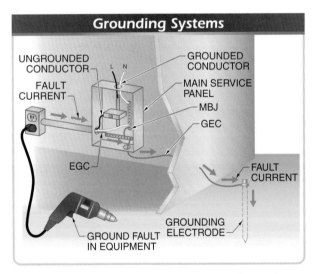

Figure 5-18. The grounding system consists of several parts that are connected together to provide a low-impedance path to ground for unwanted fault currents.

A *grounded conductor* is a current-carrying conductor that has been intentionally grounded. The neutral-to-ground connection is made by connecting the neutral bus to the ground bus at the main service panel with a main bonding jumper. A *main bonding jumper (MBJ)* is a connection at service equipment that connects an equipment grounding conductor, grounding electrode conductor, and grounded conductor (neutral conductor).

Overcurrent Protective Devices

In a properly operating circuit, current is confined to the conductive paths provided by conductors and other components when a load is turned on. Every load draws a normal amount of current when switched on. This normal amount of current is the current level for which the load, conductors, switches, and system are designed to safely carry. Under normal operating conditions, the current in a circuit must be equal to or less than the normal current level. However, sometimes an electrical circuit may have an excessive current flow.

Electrical equipment must be protected from excessive current flow. An *overcurrent protective device (OCPD)* is a device that prevents conductors or devices from reaching excessively high temperatures from high currents by opening the circuit. High temperatures can damage components and conductor insulation, which can cause electrical shock and create fire hazards. An overcurrent condition can be the result of an overload, ground fault, or short circuit.

OCPDs include fuses and circuit breakers that are usually located in panelboards. **See Figure 5-19.** A *fuse* is an overcurrent protective device with a fusible link that melts and opens a circuit when an overcurrent condition occurs. A *circuit breaker* is an overcurrent protective device with a mechanism that automatically opens a switch in a circuit when an overcurrent condition occurs. Two advantages of circuit breakers are that they are resettable after an overcurrent trip and, in certain circumstances, can be used as disconnects for installation and maintenance.

An *overcurrent* is electrical current in excess of the equipment limit, total amperage load of the circuit, or conductor or equipment rating. An overcurrent may be a short circuit current or an overload current.

A *short circuit* is a circuit in which current takes a low-impedance path around the normal path of current flow. A short circuit causes current to rise hundreds of times higher than normal at a fast rate. Short circuits may cause a fire, shock, or explosion and may damage equipment. All circuits must be protected from short circuits.

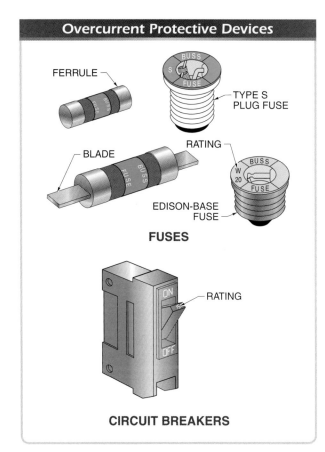

Figure 5-19. Fuses and circuit breakers protect electrical systems from overcurrent conditions that cause equipment damage, can result in electrical shock, and create fire hazards.

An *overload* is a small-magnitude overcurrent that, over a period of time, may trip the fuse or circuit breaker. Overloads are caused by defective equipment, overloaded equipment, or excessive loads on one circuit. Overloaded equipment draws a higher-than-normal current based on the degree to which the equipment is overloaded. The more overloaded the equipment, the higher the current draw. As with short circuits, overloads must be removed from the system.

Fuses and circuit breakers are installed at various points in a power distribution system and in individual pieces of equipment. Fuses and circuit breakers include current and voltage ratings. The current rating is the maximum amount of current the OCPD can carry without opening or tripping. OCPD current ratings are determined by the size and type of conductors, control devices used, and loads connected to the circuit. The voltage rating is the maximum amount of voltage that can be applied to an OCPD.

Every ungrounded (hot) power line must be protected from short circuits and overloads. Therefore, a fuse or circuit breaker is installed in every ungrounded power line. One OCPD is required for low-voltage single-phase circuits (120 V or less) and all DC circuits. The neutral line in AC circuits or the negative line in DC circuits does not include an OCPD. Two OCPDs are required for high-voltage single-phase circuits (208 V or 240 V). Both ungrounded power lines include an OCPD. Three OCPDs are required for all three-phase circuits regardless of the voltage level. Each of the three ungrounded power lines includes an OCPD. **See Figure 5-20.**

ELECTRICITY

Electricity is a natural phenomenon that has been developed as a technology to easily store and distribute energy with the potential to do useful work when used by electrical appliances. Electricity is monitored and controlled to make the most efficient use of this energy.

Electricity Switching

The switching of electricity is the alternation between energized and deenergized states by opening or closing ungrounded (hot) conductor(s) in a circuit. *Switching* is the complete interruption or resumption of electrical power to a device. Switching can turn individual electrical appliances or entire circuits on or off. Switching can also be used to alternate between separate power supplies, which can be an important part of managing a reliable electrical system for critical loads.

Circuit breakers are used to protect a circuit from overloads.

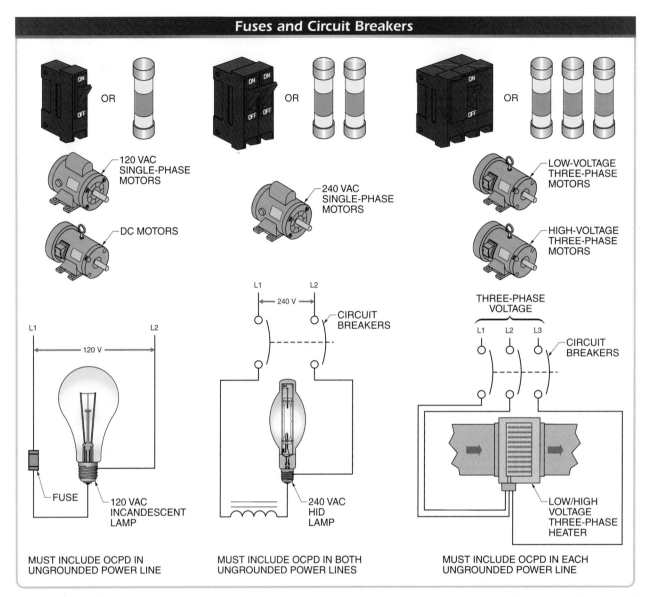

Fuses and Circuit Breakers

120 VAC SINGLE-PHASE MOTORS

DC MOTORS

240 VAC SINGLE-PHASE MOTORS

LOW-VOLTAGE THREE-PHASE MOTORS

HIGH-VOLTAGE THREE-PHASE MOTORS

FUSE

120 VAC INCANDESCENT LAMP

240 VAC HID LAMP

LOW/HIGH VOLTAGE THREE-PHASE HEATER

MUST INCLUDE OCPD IN UNGROUNDED POWER LINE

MUST INCLUDE OCPD IN BOTH UNGROUNDED POWER LINES

MUST INCLUDE OCPD IN EACH UNGROUNDED POWER LINE

Figure 5-20. Fuses and circuit breakers protect a distribution system from overcurrents that can damage equipment and injure personnel.

Switching is accomplished with sets of contacts that make or break the electrical continuity of a circuit. A switch can make or break multiple contacts simultaneously, so there are many possible physical arrangements of contacts. **See Figure 5-21.** The symbols for contact types are similar to these arrangements. The terms "pole" and "throw" are used to describe these arrangements. A *pole* is a set of contacts that belong to a single circuit. A *throw* is a position that a switch can adopt. Contacts may also be either single-break (opening the circuit in one place) or double-break (opening the circuit in two places). The most common types of switches are single-pole, single-throw switches.

Switch description abbreviations refer to these terms in a shorthand style. For example, an SPDT-SB switch has a single-pole, double-throw, single-break set of contacts. Some contacts may be normally open (NO) or normally closed (NC) so that when not actuated, they return to an open or closed position, respectively.

Electromechanical and solid-state relays are used in most control circuits. For example, an HVAC circuit includes an electromechanical relay "CHR" that controls power to the liquid line solenoid (LLS). The CHR relay coil may require only 20 mA to 100 mA to operate, but the relay contacts can be rated at 10,000 mA or more. **See Figure 5-22.**

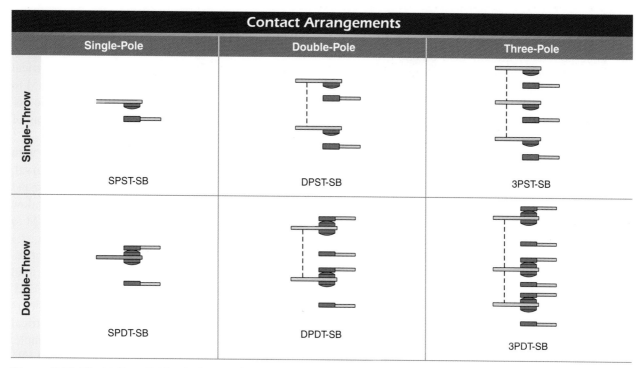

Figure 5-21. Electricity switching is the opening or closing of ungrounded (hot) conductors in one or more circuits with an arrangement of contacts.

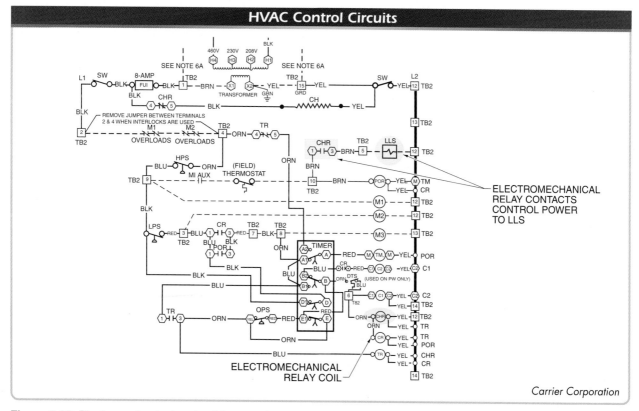

Figure 5-22. Electromechanical and solid-state relays are used in most control circuits.

Electrical Demand

Consumers commonly use more electricity during some periods of the day than others. *Electrical demand* is the amount of electrical power drawn by a load at a specific moment. Electrical demand is measured in kilowatts (kW). High electrical demands make it more difficult for electric utilities to supply power to all of their customers, so the utilities sometimes change their rates based on demand. Electricity during high-demand periods is more valuable; therefore, the rate is often higher. An electric bill for a commercial or industrial building may reflect changing rates for electricity during the month or include extra charges for high electrical demand. In some cases, the bill for an entire month or an entire year may be based on the rate for the highest demand during that period.

Electrical demand is an instantaneous value, analogous to speed. However, for billing purposes, utilities typically average the demand over a time interval of 15 min or 30 min. **See Figure 5-23.** For example, over a 15 min period, the average demand may be 40 kW, though the actual demand ranges from 30 kW to 50 kW. Utilities monitor the demand from commercial and industrial facilities and charge for the demand during a certain time interval each monthly billing period. Methods used to determine the time interval can vary.

The fixed interval method uses a set period to determine electrical demand. For example, electrical demand may always be monitored from 10:00 AM to 10:15 AM. Some utilities send signals on the incoming power service lines that indicate the beginning of each new period.

The sliding window method uses the interval with the highest demand. Any 15 min or 30 min interval, regardless of when it occurs, can be a new peak demand period for the month. For example, a sliding window can be from 10:01 AM to 10:16 AM or 10:02 AM to 10:17 AM. The interval may change each month.

Electrical demand over time results in electricity consumption. *Electricity consumption* is the total amount of electricity used during a billing period. Electricity consumption is measured in kilowatt-hours (kWh) and is calculated by multiplying the electrical demand by the amount of time at that rate. For example, if the daily average demand in a building is 40 kW, then over a 24 hr period, the loads consume 960 kWh (40 kW × 24 hr = 960 kWh).

Electricity consumption is metered for the purposes of determining the amount of electricity delivered to (or from) a customer's facility for billing purposes. Meters are installed at the service entrance and establish the transition between utility and customer-owned equipment. Both electrical demand and electrical consumption are important for supplying commercial and industrial facilities.

Power Quality

Power quality is a measure of how closely the power in an electrical system matches the nominal (ideal) characteristics. It is common for actual electrical parameters to vary somewhat, but allowable ranges are typically very small. Good power quality means that the parameters are within acceptable limits for the electrical system. Poor power quality has excessive variations in the parameters that can cause damage to loads and circuit equipment. Power quality is influenced by the performance of the electrical generation and distribution equipment, as well as electrical loads operating on the system.

Power quality can involve many characteristics of the electrical supply, such as voltage, frequency, harmonic distortion, noise, power factor, and unbalanced conditions in 3ϕ power supplies. These can all

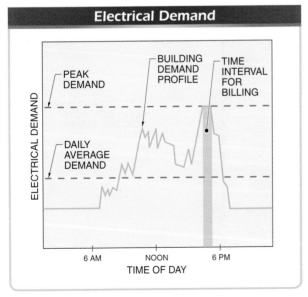

Figure 5-23. Utility billing for electrical demand throughout the billing period is sometimes based on the highest demand interval rather than on daily average demand.

be monitored and conditioned if improvement is necessary. The most common power quality parameters that are controlled by automated systems are voltage, current, and frequency.

Voltage. *Voltage* is the difference in electrical potential between two points in an electrical circuit. Voltage is the electrical pressure that causes current to flow when the two points are connected by a conductor. Common nominal voltage levels for electrical systems within buildings include 120 V, 208 V, 240 V, 277 V, 480 V, and 575 V.

Most electrical devices tolerate some degree of voltage fluctuation, but excessive deviations can cause circuit and load problems. Voltages are typically acceptable within the range of +5% to –10% from the nominal voltage. **See Figure 5-24.** Computer equipment is particularly sensitive. Noise or voltage fluctuations can disrupt computers, causing software errors, data loss, and communication failures.

Voltage sags are commonly caused by overloaded transformers, undersized conductors, conductor runs that are too long, too many loads on a circuit, peak power usage periods (brownouts), and high-current loads being turned on. Voltage sags are often followed by voltage swells as voltage regulators overcompensate.

Voltage swells are caused by loads near the beginning of a power distribution system, incorrectly wired transformer taps, and large loads being turned off.

Voltage swells are not as common as voltage sags but are more damaging to electrical equipment.

A *transient voltage* is a temporary, undesirable voltage in an electrical circuit, ranging from a few volts to several thousand volts and lasting from a few microseconds up to a few milliseconds. Transient voltages are caused by lightning strikes, unfiltered electrical equipment, contact bounce, arcing, and high-current loads being switched on and off. Transient voltages differ from voltage sags and swells by being larger in amplitude, shorter in duration, steeper in rise time, and erratic. High-voltage transients can permanently damage circuits or electrical equipment.

Voltage Measurement. AC voltage is stated and measured as peak, average, or rms values. The *peak value* (V_{max}) is the maximum instantaneous value of a sine wave of either the positive or negative alternation. The positive and negative alternation peak values are equal in a sinusoidal sine wave. The root-mean-square voltage, also known as rms voltage, is the AC voltage that produces the same amount of heat in a pure resistive circuit as DC voltage of the same value. The peak-to-peak value ($V_{p\text{-}p}$) is the value measured from the maximum positive alternation to the maximum negative alternation. Peak, average, and rms values are taken and compared when troubleshooting a power distribution system to determine if there are any types of problems, such as harmonics, present on the lines. **See Figure 5-25.**

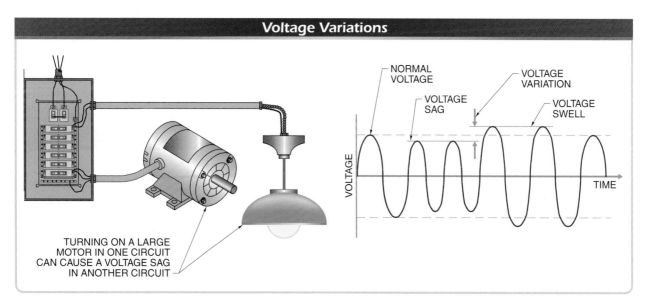

Voltage Variations

NORMAL VOLTAGE

VOLTAGE VARIATION

VOLTAGE SAG

VOLTAGE SWELL

VOLTAGE

TIME

TURNING ON A LARGE MOTOR IN ONE CIRCUIT CAN CAUSE A VOLTAGE SAG IN ANOTHER CIRCUIT

Figure 5-24. Voltage variations outside of the allowable range include voltage sags, voltage swells, and transients.

Peak-to-Peak Values

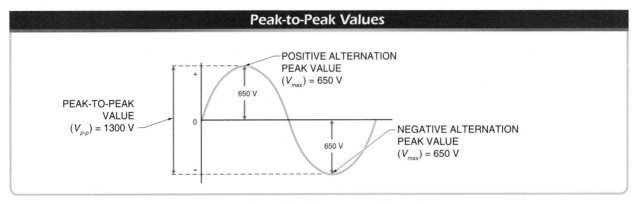

Figure 5-25. The positive and negative alternation peak values are equal in a sine wave.

To calculate peak-to-peak value, the following formula is applied:

$$V_{p\text{-}p} = 2 \times V_{max}$$

where

$V_{p\text{-}p}$ = peak-to-peak value (in V)

2 = constant (to double peak value)

V_{max} = peak value (in V)

For example, if the peak value (V_{max}) is 650 V, what is the peak-to-peak value ($V_{p\text{-}p}$)?

$$V_{p\text{-}p} = 2 \times V_{max}$$
$$V_{p\text{-}p} = 2 \times 650$$
$$V_{p\text{-}p} = \mathbf{1300\ V}$$

The *average value* (V_{avg}) is the mathematical mean of all instantaneous voltage values in a sine wave. The average value is equal to 0.637 of the peak value of a standard sine wave. To calculate average value, the following formula is applied:

$$V_{avg} = V_{max} \times 0.637$$

where

V_{avg} = average value (in V)

V_{max} = peak value (in V)

0.637 = constant (mean of instantaneous values)

For example, if the peak value (V_{max}) is 650 V, what is the average value (V_{avg})?

$$V_{avg} = V_{max} \times 0.637$$
$$V_{avg} = 650 \times 0.637$$
$$V_{avg} = \mathbf{414\ V}$$

DC Voltage. DC voltage is used in almost all portable equipment such as automobiles, golf carts, flashlights, and cameras. DC voltage is also used for the operation of most electronic circuits. DC voltage is obtained directly from batteries and photocells or is rectified to DC from an AC voltage supply. DC voltage produced from AC voltage passed through a rectifier varies from almost pure DC voltage to half-wave DC voltage. **See Figure 5-26.** DC voltage produced from rectified AC is commonly used in electronic devices such as computers, printers, photocopiers, and TVs. Power from an AC outlet is converted to DC voltage levels required.

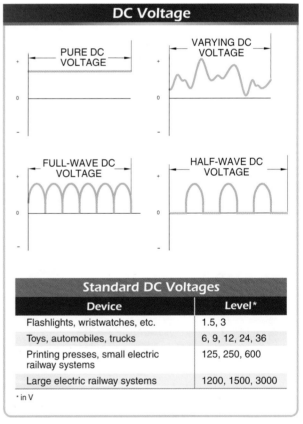

DC Voltage

Standard DC Voltages	
Device	**Level***
Flashlights, wristwatches, etc.	1.5, 3
Toys, automobiles, trucks	6, 9, 12, 24, 36
Printing presses, small electric railway systems	125, 250, 600
Large electric railway systems	1200, 1500, 3000

* in V

Figure 5-26. DC voltage varies from almost pure DC voltage to half-wave DC voltage and is commonly used in devices ranging from a wristwatch to large electric railway transportation systems.

Current. Current flows in a circuit when a source of power is connected to a device that uses electricity. *Current (I)* is a measure of the flow of charged particles flowing in a circuit and is measured in amperes (A). The more power a load requires, the larger the amount of current in a circuit. Current can be measured using a clamp-on ammeter, a DMM with a clamp-on current probe accessory, or a DMM connected as an in-line ammeter. **See Figure 5-27.**

Frequency. *Frequency* is the number of AC waveforms per interval of time. Frequency is measured in hertz (Hz) and is equivalent to cycles per second. The frequency of AC power in the United States is 60 Hz, though it is 50 Hz in many other parts of the world. These frequencies are low enough for efficient electricity transmission, but high enough that the resulting flicker of incandescent lamps is not noticeable.

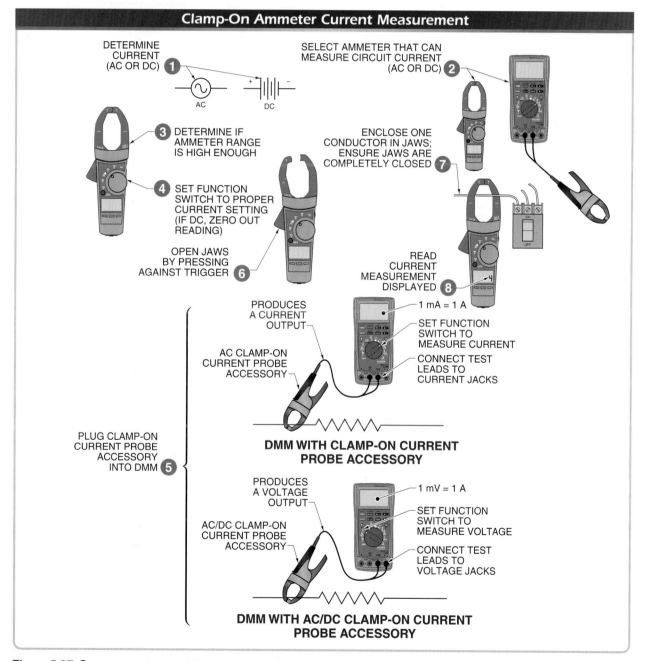

Figure 5-27. Current measurement is an important part of troubleshooting an overloaded load or power distribution system.

Frequency of the utility power supply system will vary as load or generating capacity changes. Overloading the system causes the frequency to decrease, while too much generating capacity causes the frequency to increase. Utility generators are constantly adjusted in speed so that the system frequency remains nearly constant. The frequency is so tightly regulated that it typically varies by less than ±1%. This tight tolerance is necessary to be able to synchronize the many generators supplying power to the grid. It also helps maintain accuracy for clocks and motors relying on frequency for their speed.

Electrical systems monitoring for frequency changes may automatically initiate load shedding (demand limiting) when the frequency falls in order to preserve proper functioning of critical systems. Since line frequency determines the speed of AC electric motors, frequency can also be manipulated in individual circuits within a system to control their speeds.

Electrical Noise. Electrical noise can enter the power distribution system directly on the wires or grounds or through magnetic coupling of adjacent wires. Noise is produced on power lines from two different points. *Common mode noise* is electrical noise produced between the ground and hot lines or the ground and neutral lines. *Transverse mode noise* is electrical noise produced between the hot and neutral lines. **See Figure 5-28.**

Common mode noise is also caused by arcing at motor brushes, ground faults, poor grounds, radio transmitters, ignition systems, and the opening of electrical contacts. The opening of electrical contacts produces noise because an arc is produced as they are pulled apart. The higher the current in the circuit being opened, the larger and longer the arc. Transverse mode noise is also caused by welders, switched power supplies, and the firing of silicon-controlled rectifiers (SCRs) in electrical equipment.

Noise in a system can produce false signals in electronic circuits, leading to processing errors, incorrect data transfer, and printer errors. Noise problems can be reduced by using noise suppressors at the power source.

Harmonics. A *harmonic* is voltage or current at a frequency that is an integer (whole number) multiple (2nd, 3rd, 4th, etc.) of the fundamental frequency. For example, when the power supply is 60 cycles per second (Hz) AC, the first harmonic (60 Hz) is the fundamental frequency. The second harmonic on a 60 Hz power distribution line is 120 (60 × 2) Hz. The second harmonic waveform completes two cycles during one cycle of the fundamental waveform over the same period of time.

The third harmonic is 180 (60 × 3) Hz. The third harmonic waveform completes three cycles during one cycle of the fundamental waveform over the same period of time. When these harmonics are present in a circuit, the resulting waveform consists of the sum of the fundamental and the higher harmonics at every instant. **See Figure 5-29.** The result is a distorted waveform from the contribution of the harmonics.

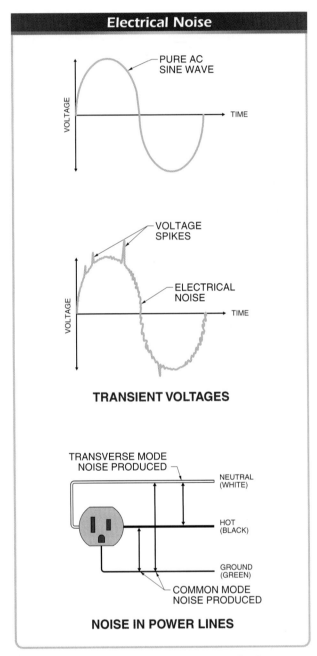

Figure 5-28. Noise can enter the power distribution system directly on the wires or grounds or through magnetic and capacitive coupling of adjacent wires.

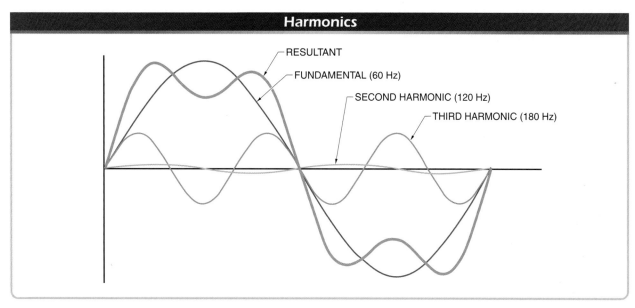

Figure 5-29. Harmonics are multiples of the fundamental sine wave.

Loads that typically produce harmonics on the power distribution system within a building are personal computers (PCs), copiers, printers, TVs, electronic lighting ballasts, microwave ovens, programmable logic controllers (PLCs), and solid-state motor drives. Harmonics can cause motors to burn out, transformers to fail, false circuit-breaker tripping (nuisance tripping), and overheating of neutral conductors and other parts of the power distribution system. Severe overheating can lead to a fire. When evaluating power quality, the incoming power, types and number of loads, and equipment used in the distribution system must all be tested.

Power Factor

Power factor (PF) is the ratio of true power used in an AC circuit to apparent power delivered to the circuit. Power factor is commonly expressed as a percentage. True power equals apparent power only when the power factor is 100% or 1. When the power factor is less than 100% or 1, the circuit is less efficient and has a higher operating cost because not all current is performing work.

True Power. The term "power" is used to express the rate of doing work or converting energy. *True power (P_T) is the* actual power used in an electrical circuit. True power is the power that is converted for use by devices into work, such as sound produced by speakers, rotary motion by motors, light by lamps, linear motion by solenoids, and heat by heating elements. True power is measured in watts (W), kilowatts (kW), or megawatts (MW). **See Figure 5-30.**

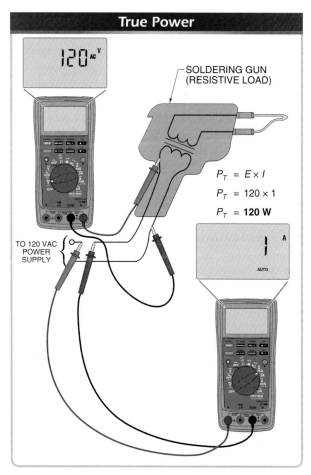

$P_T = E \times I$

$P_T = 120 \times 1$

$P_T = \textbf{120 W}$

Figure 5-30. True power is power converted into work and is measured in watts (W), kilowatts (kW), or megawatts (MW).

Reactive Power. *Reactive power (VAR)* is power supplied to a reactive load. The unit of reactive power is volt-amps reactive (VAR) instead of watts. True power represents a pure resistive component or load, and VAR represents a pure reactive (inductor or capacitor) component or load. In an AC circuit in which voltage and current are in phase, such as a resistive load, power in the circuit is true power. If all loads and circuits contained only resistance, all power would be true power. However, almost all AC circuits include impedance in the form of inductive reactance and/or capacitive reactance. Inductive reactance is by far the most common, since all motors, transformers, solenoids, and coils have inductive reactance. Inductive reactance or capacitive reactance causes voltage and current to be out of phase.

True power is different from reactive power. True power supplied to a resistive load is used to perform work or is dissipated as heat. Reactive power supplied to a reactive component such as an inductor or capacitor averages out to zero and is not converted into sound, rotary motion, light, or heat.

Calculating Power Factor. The power factor depends on the true power and the reactive power. A power triangle can be used to show the relationship between true power, apparent power, reactive power, and power factor in a circuit. **See Figure 5-31.** In a power triangle, a line indicates both magnitude and direction. Magnitude is indicated by the line length and direction is indicated by an angle of rotation from 0°. True power is represented by the horizontal vector line, apparent power is drawn lagging or leading true power by a line at angle theta (θ), and reactive power is represented by a vertical line that completes the triangle.

Power factor is lagging for an inductive load, leading for a capacitive load, and in phase for a resistive load. As circuit impedance increases, angle theta (θ) also increases. Likewise, as circuit impedance decreases, angle theta also decreases. Angle theta can be used to find the power factor because the cosine of angle theta is equal to the circuit power factor. Power factor can be found by applying either formula:

$$PF = \frac{P_T}{P_A} \times 100$$

or

$$PF = \cos \theta \times 100$$

where

PF = power factor (in %)

P_T = true power (in W)

P_A = apparent power (in VA)

$\cos \theta$ = cosine of angle θ

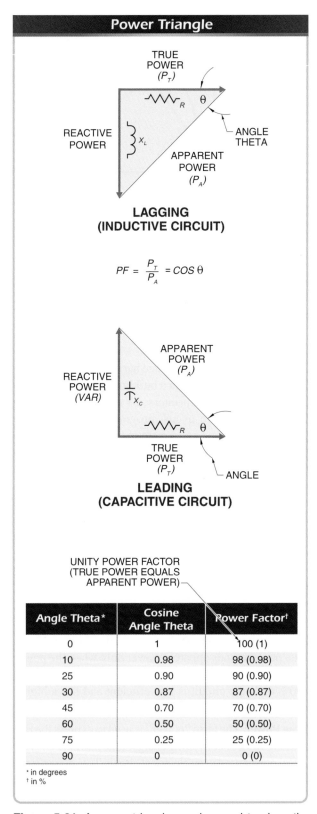

Power Triangle

$$PF = \frac{P_T}{P_A} = COS\ \theta$$

LAGGING (INDUCTIVE CIRCUIT)

LEADING (CAPACITIVE CIRCUIT)

UNITY POWER FACTOR
(TRUE POWER EQUALS
APPARENT POWER)

Angle Theta*	Cosine Angle Theta	Power Factor†
0	1	100 (1)
10	0.98	98 (0.98)
25	0.90	90 (0.90)
30	0.87	87 (0.87)
45	0.70	70 (0.70)
60	0.50	50 (0.50)
75	0.25	25 (0.25)
90	0	0 (0)

* in degrees
† in %

Figure 5-31. A power triangle can be used to show the relationship of true power, apparent power, reactive power, and power factor in a circuit.

For example, the power factor of a small 1ϕ motor is typically very poor. **See Figure 5-32.** With 186.5 W of true power and 575 VA of apparent power, the power factor of the ¼ HP, 1ϕ motor is only 32.5%. The lower the power factor, the less efficient the circuit and the higher the overall operating cost. The overall operating cost is increased because every component in the system, such as transformers and conductor sizes, must be sized for the higher current caused by a lower power factor. The power factor varies with different load types.

If the power factor is unity ($cos\ \theta = 1$), true power is equal to apparent power. In most AC circuits, power factor is not equal to unity (1) because there is always some impedance on the power lines. The $cos\ \theta$ varies between 0 and 1, and power factor varies between 0 and 1 for most circuits.

Power Factor Measurement

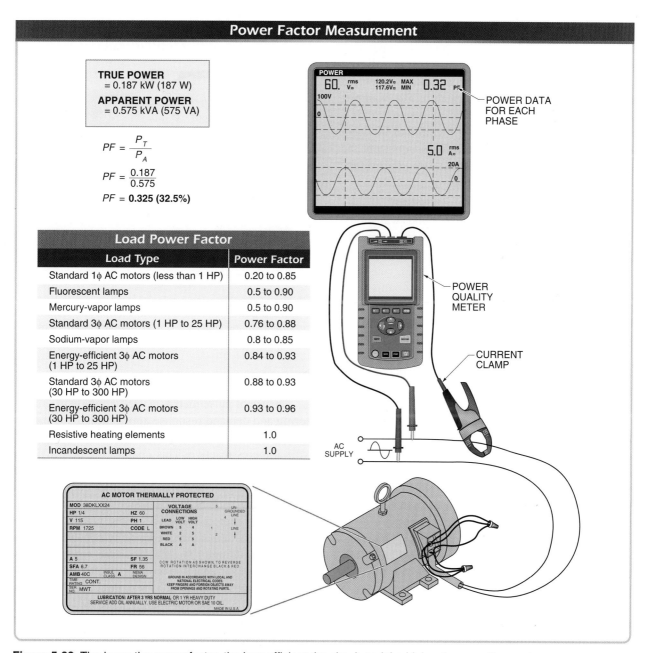

TRUE POWER
= 0.187 kW (187 W)
APPARENT POWER
= 0.575 kVA (575 VA)

$$PF = \frac{P_T}{P_A}$$

$$PF = \frac{0.187}{0.575}$$

$$PF = \textbf{0.325 (32.5\%)}$$

POWER DATA FOR EACH PHASE

POWER QUALITY METER

CURRENT CLAMP

AC SUPPLY

Load Power Factor

Load Type	Power Factor
Standard 1ϕ AC motors (less than 1 HP)	0.20 to 0.85
Fluorescent lamps	0.5 to 0.90
Mercury-vapor lamps	0.5 to 0.90
Standard 3ϕ AC motors (1 HP to 25 HP)	0.76 to 0.88
Sodium-vapor lamps	0.8 to 0.85
Energy-efficient 3ϕ AC motors (1 HP to 25 HP)	0.84 to 0.93
Standard 3ϕ AC motors (30 HP to 300 HP)	0.88 to 0.93
Energy-efficient 3ϕ AC motors (30 HP to 300 HP)	0.93 to 0.96
Resistive heating elements	1.0
Incandescent lamps	1.0

Figure 5-32. The lower the power factor, the less efficient the circuit and the higher the operating cost.

ELECTRICAL SYSTEM CONTROL DEVICES

Electrical system control devices are primarily used to monitor and maintain a reliable and quality power supply. If necessary, the control devices can either condition the power to improve its quality or switch the electrical system over to a secondary power source. This switching ability can also be used to respond to high utility rates or requests to shed loads due to high electrical demand.

Electrical Parameter Sensors

Electrical parameter sensors monitor various aspects of electrical demand and power quality. Depending on the measurement, the sensors may use digital (contact closure) signals, analog signals, or structured network messages to share information. For example, if an application requires only verification of voltage, a voltage sensor with a digital output can be used. However, if the exact voltage level is important, an analog voltage sensor can be used so that the voltage level can be monitored at all times.

Demand Meters. Electrical demand and electricity consumption are measured with demand (watt-hour) meters. Electromechanical induction demand meters are conceptually similar to motors. Electricity passing through coils creates a magnetic field, causing a metal disk to rotate. Each revolution accounts for a certain amount of energy transfer, and the rate of disk rotation is proportional to the amount of electrical power passing through the meter. The number of disk revolutions is counted mechanically or electronically, and the corresponding energy use is displayed on a register in digits or dials or shared with other devices via electronic signals.

Today, many electromechanical demand meters are being replaced by electronic demand meters that use current and voltage transformers and microprocessors to measure, process, and record data. **See Figure 5-33.** Some can record other electrical service information, such as peak power demand, power factor, reactive energy, time-of-use consumption, and, in the case of distributed generation, exported energy. Many electronic meters allow the meter data to be read remotely by infrared, radio frequency, telephone, network, or power-line carrier signals.

One type of electronic demand meter is a pulse meter. A *pulse meter* is a meter that outputs a pulse for every predetermined amount of flow in a circuit or pipe. The pulse can be either dry contact closures or a short application of voltage (typically 5 VDC). Flow is measured at a register by counting the pulses. A flow unit per pulse is set at the factory. The accumulated number of pulses represents the electricity consumption, and the pulse rate represents the electrical demand. For example, if each pulse corresponds to 1 kWh, 40 pulses means that 40 kWh of electricity were consumed. If the 40 pulses accumulated within an hour, then the average electrical demand during that hour was 40 kW (40 kWh ÷ 1 hr = 40 kW).

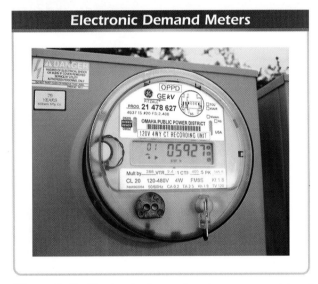

Figure 5-33. Electronic demand meters are capable of recording many different power supply parameters and allow remote access to the data.

The use of pulse meters is very common in submetering and demand response applications. Some electric pulse meters can output pulses for both watts and volt-amperes reactive. This enables the power factor to be calculated and monitored. Meters using a similar pulse-counting method are also used for counting units of fluid flow, such as in water meters.

Demand meters are used by the utility to record the electrical demand and electricity consumption for an entire building or facility. However, some facilities may use similar meters to monitor electricity usage for electrical subsystems, such as certain building areas or production lines, for maintenance or accounting purposes. Additional demand meters can be installed temporarily or permanently throughout the building to provide this data to a building automation system.

Power Quality Sensors. A variety of sensors can be used to monitor the power quality of an electrical power source. Such sensors measure current, voltage, frequency, harmonics, noise, or unbalance between 3ϕ power supplies. **See Figure 5-34.** Many of these sensors are similar to handheld and bench-top test instruments, except that instead of displaying the measured value, they output this information via electrical signals to other devices. They also may be permanently installed in the circuit to measure the value continuously.

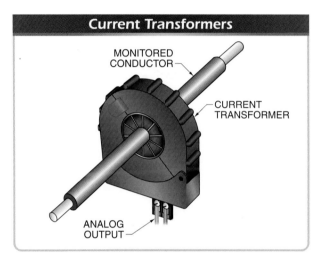

Figure 5-34. Current transformers are electrical sensors used to measure the current flow through conductors.

Most equipment will only operate properly at the frequency for which it was designed. Therefore, most frequency sensors have digital (on/off) outputs because it is not necessary to know the exact frequency value, only if the supplied power is or is not at the correct frequency.

Relays

Electrical circuits can be automatically switched using relays. A *relay* is an electrical switch that is actuated by a separate electrical circuit. The load in the control circuit is the relay coil, which is a winding of a conductor. When energized, the coil becomes an electromagnet that causes an armature to move and switch the state of a set of contacts. **See Figure 5-35.**

The control circuit commonly operates at voltages of 24 V or less, with a current draw of less than 1 A. The load circuit can be hundreds of volts and many tens of amperes or more. The two circuits are electrically separate, so they can also be any combination of AC and

DC circuits. This flexibility allows a relay to control a variety of load circuits, such as motors, lights, or heating elements, with a low-voltage control circuit.

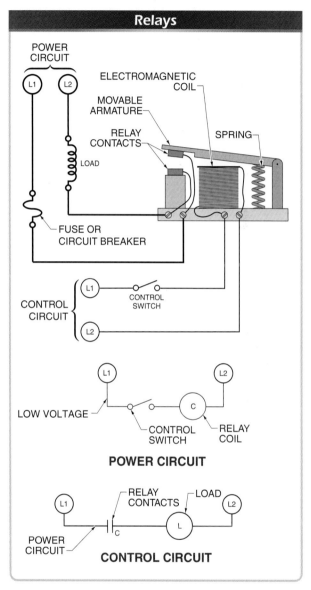

Figure 5-35. A relay allows a low-voltage control circuit to energize or deenergize loads on a higher-voltage power circuit.

Relays can include any arrangement of switch contacts. A common type is a single-pole, double-throw arrangement that allows for both a normally open (NO) and normally closed (NC) set of contacts, sharing a common (COM) third terminal. When deenergized, the NC and COM set is closed, and the NO and COM set is open. When energized, the NO and COM set is closed, and the NC and COM set is open.

Once energized, relays may hold the contacts in position either electrically or mechanically. **See Figure 5-36.** Electrically held relays require a constant application of the control voltage in order to remain in the energized position. This consumes an appreciable amount of power and creates heat, requiring large heat sinks. Mechanically held relays include a latching mechanism that holds the contacts in either position without constant voltage. Momentary applications of voltage can either open or close the contacts, making mechanically held relays more energy efficient.

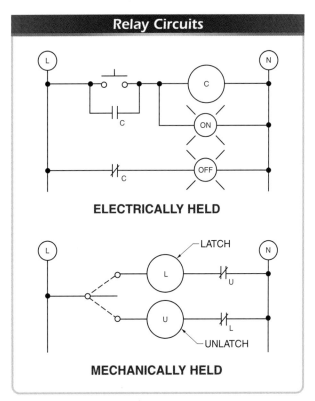

Figure 5-36. Relays may be held in the energized position either electrically or mechanically.

General-purpose relays are available for many combinations of voltage types, voltage levels, and contact arrangements. Many are designed for the circuit requirements of certain applications, such as controlling motors or lighting circuits. These may be known by different names but are really just specialized types of relays.

A *contactor* is a heavy-duty relay for switching circuits with high-power loads. Contactors use contacts made from pure silver, which remains a good electrical conductor even after oxidizing from the arcing of switching high currents.

A *magnetic motor starter* is a specialized contactor used for switching electrical power to a motor and includes overload protection. Magnetic motor starters commonly control 3ϕ power to electric motors. **See Figure 5-37.** Each output terminal contains an overload contact operated by a heater unit that responds to the heat generated by current flow. Excessive current in a conductor for more than a few minutes causes the heater to trip (open) the NC overload contact in the control circuit, deenergizing the coil and opening the load circuit contacts.

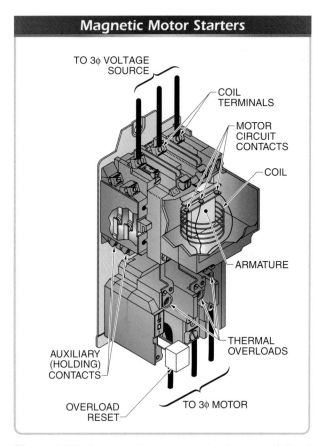

Figure 5-37. A magnetic motor starter is a specialized contactor that directly energizes or deenergizes an electric motor.

Variable-Frequency Drives

The frequency of a power supply is changed in order to increase or decrease the speed of AC electric motors. A *variable-frequency drive (VFD)* is a motor controller that is used to change the speed of an AC motor by changing the frequency of the supply voltage. In addition to controlling motor speed, variable-frequency drives can control motor acceleration time, deceleration time, torque, and braking.

A motor drive changes the frequency of the voltage applied to a motor by converting the incoming AC voltage to a DC voltage and inverting it back to an AC voltage that simulates the desired fundamental frequency. **See Figure 5-38.** The fundamental frequency is the voltage frequency simulated by the changing pulse widths of the carrier frequency. Fundamental frequencies typically range from nearly 0 Hz to over 60 Hz and directly determine the motor speed.

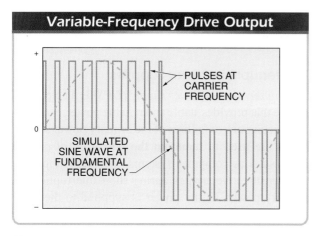

Figure 5-38. Variable-frequency drives simulate fundamental frequencies with pulses at a carrier frequency.

Variable-frequency drives can accept analog signals, commonly 0 VDC to 10 VDC, that are proportional to the desired motor speed. The carrier frequency is the frequency of the ON/OFF voltage pulses that simulate the fundamental frequency. The carrier frequencies of most variable-frequency drives range from 1 kHz to about 20 kHz.

Carrier frequency of a variable-frequency drive can be changed to meet particular load requirements. Higher carrier frequencies reduce heat in the motor because the voltage more closely simulates a pure sine wave. This is because higher carrier frequencies have more individual pulses to reproduce each cycle of the fundamental frequency. Higher carrier frequencies also reduce audible noise because the 1 kHz to 2 kHz range is within the range of human hearing. However, the resulting fast switching of the inverter section of the variable-frequency drive produces large voltage spikes that can damage motor insulation over time.

Transfer Switches

A *transfer switch* is a switch that allows an electrical system to be switched between two power sources.

This is most commonly implemented in order to keep critical loads operating during utility outages with a secondary power source, such as an emergency engine generator. **See Figure 5-39.**

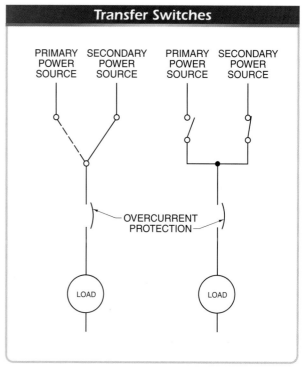

Figure 5-39. Transfer switches are configured to transition from one power source to another.

Transfer switches are manual or automatic. With a manual transfer switch, the transfer to secondary power is done by manually actuating a switch. Automatic transfer switches do not require manual intervention. These switches continually monitor the primary power source and, when the voltage falls below a certain level for a predetermined amount of time, the switch automatically initiates a power transfer operation.

The transfer operation may include more than just switching power sources. Some secondary power sources, such as engine generators, require time to complete a start-up sequence before they can be loaded. When a power transfer is needed, the automatic transfer switch sends an electronic start-up command to the secondary power source and monitors the resulting voltage and frequency. Only when the secondary power is stable does it complete the transfer by switching the connections. The transfer switch also manages the transfer back to primary power when appropriate.

Automatic transfer switches deal with the resulting transition period between power sources in different ways. In an open-transition transfer (OTT), one power source must be completely disconnected from the electrical system before the other power source is connected. This is known as a "break-before-make" transfer, which prevents the power from back-feeding into the primary system but results in a short power interruption. Depending on the system, the interruption may last from a fraction of a second to several seconds. Sometimes the system's critical loads are supported by power from a UPS battery during the transfer.

In a closed-transition transfer (CTT), however, the transition is smoothed by slightly overlapping the connection of the two power sources. This is a "make-before-break" transition. The transfer switch monitors the frequency and phase of each power source. When they are synchronized, the second power source is connected and then the first is disconnected. There is a brief period, typically less than a second, when they are connected in parallel and both are supplying power to the loads. CTT switches used with engine generators require isochronous governors, which keep the generator output frequency constant under any load conditions.

A paralleling load transfer is very similar to a CTT, but it parallels the power sources for a longer period, allowing a smoother transfer of loads from one source to the other. These paralleling load transfer systems require load sharing and continuous synchronization equipment and are most often used in large applications paralleling multiple engine generators for "peak shaving" or cogeneration with the utility.

Distributed Generator Interconnection Relays

Distributed generation involves exporting electricity generated with a secondary power source to the utility grid. This can be useful to both the utility and the customer. The utility adds electrical demand capacity to its network and the customer sells excess electricity. However, there are concerns with paralleling two power sources. First, the power sources must be carefully synchronized. Second, if the utility fails while connected, the secondary power source may back-feed into the utility's system, potentially causing equipment damage and safety hazards. This is of particular concern to the utility. Distributed generators must immediately disconnect from the grid if there is a utility outage. Otherwise, utility workers can be seriously injured by unexpectedly energized equipment.

To avoid these problems, distributed generators must interface with the grid through a distributed generator interconnection relay. A *distributed generator interconnection relay* is a specialized relay that monitors both a primary power source and a secondary power source for the purpose of paralleling systems. The interconnection relay includes sensors for voltage, current, frequency, direction of power flow, and other parameters. The interconnection relay only allows the paralleling of connections if the two power sources are in phase and immediately opens the connection if they become out of phase.

Uninterruptible Power Supplies

An *uninterruptible power supply (UPS)* is an electrical device that provides stable and reliable power, even during fluctuations or failures of the primary power source. UPSs are used to maintain the operation of critical loads such as computer systems, medical equipment, and sensitive electronics during short interruptions in primary power. Utility power is supplied to the UPS, which supplies conditioned, stable, and disturbance-free output power to the critical loads.

There are several different ways in which UPSs can be used. A small UPS can be a stand-alone device installed at the point of use or a larger piece of equipment integrated into a critical power infrastructure system. Medium-size UPSs protect clusters of sensitive equipment, and centrally located UPSs protect larger electrical systems.

Batteries are the most commonly used power source for a UPS, although flywheels and fuel cells are becoming more common for certain applications. The overall functionality of the UPS is the same, regardless of the power source. When utility power is interrupted or the quality is poor, current is drawn from the back-up power source to continue supplying adequate power from the output section of the UPS. The loads continue operating without interruption because they are typically always connected to both power sources. The UPS includes an inverter to convert the DC power to AC power. When acceptable utility power returns, the UPS input returns to using utility power to provide power to the load while also charging the back-up power source.

Many UPSs have an internal bypass circuit that can connect the utility input power directly to the load. The bypass circuit automatically activates if there is an internal failure or the load draws too much power. The bypass typically operates within 10 ms so that the critical load continues to operate without interruption. The bypass will stay on until the fault or overload is cleared. It can also be manually operated as required to isolate the back-up power source for service or tests. An external bypass circuit can also be incorporated to isolate the entire UPS system while maintaining load operation.

There are several different designs of UPSs, each with relative advantages and disadvantages. The most common designs are standby, line-interactive, rotary, double-conversion on-line, and delta-conversion on-line UPSs.

Standby UPSs. Standby UPSs are typically used for small applications (less than about 600 VA) and are commonly used to protect personal computers. This design is also known as an "off-line" UPS because, under normal conditions, it simply passes utility power through to the output side and the inverter does not operate. In this case, the back-up power system inverter is in standby or off-line mode. The UPS provides some power conditioning to suppress surges or filter noise, but only if the utility power falls below minimum requirements does the UPS use back-up power. Then a transfer switch changes completely over to the back-up source. **See Figure 5-40.**

High efficiency, small size, and low cost are the main benefits of this design. However, this type of UPS will switch completely over to the back-up mode for nearly any power quality problem, which can quickly drain the battery. For example, a modest voltage sag can trigger back-up operation, even though some line voltage is present. For this reason, a standby UPS is rarely used for critical commercial loads.

Line-Interactive UPSs. The line-interactive UPS design incorporates a transformer and an inverter/rectifier unit that are always connected to the output of the UPS. **See Figure 5-41.** During normal conditions, the utility power flows through both devices, and a small amount is rectified to charge the back-up power source. When the input power fails, a transfer switch opens and DC power flows from the back-up power source, through the inverter to convert it to AC, and to the UPS output. With the inverter always connected to the output, this design provides additional power filtering and yields fewer switching transients than standby UPS types.

> **Facts**
>
> *Reactive power is present in a circuit when the current and voltage are out of phase. This happens when inductive loads are present.*

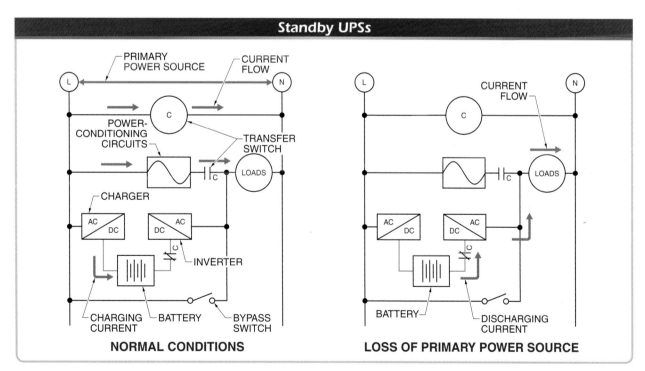

Standby UPSs

NORMAL CONDITIONS

LOSS OF PRIMARY POWER SOURCE

Figure 5-40. The inverter in a standby UPS system does not operate under normal conditions. Therefore, this system is known as an off-line UPS.

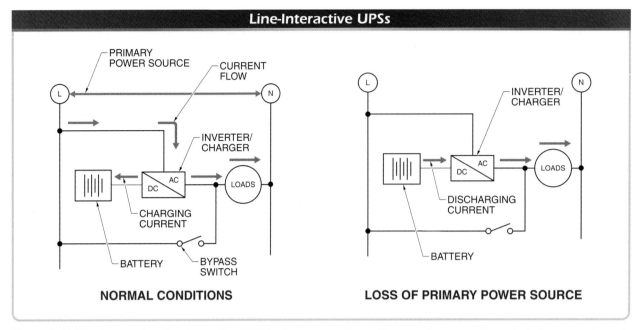

Figure 5-41. The inverter in a line-interactive UPS is always operating, either passing power through during normal conditions or converting DC power to AC while in the back-up mode.

A feature of line-interactive UPSs is the variable transformer that automatically adjusts the secondary winding taps as the input voltage varies. This voltage regulation feature allows the UPS to adapt to a wide range of high- or low-voltage conditions without switching completely over to the back-up power source. This reduces the charge/discharge cycling of back-up batteries, thus extending their useful life.

High efficiency, small size, low cost, and high reliability, coupled with the ability to correct low or high line voltage conditions, make this the dominant type of UPS in the 0.5 kVA to 5 kVA power range. A line-interactive UPS can be either a stand-alone or rack-mounted unit and is common for small business and departmental servers.

Rotary UPSs. The rotary UPS is one of the oldest UPS system designs. There are a few variations of this system, but all include a motor-generator combination. This system uses a rectifier to convert all AC input power to DC. **See Figure 5-42.** This DC power is then used to charge the back-up power source and operate the DC motor. The generator coupled to the motor converts the mechanical rotating energy into AC power. The motor and generator combination operates at all times. If the utility AC input power fails, the DC motor runs on battery power instead.

A rotary UPS provides excellent isolation of the load from utility power and has a high overload and fault-clearing capability. Also, the output from the generator is a perfect sine wave. No other power quality conditioning is needed.

This design, however, includes many moving parts that are subject to wear and mechanical failure, so significant maintenance is required. Therefore, rotary UPSs are most common for very large installations (>200 kVA) with the personnel to accommodate the maintenance requirements.

Double-Conversion On-Line UPSs. A double-conversion on-line UPS is similar to the rotary UPS design in that both perform two conversions of all the power flowing through the UPS: first AC to DC and then DC back to AC. The double-conversion on-line UPS, however, uses an electronic inverter instead of a motor-generator combination.

A rectifier changes all the AC input power to DC power. Some DC power is used to charge the back-up power supply, while the majority goes to the inverter to create AC power output. When the input AC power fails, there is no interruption in the output power because the back-up power source is always on-line and ready to supply power to the DC bus. Since all incoming power is rectified to DC, most AC power quality problems become nonissues.

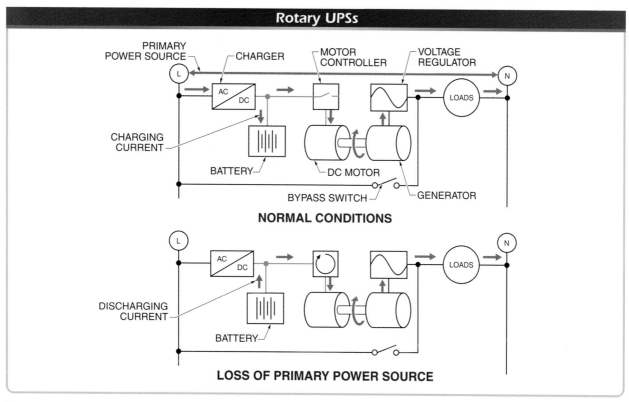

Figure 5-42. A rotary UPS isolates the input and output power by way of a mechanical coupling between a DC motor and an AC generator.

The double-conversion process allows for excellent regulation of the output voltage and frequency. **See Figure 5-43.** The rectifier may operate as a nonlinear load and, due to the multiple power conversions, each of which can contribute power losses, the efficiency is somewhat lower than other UPS designs. This type of UPS is most common for systems of 5 kVA and larger.

Delta-Conversion On-Line UPSs. The delta-conversion on-line UPS is the most recent UPS design. It includes a circuit that is similar to a double-conversion on-line UPS, but instead of processing all of the incoming power, this circuit only contributes the difference needed to maintain a steady output. For example, if the voltage of the incoming utility power is lower than the minimum level, the inverter delivers extra voltage to the output to make up the difference. **See Figure 5-44.** If the utility power is normal, it passes through the UPS with no further processing other than basic power conditioning for surges and noise.

The inverter is always ON and maintains the load voltage, while the delta converter controls the current to the output. If the utility power completely fails, the back-up power source supplies the inverter with power without interruption.

The delta-conversion system provides excellent voltage regulation and is highly efficient. The UPS operates as a linear device so that there are no harmonic problems related to its operation. However, it cannot provide frequency conversion. This UPS is common for systems 5 kVA and larger.

ELECTRICAL SYSTEM CONTROL APPLICATIONS

There are numerous examples of electrical system control applications in which electrical loads are switched by relays or electric motors are controlled with variable-frequency drives. These applications are very simple actions, typically embedded, and one control step within a more sophisticated system application, such as the control of an HVAC air-handling unit. For example, relays are used to energize heating or cooling functions, and drives control the speed of blower motors.

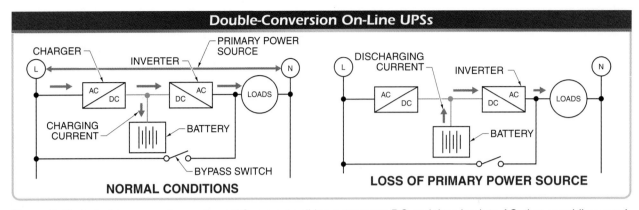

Figure 5-43. A double-conversion on-line UPS converts all input power to DC and then back to AC, thus providing excellent voltage and frequency regulation.

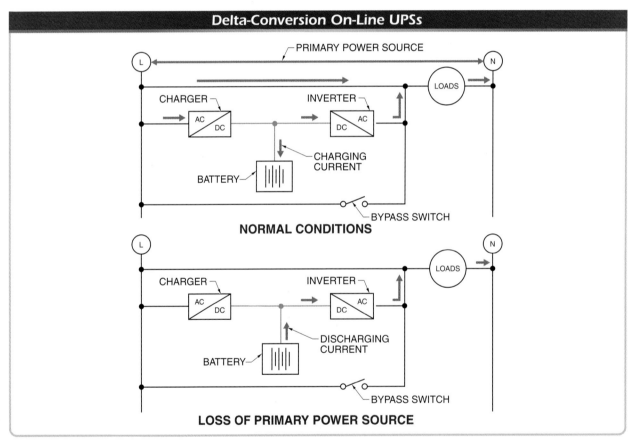

Figure 5-44. During normal operation, a delta-conversion on-line UPS system contributes the voltage necessary to bring the line voltage within the acceptable range.

Most electrical system control applications deal primarily with the management of the electrical power supply to ensure reliable and adequate electrical power to all the building systems, particularly critical loads. Electrical system controls can also be used to reduce energy costs. These goals can involve reducing the overall building electrical demand, switching to back-up power sources during utility outages, and operating secondary power sources in parallel with the utility to supplement the energy supply.

Demand Limiting

Electrical demand can have a big impact on utility energy costs, but building automation systems can be used to help avoid high rates or extra charges by limiting the building's overall electrical demand. This is accomplished by temporarily reducing the noncritical electrical loads in the building until overall demand requirements fall. **See Figure 5-45.** For example, if the midday electrical demand for a building during a hot summer day rises too high, the system can be programmed to reduce the power consumption by some of the largest system loads, usually lighting and HVAC systems. For example, the HVAC temperature setpoints may be raised slightly to reduce the cooling loads and the lighting in certain areas may be dimmed.

Operating Conditions. Demand limiting requires the involvement and automation of other building systems in order to reduce the electrical demand. The most common systems to involve are lighting and HVAC, since these are typically the largest consumers of energy and can usually tolerate some operating reductions. Lighting in certain areas of the building may be switched off or dimmed, temperature setpoints may be adjusted by a few degrees to reduce heating or cooling loads, and blowers and pumps may be cycled on and off for limited operation. It is common for these small changes to make a significant difference in the electrical demand and yet go largely unnoticed by the building occupants for short periods.

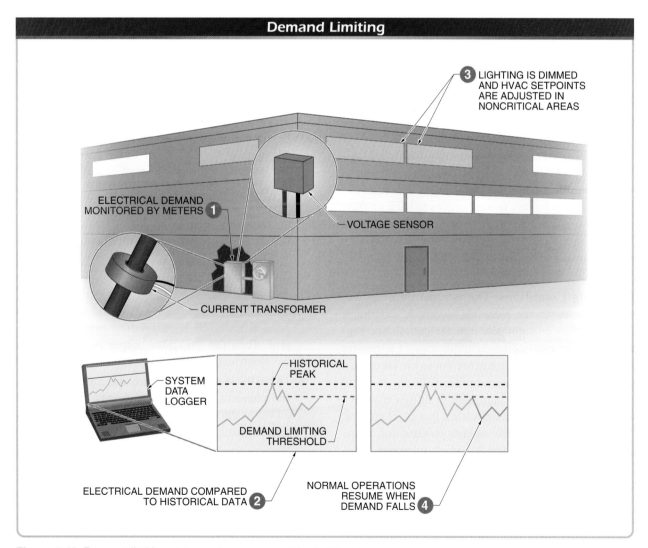

Figure 5-45. Demand limiting reduces the power used by building systems.

Control Sequence. The control sequence for demand limiting with electrical system controls is as follows:

1. Electrical demand is monitored by meters or electricity sensors, such as current transformers and voltage sensors.

2. The electrical demand is continuously compared to the historical demand data. If the demand exceeds a certain threshold (typically 85% to 90%) of the peak demand, the system initiates a demand-limiting program.

3. An electrical system controller signals the appropriate lighting and HVAC controllers to override their normal operating program. Some lighting is switched off or dimmed and HVAC temperature setpoints are adjusted.

4. When the electrical demand falls below a preconfigured percentage of the historical peak demand (typically 70% to 80%), the lighting and HVAC systems are released from the demand-limiting override and allowed to return to their normal operations.

Emergency Back-Up Power Source Transfer

Complete power outages or power quality problems, such as very low voltage, of the utility-supplied power are uncommon and typically short in duration. However, for buildings with critical loads, even this small risk for problems is intolerable. These facilities use one or more secondary power sources to back up their operations in the event of a utility power problem. These systems must constantly monitor the primary power supply and be ready to switch over to secondary power sources at any time.

Operating Conditions. The alternation of power sources to supply the electrical system of a building is managed by an automatic transfer switch. **See Figure 5-46.** The most common application of a transfer switch uses a diesel engine generator as the secondary power source. An engine requires time to start and come up to speed before being able to sufficiently drive the generator and produce electrical power. Therefore, a UPS system is typically used to power some of the critical building loads for this short transition time. The UPS returns to battery charging when the generator is running.

It is usually not feasible to supply an entire building's electrical system with back-up power because the necessary secondary power source would be too large. Instead, the electrical system is divided into critical and noncritical load circuits. The primary power source would supply electricity to all loads, but the secondary power source would supply electricity to only the critical loads. The critical loads must be grouped together in common switchboards or panelboards. Therefore, this type of arrangement must be planned and executed during the design and construction of the building.

To avoid initiating engine startup for very short voltage dips or interruptions, the transfer switch delays the start-up signal for a predetermined period. This time delay is at least 1 sec and may be up to 30 sec. If adequate utility power returns within the delay period, the transfer switch returns to the monitoring mode. Otherwise, it continues with the power source transfer operation.

Control Sequence. The control sequence for emergency back-up power transfer with electrical system controls is as follows:

1. The transfer switch continuously monitors incoming utility power for outages or poor power quality.

2. When the utility power is inadequate, the UPS system provides or conditions power to some of the critical loads, while the transfer switch waits for the set delay period. If the power remains inadequate at the end of the delay period, it sends a start-up signal to the engine generator.

3. The transfer switch monitors the power output from the generator. When the engine comes up to speed, the transfer switch senses that the voltage and frequency stabilize and initiates the transfer operation.

4. The transfer switch changes the building electrical system over to the secondary power source, the engine generator. First, the connection between the building electrical system and the utility is opened to avoid back-feeding power onto the grid. Then the generator output is connected to the electrical system.

5. The UPS synchronizes with the input frequency and phase of the generator, maintaining constant power to the load, and then deactivates. The UPS then begins charging its batteries.

6. Upon the restoration of adequate utility power, the transfer switch initiates the process of retransferring the electrical system back to the utility. Another short time delay ensures that the utility power source is stable before the retransfer. Then the generator is disconnected and the utility is reconnected.

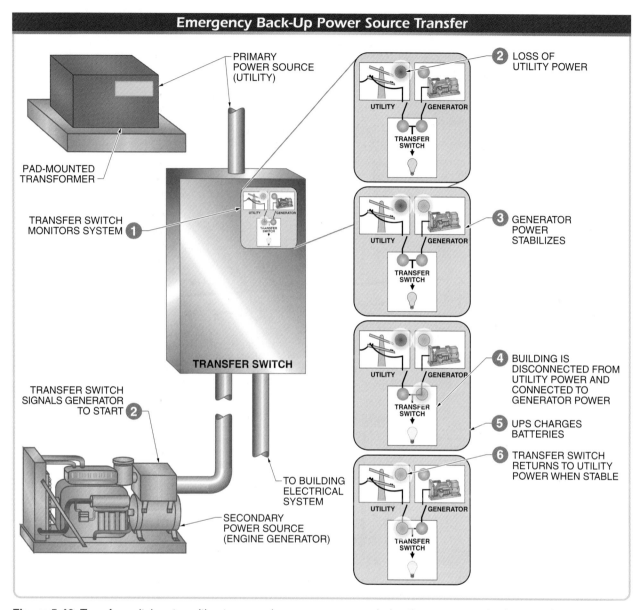

Figure 5-46. Transfer switches transition to secondary power sources during the emergency back-up mode.

On-Demand Distributed Generation

Besides the transfer of power sources for emergency reasons (utility power outage or poor power quality), an automated electrical system can use secondary power sources for on-demand distributed generation.

Facilities with a significant secondary power source capability may be able to contract with the utility to be an on-demand distributed generator. This scenario involves the utility signaling to the energy management system at the facility that the overall electrical demand on the grid is high and

extra electricity is requested to avoid brownouts. **See Figure 5-47.** Even if the utility-supplied power to the facility is still adequate, the generator starts up, synchronizes with the grid and the building electrical system, and begins contributing electricity when the switch closes. This immediately reduces the building's electrical demand, which helps relieve some of the high-demand problem. If the generator produces enough electricity to cover the building's usage and still export excess electricity, this further helps relieve the high-demand problem.

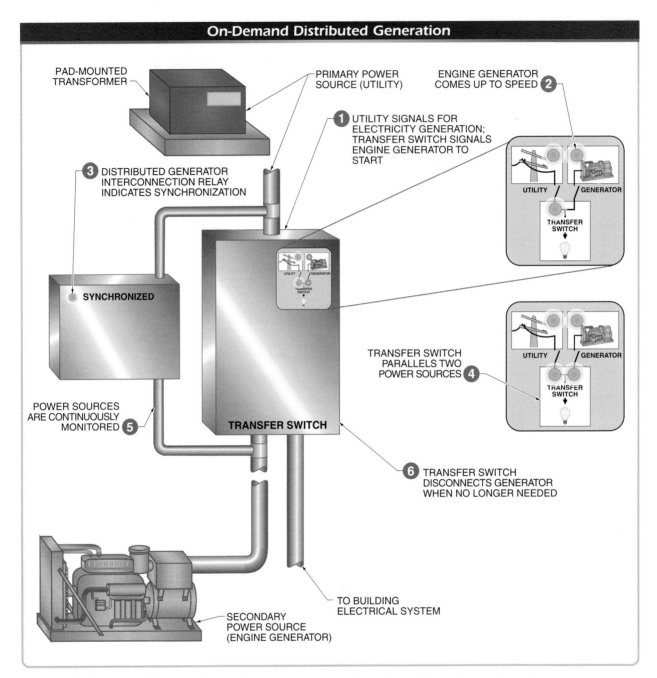

On-Demand Distributed Generation

PAD-MOUNTED TRANSFORMER

PRIMARY POWER SOURCE (UTILITY)

ENGINE GENERATOR COMES UP TO SPEED **2**

1 UTILITY SIGNALS FOR ELECTRICITY GENERATION; TRANSFER SWITCH SIGNALS ENGINE GENERATOR TO START

3 DISTRIBUTED GENERATOR INTERCONNECTION RELAY INDICATES SYNCHRONIZATION

UTILITY   GENERATOR

TRANSFER SWITCH

SYNCHRONIZED

UTILITY   GENERATOR

TRANSFER SWITCH PARALLELS TWO POWER SOURCES **4**

TRANSFER SWITCH

POWER SOURCES ARE CONTINUOUSLY MONITORED **5**

TRANSFER SWITCH

6 TRANSFER SWITCH DISCONNECTS GENERATOR WHEN NO LONGER NEEDED

TO BUILDING ELECTRICAL SYSTEM

SECONDARY POWER SOURCE (ENGINE GENERATOR)

Figure 5-47. On-demand distributed generation exports power to the utility grid through transfer switches.

Operating Conditions. Like emergency back-up applications, this scenario requires an automatic transfer switch and typically involves diesel engine generators, though some other secondary power sources are also compatible. Since this application also tends to activate a secondary power source more often than is required for only emergency needs, it helps keep the systems exercised, which improves reliability.

Control Sequence. The control sequence for on-demand distributed generation with electrical system controls is as follows:

1. After receiving the signal to start distributed generation, a controller signals the engine generator to begin its start-up sequence.

2. The engine starts up and the power output from the generator stabilizes.

3. The distributed generator interconnection relay verifies the voltage, frequency, and phase synchronization of the generator to the utility source.

4. The transfer switch connects both the generator and the utility to the building electrical system.

5. While the two power sources are paralleled, the utility feed is continuously monitored. If there is a problem with the utility feed, the transfer switch opens the utility circuit and the building operates from generator power alone.

6. When the utility no longer needs the distributed generation, it signals the electrical system controllers of the building. The transfer switch closes the connection to the utility power if it is not already closed and the generator connection is opened. This returns the system to normal operation. The engine generator begins its shutdown sequence.

Large industrial processes can be good candidates for electrical demand limiting if the processes can be easily shut down and started up again.

REVIEW QUESTIONS

1. What are the advantages of having a secondary power source available?

2. What is the primary difference between centralized generation and distributed generation?

3. Briefly explain how electricity is distributed within a building from the electrical service to a receptacle.

4. How does electrical demand affect the billing for electricity usage to some customers?

5. With respect to voltage and frequency, how does power quality affect electrical systems and their loads?

6. Briefly describe two examples of general-purpose relays.

7. How do variable-frequency drives change the frequency of power supplied to electric AC motors?

8. Compare transfer switches that use open-transition transfer (OTT) and closed-transition transfer (CTT).

9. Explain the overall functionality of an uninterruptible power supply (UPS).

10. How can building automation systems be used to limit electrical demand?

Chapter 5 Review and Resources

HVAC System Energy Sources 6

The purpose of an HVAC system is to provide comfort to the occupants of a building space. HVAC systems require energy to provide heating, cooling, humidification, and air circulation in building spaces. The most common heating system energy sources are electricity, natural gas, fuel oil, solar energy, and heat pumps. The most common cooling system energy sources are outside air, electricity, cold water, steam and hot water, and heat pumps. The energy source used for an HVAC system depends on cost and availability.

HEATING SYSTEM ENERGY SOURCES

Commercial HVAC systems must provide adequate heat to building occupants for comfort in the winter months. Commercial building heating system failure results in occupant discomfort and sometimes building damage such as frozen pipes.

Commercial building heating system energy sources include electricity, natural gas, fuel oil, solar energy, and mechanical system heat transfer. Factors considered when choosing a heating system energy source include installation cost, energy cost per unit of energy used, and local climate.

Electricity

Electricity is an energy source used to heat commercial buildings. Electricity is normally an existing energy source within a building, which minimizes the installation cost of piping distribution systems required for natural gas and fuel oil energy sources. Traditionally, the cost of electricity as a heating source has been higher than natural gas, which discouraged its use in heating. However, the current volatile pricing of electricity and natural gas varies the cost effectiveness of electricity systems in comparison to natural gas systems.

Electric heating systems commonly contain electric heating elements. An *electric heating element* is a device that consists of wire coils that become hot when energized. When placed in building ductwork, an electric heating element heats the air flowing through the ductwork. The heated air is then delivered to the building space. **See Figure 6-1.** Electric heating elements are used in electric baseboard heaters, radiant heat panels, air-handling units (AHUs), and variable-air-volume (VAV) terminal boxes.

Electric heat is widely used even though it is sometimes more expensive to heat a building with electricity than with other heating sources. The greatest advantage of electric heat is that no extensive piping or plumbing system is required. In addition, radiant heat panels can be designed for easy movement to new locations in a building. These advantages may outweigh the disadvantages for a specific application.

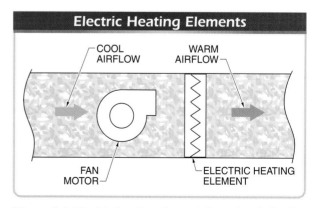

Electric Heating Elements

COOL AIRFLOW — WARM AIRFLOW —

FAN MOTOR — — ELECTRIC HEATING ELEMENT

Figure 6-1. Electric heating elements heat the air flowing through ductwork.

Natural Gas

Natural gas is a colorless, odorless fossil fuel. Natural gas is commonly used as an energy source for heating commercial buildings because it is plentiful and relatively inexpensive. Natural gas is also clean burning, which aids in meeting air pollution standards.

Natural gas heating applications generate heat through combustion. *Combustion* is the chemical reaction that occurs when oxygen (O) reacts with the hydrogen (H) and carbon (C) present in a fuel at ignition temperature. *Ignition temperature* is the intensity of heat required to start a chemical reaction. Fuel, oxygen, and ignition temperature (heat) are the three requirements for combustion. **See Figure 6-2.**

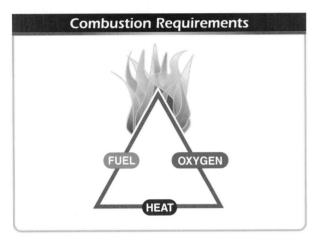

Figure 6-2. Fuel, oxygen, and ignition temperature (heat) are the three requirements for combustion.

Natural gas provides chemical energy, oxygen reacts with the fuel, and the ignition temperature starts the reaction. Ignition temperature may be provided by a pilot light, electric spark, or some other means. When combustion has started, the temperature must remain at or above the ignition temperature or combustion stops. The heat produced by combustion maintains the ignition temperature. Commonly, the hot gases of combustion are used to heat the air in the building space. An air-to-air heat exchanger is used in heating units such as furnaces.

If large amounts of heat are needed in a cold climate, burning natural gas to create hot water or steam is efficient because hot water and steam carry a large amount of heat energy (over 1000 Btu/lb). **See Figure 6-3.** In buildings such as hospitals, steam may also be needed for sterilization and laundry purposes. In facilities that use large amounts of natural gas, long-term contracts may be negotiated with the local utility, further reducing the cost.

A disadvantage of using natural gas as a heating source is the cost of installing natural gas piping in a building. Piping system integrity must be maintained to prevent contamination of the building spaces. Safety controls must be included to stop the flow of gas if improper conditions occur. In addition, the hazardous by-products of the combustion process, such as carbon monoxide, must be vented outside the building to prevent exposure to the occupants.

Fuel oil is normally stored in tanks located near the point of use.

Fuel Oil

Fuel oil is a petroleum-based product made from crude oil. Fuel oil is a common energy source for heating in the northeast U.S. Different grades of fuel oil are available. The grade of a fuel oil is based on the weight and viscosity of the oil. *Viscosity* is the ability of a liquid to resist flow. The viscosity of fuel oil is lowered by raising its temperature. For example, for pumping some fuel oil, the temperature must be raised by heating it in a fuel oil heater.

Heating value is the amount of British thermal units (Btu) per pound or gallon of fuel. A *British thermal unit (Btu)* is the amount of heat energy required to raise the temperature of 1 lb of water 1°F. The Btu rating of a fuel indicates how much heat it can produce.

The American Society for Testing and Materials (ASTM) has established standards for grading fuel oil. Each grade of oil has properties required for specific applications. The four grades of fuel oil used in boilers are No. 2 fuel oil, No. 4 fuel oil, No. 5 fuel oil, and No. 6 fuel oil. **See Figure 6-4.** A No. 2 fuel oil has a heating value of approximately 141,000 Btu/gal. No. 2 fuel oil does not have to be preheated. Almost all oil-fired HVAC systems use No. 2 fuel oil. It is rare to use other fuel oil grades for these systems.

Gas-Fired Boilers

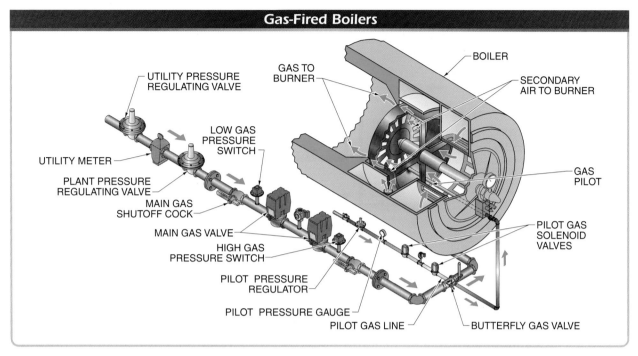

Figure 6-3. In a gas-fired boiler application, natural gas is piped from a utility to a gas burner.

Fuel Oil Grades				
Characteristics	**No. 2**	**No. 4**	**No. 5**	**No. 6**
Type	light distillate	light distillate or blend	light residual	residual
Color	amber	black	black	black
Specific gravity	0.8654	0.9279	0.9529	0.9861
Btu/gal.	141,000	146,000	148,000	150,000
Btu/lb	19,500	19,100	18,950	18,750

Figure 6-4. ASTM has established standards for grading fuel oils based on their characteristics.

No. 4 fuel oil is heavier than No. 2 fuel oil and has a heating value of approximately 146,000 Btu/gal. No. 4 fuel oil is most commonly a blend of No. 2 fuel oil and heavier fuel oils. In colder climates, No. 4 fuel oil may require preheating to reduce its viscosity for pumping.

No. 5 fuel oil has a heating value of approximately 148,000 Btu/gal. Preheating may be required in cold climates to reduce the viscosity to facilitate pumping. No. 6 fuel oil (bunker C) contains heavy elements from the distillation process. No. 6 fuel oil has a heating value of approximately 150,000 Btu/gal. Fuel oil tank heaters and line heaters must be used to heat No. 6 fuel oil to the required temperature for transport and combustion. The fuel oil temperature required depends on the type of burner and whether a straight distillate fuel oil or a blend of fuel oils is used.

Fuel oil systems include the accessories required to safely and efficiently operate the fuel oil burner. **See Figure 6-5.** Fuel oil accessories clean, control the temperature of, and regulate the pressure of fuel oil. In a fuel oil system, fuel oil is pumped through pipes from a storage tank to a furnace or boiler. Fuel oil may require filtering to remove impurities.

In HVAC applications, fuel oil and natural gas are interchangeable energy sources for heating. This allows flexibility in choosing which source to use. Common fuel oil applications include fuel oil-fired rooftop units and fuel oil-fired boilers. Advantages of using fuel oil include possible lower cost and easier availability than other energy sources. For example, fuel oil may be cost-effective in areas with exorbitant electrical costs and little availability of natural gas. In addition, because fuel oil is stored in large tanks close to the facility, fuel oil may be purchased in large quantities when costs are low and stored until needed.

Disadvantages of using fuel oil include increased environmental pollution compared to natural gas and costs associated with leak prevention and cleanup requirements. In addition, the capital costs of storage tanks, piping, and winter heating of the fuel oil also increase the cost of using fuel oil.

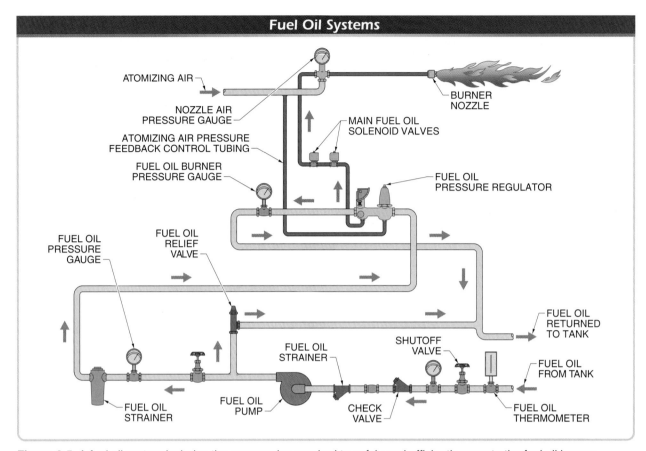

Figure 6-5. A fuel oil system includes the accessories required to safely and efficiently operate the fuel oil burner.

Solar Energy

Solar energy is energy (radiant heat) transmitted from the sun. Radiant heat travels through space to Earth and can be captured and used for heating.

Some HVAC systems use solar energy as the primary energy source and others use it as a backup source to complement an existing system. Solar energy is plentiful and can provide substantial amounts of energy that can be used to replace more expensive or less available fuels. Most solar systems heat water for small commercial and residential applications. The water is heated by the solar rays and stored in tanks until needed. **See Figure 6-6.**

The application of solar power systems is limited by the climate, the amount of available sunlight, and the size of the installed equipment. Solar energy collection and storage systems are built to collect and store the maximum amount of solar energy. The amount of energy received at the surface of the Earth can exceed 200 Btu/hr per sq ft of surface, depending on the angle of the sun's rays and the position of the solar collector.

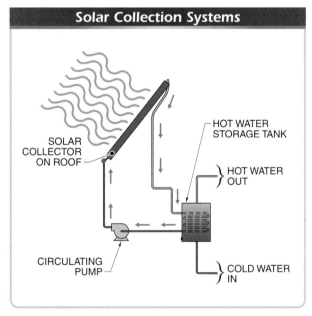

Figure 6-6. Solar collection systems may be used to heat water.

Heat Pump Heating Cycle

All air, even air at a moderate temperature, contains some heat. For example, when the outside air temperature is 45°F, it can still be used in a heat pump system to vaporize refrigerant that is at a temperature of 35°F. A *heat pump* is a mechanical compression refrigeration system that moves heat from one area to another area. **See Figure 6-7.** Heat pumps are used in residential and commercial applications.

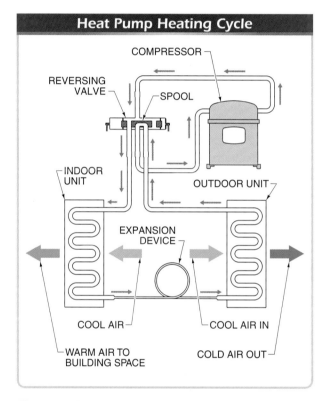

Heat Pump Heating Cycle

COMPRESSOR

REVERSING VALVE

SPOOL

INDOOR UNIT

OUTDOOR UNIT

EXPANSION DEVICE

COOL AIR

COOL AIR IN

WARM AIR TO BUILDING SPACE

COLD AIR OUT

Figure 6-7. During the heat pump heating cycle, heat is collected from the outside air at the outdoor unit and released to the building space at the indoor unit.

During a heat pump heating cycle, the balance point of a heat pump is reached when the heat losses through a building are the same as the amount of heat provided by the heat pump. This normally occurs at a temperature of approximately 40°F. For this reason, residential heat pumps are not widely used in northern climates. Heat pumps reduce or eliminate the need for natural gas or fuel oil piping and controls, while still being relatively efficient. The disadvantage of heat pumps is that they become less efficient as the outside air temperature drops.

Supplemental heat is provided if the outside air temperature drops below a set temperature or if a heat pump fails. Supplemental heat is normally provided by electric heating elements. The use of electric heating elements is normally more expensive than the operation of the normal heating system. The cost of supplemental heat from electric heating elements may be prohibitive in areas where a large number of hours of supplemental heat are required. While the use of supplemental heat is discouraged, it provides heat if a heat pump fails or if the outside temperature is too low. In some cases, an alarm light is attached to the thermostat to indicate heat pump failure. Normally, a technician is called to diagnose and repair the heat pump when the light is ON.

COOLING SYSTEM ENERGY SOURCES

Air conditioning is the process of cooling the air in building spaces to provide a comfortable temperature. Air conditioning systems are rated in terms of the available cooling capacity. Air conditioning system capacity is designed to deliver the cooling needs based on the requirements of the building space. Large building spaces require a system with a greater capacity than small building spaces. Energy sources for HVAC cooling systems include outside air, electricity, cold water, steam and hot water, and mechanical system heat transfer.

> **Facts**
>
> *In order for fuel to burn, an adequate amount of air is required to support combustion. If there is too much or too little air, the fuel will not burn. The range between too much and too little is referred to as the "flammable limits" of the particular fuel. This is the range in which combustion will be self-supporting. When the mixture of air and fuel is outside of this range, the fire will go out.*

Outside Air

A simple and inexpensive energy source used to cool a building is outside air. Outside air is brought into a building space from outside the building. The use of outside air for cooling is referred to as free cooling. **See Figure 6-8.** Outside air used for cooling purposes is limited by the temperature and humidity of the outside air. When the outside-air temperature or humidity is excessively high, the outside air is unsuitable as a cooling system energy source.

Free Cooling

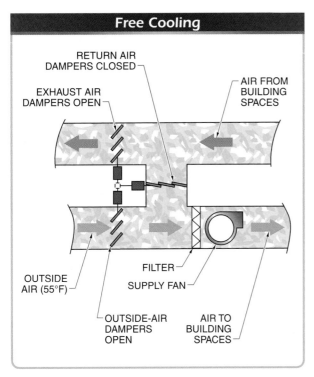

RETURN AIR
DAMPERS CLOSED

EXHAUST AIR
DAMPERS OPEN

AIR FROM
BUILDING
SPACES

OUTSIDE
AIR (55°F)

FILTER

SUPPLY FAN

OUTSIDE-AIR
DAMPERS
OPEN

AIR TO
BUILDING
SPACES

Figure 6-8. Free cooling uses outside air to cool building spaces.

Facts

Most pneumatic fan damper actuators have a mechanism for removing the pneumatic signal and manually locking the damper in one position when necessary.

Electricity

Electricity is a common energy source used to cool commercial buildings. Electricity is used in air conditioning systems to provide power for electric motors. The electric motors are used to operate fans and compressors. A compressor is a mechanical device that compresses refrigerant or other fluid. Refrigerant is fluid used for transferring heat (energy) in a refrigeration system. The refrigerant changes from a liquid to gas and back to liquid in the refrigeration cycle. **See Figure 6-9.**

When using electricity as a cooling energy source, its cost effectiveness depends on the cost per unit of electricity. For example, electrical usage costs in some parts of the country are $0.10/kWh or less. This low cost may be due to proximity to electrical power sources such as hydroelectric dams. In other parts of the country, the cost of electricity may be $0.20/kWh or more.

Refrigeration Systems

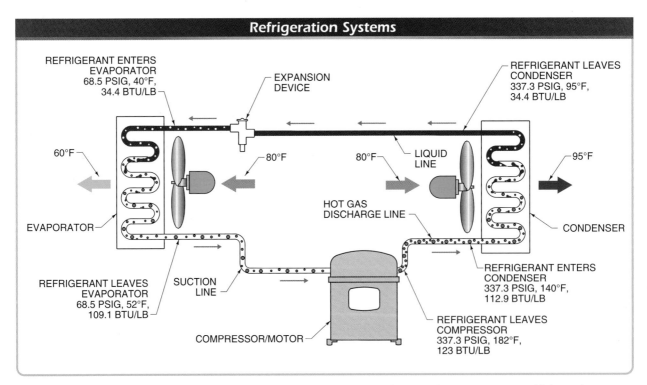

REFRIGERANT ENTERS
EVAPORATOR
68.5 PSIG, 40°F,
34.4 BTU/LB

EXPANSION
DEVICE

REFRIGERANT LEAVES
CONDENSER
337.3 PSIG, 95°F,
34.4 BTU/LB

60°F

80°F

80°F

LIQUID
LINE

95°F

EVAPORATOR

HOT GAS
DISCHARGE LINE

CONDENSER

REFRIGERANT LEAVES
EVAPORATOR
68.5 PSIG, 52°F,
109.1 BTU/LB

SUCTION
LINE

REFRIGERANT ENTERS
CONDENSER
337.3 PSIG, 140°F,
112.9 BTU/LB

REFRIGERANT LEAVES
COMPRESSOR
337.3 PSIG, 182°F,
123 BTU/LB

COMPRESSOR/MOTOR

Figure 6-9. Electricity is used in a refrigeration system to power an electric motor in a compressor, which produces a refrigeration effect used to cool building spaces.

Cold Water

Cold water is a common energy source used to cool commercial buildings. Cold water is supplied from holding ponds, cooling towers, or liquid chillers. A *liquid chiller* is a system that uses a liquid (normally water) to cool building spaces. Liquid chillers contain a compressor, expansion device, condenser, and evaporator, similar to a refrigeration system. In a liquid chiller, however, the evaporator and condenser consist of tube-in-shell heat exchangers. These heat exchangers transfer heat to and from water that contacts the refrigerant-filled tubes. The resulting warm water is pumped to a cooling tower to give up heat to the atmosphere. The cool water is pumped to building spaces to provide cooling. **See Figure 6-10.**

Steam and Hot Water

Steam and hot water are other common energy sources used to cool commercial buildings. While the majority of air conditioning systems are driven by electric motors, steam and hot water are energy sources that do not require the use of electric motor-compressors. Costs are lower when using steam or hot water because these energy sources are generated within the building space by a boiler and are not totally dependent on electricity. Steam or hot water is used to provide cooling in absorption refrigeration systems. An *absorption refrigeration system* is a nonmechanical refrigeration system that uses a fluid with the ability to absorb a vapor when it is cool and release a vapor when heated. Absorption refrigeration systems have a generator and absorber in place of the compressor to raise system pressure. A *generator* is an absorption refrigeration system component that vaporizes and separates the refrigerant from the absorbent. An *absorber* is an absorption refrigeration system component in which refrigerant is absorbed by the absorbent. An *absorbent* is a fluid that has a strong attraction for another fluid.

Absorption refrigeration systems are commonly used for large commercial or industrial applications in which mechanical compression systems are not as efficient. A refrigerant and absorbent are required in absorption refrigeration systems. A common refrigerant used is ammonia. Some absorption refrigeration systems may use combinations of absorbents and refrigerants, such as lithium bromide and water or lithium chloride and water.

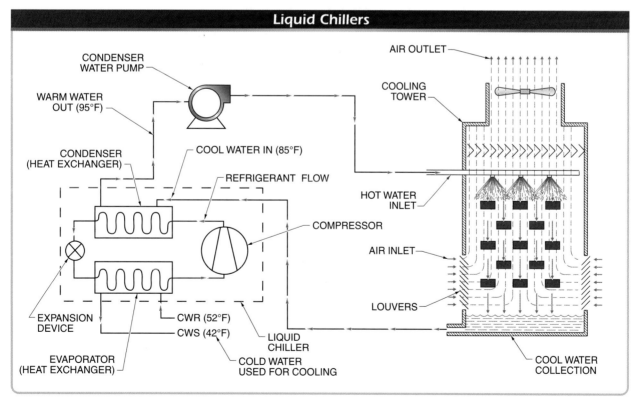

Figure 6-10. A liquid chiller contains two heat exchangers that transfer heat to and from water.

Lithium bromide-water absorption refrigeration systems operate using a generator, condenser, chiller, and absorber. A strong lithium bromide solution absorbs water vapor, making it weaker. When heated, the weak solution releases water. **See Figure 6-11.**

Heat is applied to the lithium bromide and water solution in the generator using heat from a source such as a steam coil. At the separator, the water vapor is separated from the lithium bromide solution. The heated water vapor is directed to the condenser. In the condenser, the heated water vapor is condensed into water. The water passes through an expansion valve into the chiller where it flashes (absorbs heat) because of the low pressure in the system after the expansion valve. An evaporator coil inside the chiller absorbs heat and chills water to be sent to the building space. Water vapor coming from the chiller is absorbed by the lithium bromide solution in the absorber. Cooling water is directed through condenser coils in the absorber and condenser to remove heat. The cooled lithium bromide solution in the absorber cools the solution received from the generator.

In many commercial buildings, substantial investments are made in boiler equipment. In many cases, the boiler equipment is operated year-round due to demands for hot water and steam. Air conditioning systems that use steam or hot water for cooling avoid problems with systems that use electric motors. For example, no large compressor motor with complicated electrical controls is required. Also, steam or hot water cooling systems can be operated off of a backup generator if power is lost. This feature is often not possible with large electrically driven motor-compressors. Also, hot water and steam systems do not use electricity to turn compressors, so changes in the price or availability of electricity have less economic effect.

Most motor-driven systems in the past used CFC refrigerants that are believed to damage the ozone layer when inadvertently released into the atmosphere. The change out of these refrigerants has proven to be expensive. Steam and hot water cooling systems normally do not use CFC refrigerants, avoiding the problem of refrigerant replacement.

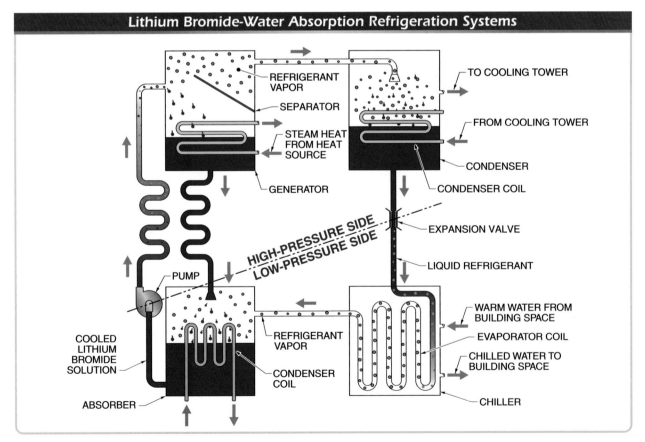

Figure 6-11. Absorption refrigeration systems are commonly used for large commercial or industrial applications where mechanical compression systems are not as efficient.

Heat Pump Cooling Cycle

The mechanical equipment of a heat pump system can be used to transfer heat from the air inside a building to the air outside a building, producing a cooling effect. The refrigerant cools as it flows through the expansion device. The heat from the building space is absorbed into the refrigerant inside the indoor coil (evaporator). The vaporized refrigerant flows to the compressor, where it is compressed. The hot, high-pressure refrigerant vapor is pumped to an outdoor coil (condenser), where it releases heat to the outside air. **See Figure 6-12.**

ALTERNATIVE HVAC SYSTEM ENERGY SOURCES

A variety of HVAC system energy sources are in use today. Ponds or lakes located near buildings may be used as sources of cold water. In heavily forested areas, wood by-products may be burned to generate steam. In some plants, coal is used as a heating source. Geothermal energy is also available underground for use in some HVAC systems. The applications of these alternative energy sources depend largely on local availability of fuel, construction codes, and energy efficiency standards.

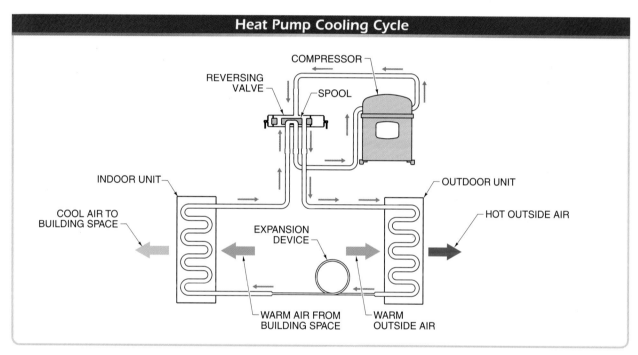

Figure 6-12. During the heat pump cooling cycle, heat is collected from the inside air at the inside unit and released to the outside air at the outdoor unit.

REVIEW QUESTIONS

1. List five energy sources for commercial building heating systems.

2. Describe electric heating elements.

3. Explain how natural gas heating applications generate heat.

4. Describe solar energy.

5. What is a heat pump?

6. What is supplemental heat?

7. How is electricity used in air conditioning systems?

8. Describe a liquid chiller.

9. What is an absorption refrigeration system?

10. Explain mechanical system heat transfer.

Chapter 6 Review and Resources

HVAC System Control Devices 7

HVAC systems are the most commonly automated building systems. They are also perhaps the largest and most complicated automated building systems because there are a large number of possible control devices available to manage them. These control devices can affect the systems in multiple ways. For example, a cooling device affects not only the temperature of the air within a building space but also the humidity. A thorough knowledge of all the application control devices and their interrelationships is vital to the successful automation of HVAC systems.

HVAC SYSTEMS

A *heating, ventilating, and air conditioning (HVAC) system* is a building system that controls the indoor climate of a building. HVAC functions are also often called "climate control." The primary functions of an HVAC system are heating, ventilating, and air conditioning (cooling), which give the system its name. **See Figure 7-1.** Additional functions include air filtration and humidity control. These functions are closely interrelated and use some of the same components.

Precise control of HVAC systems maintains a comfortable indoor climate for the building occupants, optimizes the indoor conditions for operating equipment or stored inventory, and performs these functions with minimum energy use during all times of the year. Automated HVAC systems also reduce air infiltration and maintain pressure relationships between spaces. HVAC systems are used to control the temperature of building spaces and rely on thermodynamics to function. In addition, HVAC systems are classified into two groups based on the method used to distribute heat energy throughout the building: forced-air and hydronic systems.

Facts

The laws of thermodynamics are rules that describe the science of heat and energy and involve measuring and controlling the temperatures of materials.

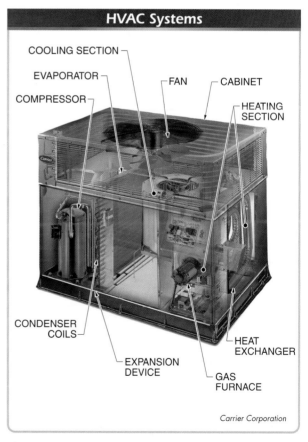

Carrier Corporation

Figure 7-1. Many commercial HVAC systems are package units that provide all of the primary HVAC functions.

Thermodynamics

Thermodynamics is the science of thermal energy (heat) and how it transforms to and from other forms of energy. **See Figure 7-2.** HVAC systems, particularly the heating and cooling functions, rely on the basic principles of thermodynamics and heat transfer. HVAC systems transform and redistribute heat energy to create the most desirable indoor climates. The two laws of thermodynamics apply to the heating and cooling of air in a building space.

First Law of Thermodynamics. The first law of thermodynamics states that energy cannot be created or destroyed but may be changed from one form to another. This is also known as the law of conservation of energy. An example of this law is the combustion process. As fuel burns, the hydrogen and carbon, which are found in most fuels, combine with oxygen in the air. This chemical reaction releases some of the chemical energy in the elements in the form of thermal energy. The hydrogen, carbon, and oxygen recombine to form new compounds that are the products of combustion. Two of the new compounds are carbon dioxide (CO_2) and water vapor (H_2O). During the process, energy has changed from chemical energy to thermal energy, but no energy has been created or destroyed.

Another example is within an air conditioning system. An electric motor uses electrical energy to drive a compressor, converting electrical energy into mechanical energy. The mechanical energy compresses the refrigerant in the system. As the compressed refrigerant expands, it produces a cooling effect. Most of the electrical energy used to drive the compressor results in the cooling effect of the system.

Some of the mechanical energy is converted to thermal energy because of friction, but no energy is created or destroyed in the process.

Second Law of Thermodynamics. The second law of thermodynamics states that heat always flows from a material at a higher temperature to a material at a lower temperature. This flow of heat is natural and does not require outside energy to facilitate the process. The second law of thermodynamics applies to all cases of heat transfer. *Heat transfer* is the movement of heat from one material to another. The rate of heat transfer increases with the temperature difference between two substances. For example, air in a furnace is heated by the products of combustion. Heat flows from the hot burner flame to the cool air.

Air conditioning systems use energy to control the movement of heat. Heat flows from warm room air into the cold refrigerant, cooling the air. Then, the now-hot refrigerant gives up this heat to the warm outside air. The refrigerant flow is driven by a compressor, but the flow of heat from the room to outside, via the medium of the refrigerant, is a natural example of the second law of thermodynamics.

The three methods of heat transfer are conduction, convection, and radiation. *Conduction* is the transfer of heat from molecule to molecule through a material. For example, if one end of a metal rod is heated, heat is transferred by conduction to the other end. *Convection* is the transfer of heat from warm to cool regions of a fluid from the circulation of currents. For example, as air is warmed by a fire, the warm air rises and is replaced by cool air. The movement of air creates a current that continues as long as heat is applied.

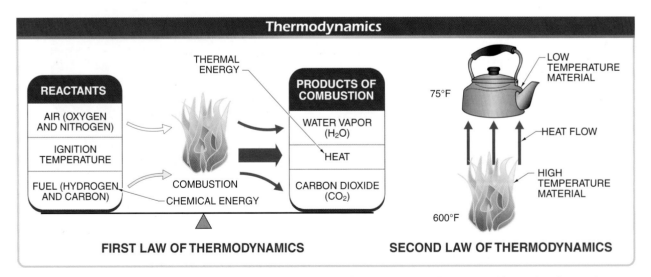

Figure 7-2. Thermodynamics is the science of thermal energy (heat) and how it transforms to and from other forms of energy.

Radiation is the transfer of heat between nontouching objects through radiant energy (electromagnetic waves). Radiant energy waves move through space, but they only produce heat when they contact an opaque object. For example, the radiant energy from an electric heating element passes through the air without heating it. However, this energy heats a person or object that comes into contact with it.

Forced-Air HVAC Systems

A *forced-air HVAC system* is a system that distributes conditioned air throughout a building in order to maintain the desired conditions. Forced-air HVAC systems use a system of ductwork, dampers, and fans to move conditioned air into the building areas where it is needed. *Conditioned air* is indoor air that has been given desirable qualities by the HVAC system. (This should not be confused with "air conditioning," which refers only to cooling the air.) For example, during the winter, the indoor air is conditioned by the heating functions of

the HVAC system. Hot air is introduced into a building space where the air is cool. The net result is a warm indoor climate.

Air-Handling Units. The hub of a forced-air HVAC system is an air-handling unit. This device combines many HVAC control devices together to fully condition the air at a central location before it is distributed throughout the building. An *air-handling unit (AHU)* is a forced-air HVAC system device consisting of some combination of fans, ductwork, filters, dampers, heating coils, cooling coils, humidifiers, dehumidifiers, sensors, and controls to condition and distribute supply air. **See Figure 7-3.** Depending on the application, AHUs may vary in size and configuration, but most are arranged similarly.

> **Facts**
>
> *A minimum amount of outdoor air must be introduced into most AHUs to ensure good IAQ and to pressurize a building.*

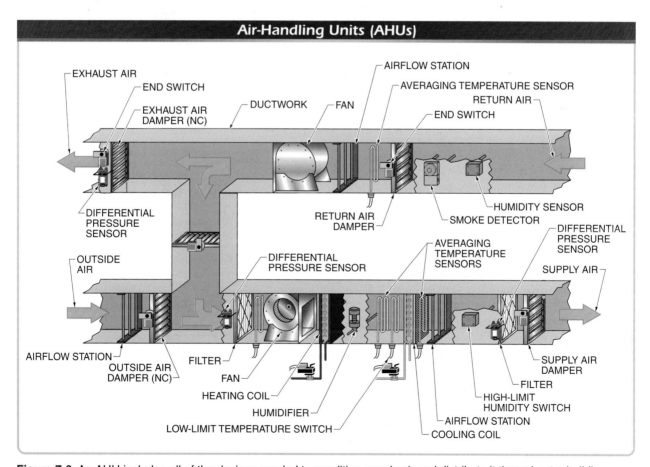

Figure 7-3. An AHU includes all of the devices needed to condition supply air and distribute it throughout a building.

AHUs manage five types of air. Entering the AHU are the return air and outside air. *Return air* is the air from within a building space that is drawn back into a forced-air HVAC system to be exhausted or reconditioned. *Outside air* is fresh air from outside a building that is incorporated into a forced-air HVAC system. Outside air is also known as makeup air, since it replaces air exhausted from the AHU. *Mixed air* is the blend of return air and outside air that is combined inside an AHU and goes on to be conditioned. Exhaust air and supply air leave the AHU. *Exhaust air,* also known as relief air, is the air that is ejected from a forced-air HVAC system. *Supply air* is the newly conditioned mixed air that is distributed from a forced-air HVAC system to a building space.

One or more fans draw air through the AHU. Several sets of dampers control the relative proportions and mixing of the outside air and return air in the mixed air plenum (duct section). Incorporating outside air is a way to supply fresh air into the building to prevent the air from becoming stale and to permit natural heating and cooling. Mixed air is filtered and passes across heating coils and cooling coils, as well as humidifiers, dehumidifiers, and/or sensors, and enters the building spaces as conditioned supply air. By controlling the fans, dampers, coils, and other devices within an AHU, the unit can be used to heat, cool, humidify, dehumidify, or simply ventilate a building space.

The devices within an AHU can be controlled individually, but it is more common in building automation systems to use a controller that includes most or all the functions of the AHU. The controller includes input connections for temperature, pressure, and other sensors and output connections for fan and damper control devices. The logic of how to change the outputs based on the different air conditions indicated from the input information may already be programmed into the controller. This is a much simpler way to interface with all of these devices, but the controller must be compatible with the specific AHU and its functions. For example, not all AHUs include heating functions, and some supply multiple building spaces with different climate requirements.

After being distributed from the AHU, air enters the building space through a terminal unit. A *terminal unit* is the end point in an HVAC distribution system where the conditioned medium (air, water, or steam) is added to or directly influences the environment of the conditioned building space. Forced-air terminal units include dampers to modulate the amount of conditioned supply

air into the space and may also include other devices to further condition the supply air. There are two ways that AHUs add conditioned air to the indoor building spaces in order to achieve climate control: the constant-volume method and the variable-volume method. **See Figure 7-4.**

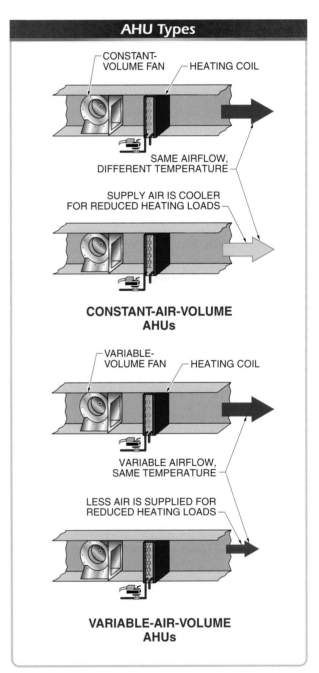

Figure 7-4. The two ways that AHUs add conditioned air to the indoor building spaces are the constant-volume method and the variable-volume method.

Constant-Air-Volume AHUs. A *constant-air-volume AHU* is an AHU that provides a steady supply of air and varies the heating, cooling, or other conditioning functions as necessary to maintain the desired setpoints within a building zone. Constant-air-volume AHUs operate at their rated airflow capacity (in cubic feet per minute) at all times. Many AHUs found in existing light commercial buildings are constant-air-volume AHUs. The disadvantage of constant-air-volume AHUs is that the fans operate at maximum power 100% of the time, which uses a great deal of energy.

Variable-Air-Volume AHUs. A *variable-air-volume AHU* is an AHU that provides air at a constant air temperature but varies the amount of supply air in order to maintain the desired setpoints within a building zone. Many variable-air-volume AHUs are used for cooling only. When heating is provided, it is by reheat coils in terminal units. Variable-air-volume AHUs are the most common AHUs installed in new commercial buildings. They generally use less energy and are quieter than constant-air-volume types because the fans operate at lower speeds.

A variable-air-volume AHU provides air at a constant air temperature.

Supply airflow is modulated either by controlling supply fan output or managing airflow within the AHU. For example, bypass dampers (common for small rooftop units) can reduce the supply airflow by allowing excess supply air to flow directly into the return duct.

A building may employ a small group of centralized variable-air-volume AHUs, or it may use a system that incorporates many smaller units located throughout

the building. A *variable-air-volume (VAV) terminal box* is a device located at a building zone that provides heating and airflow as needed in order to maintain the desired setpoints within the building zone. **See Figure 7-5.** As the zone temperature drops, heat is provided by a heating coil within the VAV terminal box. When the zone temperature reaches the setpoint, the heating coil is modulated to maintain the zone setpoint. Most VAV terminal boxes have a differential pressure switch that deenergizes the heating coil if airflow falls below a minimum setting, preventing damage to the VAV terminal box from overheating.

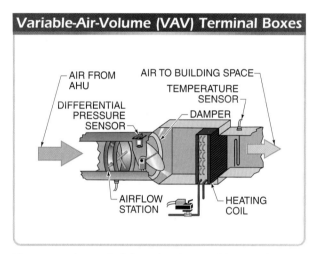

Figure 7-5. Instead of changing the conditioning of the air, VAV terminal boxes change the airflow volume of consistently conditioned air added to a building zone.

Rooftop Units. A *rooftop unit* is an HVAC package unit that provides heating and cooling to a building space but is mounted in an enclosure on the roof. **See Figure 7-6.** The mounting of the unit on the roof creates valuable free space inside the building and may make maintenance access easier. However, this ease of access may be offset by the technician's exposure to the weather during service and the possibility of having to carry heavy tools and equipment up to the roof when performing service. When unit replacement is needed, it is usually simple to use a crane to bring up one or more new units to replace the old one. Rooftop units usually have an electromechanical or electronic programmable thermostat in the zone or building. Modern rooftop units normally include direct digital control. This has allowed building management systems to monitor and control rooftop units.

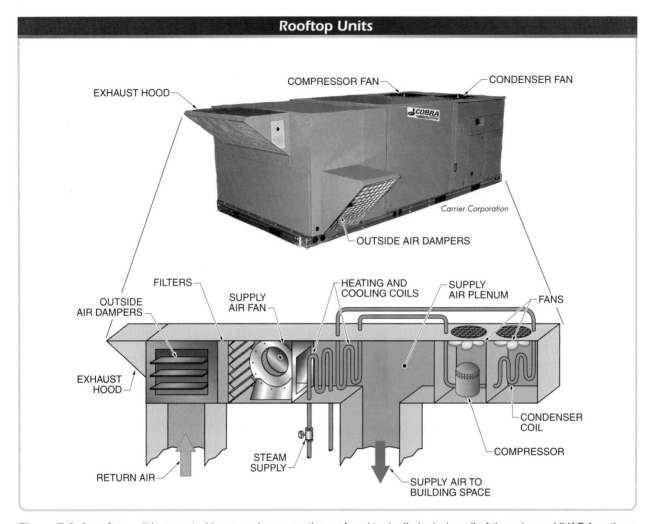

Figure 7-6. A rooftop unit is mounted in an enclosure on the roof and typically includes all of the primary HVAC functions.

Constant-Volume Variable-Temperature (CVVT) Systems. A constant-volume variable-temperature (CVVT) system is a single heating/cooling package unit that supplies air at a constant volume and uses dampers to vary the flow of conditioned (hot or cold) air to the zones. For heating, the unit switches from supplying cold air to supplying hot air to the CVVT boxes.

When both heating and cooling are needed to satisfy all the zones, a CVVT system becomes a time-share system with the package unit switching back and forth. If the package unit is in a heating mode and one of the zones requires cooling, the controls shut down the air-flow from that CVVT terminal unit until the package unit mode and zone needs match. These systems usually use little transportation energy because of their short duct runs. Also, they do not use reheat energy. They are often used with rooftop units.

> **Facts**
>
> *In cold climates, radiant floor heat has become popular in residential applications. Some homeowners prefer the gentle, even heat of a hydronic system to the heated air distributed by a forced air fan.*

Hydronic Systems

A *hydronic system* is a system that distributes water throughout a building as the heat-transfer medium for heating and cooling systems. Hydronic systems require pipes, valves, and pumps to distribute water throughout the building. **See Figure 7-7.** Hot water is used for heating, while chilled water is used for cooling. The hot water and cold water are typically distributed in two separate piping loops.

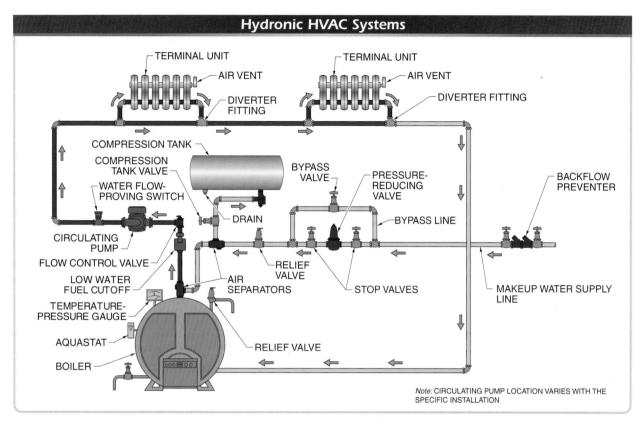

Figure 7-7. Hydronic HVAC systems consist of a device to heat or chill water and the piping and equipment to distribute that water throughout the building for heating and cooling purposes.

The water in a hydronic system can be used in two different ways. First, it can be piped directly into the building spaces, where it flows through a hydronic terminal unit that conditions the indoor air without forced airflow. In this case, the terminal unit is a heat exchanger that transfers heat between the water and the indoor air through radiative and/or convective means. A *heat exchanger* is a device that transfers heat from one fluid to another fluid without allowing the fluids to mix. **See Figure 7-8.** This type of system is primarily used for heating. Some of the oldest and most common examples of terminal units are hot-water radiators.

Alternatively, the water can flow through a coil within the ductwork of a forced-air HVAC system, transferring heat between the air and the water. Moving air provides a more efficient method of heat transfer. This arrangement can be used for both heating and cooling. Many commercial HVAC systems incorporate aspects of both forced-air and hydronic systems in this way, such as AHUs that use hot water from a boiler in the heating coils and chilled water from a chiller in the cooling coils.

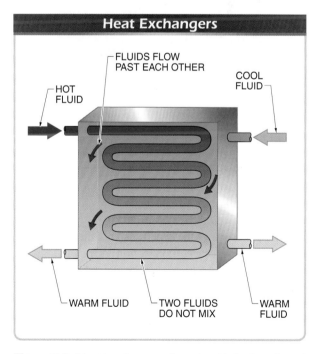

Figure 7-8. A heat exchanger allows heat to be transferred between two fluids without the fluids mixing.

Heating and Cooling Coils. Heating and cooling coils add or remove heat from indoor air. A *heating coil* is a heat exchanger that transfers heat to the air surrounding or flowing through it. Heating coils typically use hot water to heat the air. A *cooling coil* is a heat exchanger that transfers heat from the air surrounding or flowing through it. Cooling coils use chilled water or refrigerant vapor to cool the air. AHUs may contain one or more of each type of coil to provide the required heating or cooling capacity.

The two types of coils are constructed in much the same way. Most consist of a long length of tubing shaped to maximize the surface area exposed to the air. Fins of sheet aluminum may be attached to the tubing to add more surface area for heat conduction. **See Figure 7-9.** Either hot or cold fluid flows through the tubing, which acts as a heat exchanger, transferring heat to or from the air.

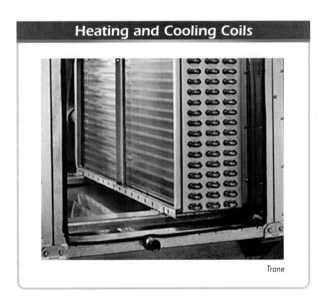

Heating and Cooling Coils

Trane

Figure 7-9. Heating and cooling coils are heat exchangers made from folded lengths of pipe with tabs of sheet metal added to increase the surface area for heat conduction.

Heating coils may also be made from electric heating elements. These are shaped similarly to the fluid versions in order to maximize the heat transfer to the air, but they require only an electrical connection and no fluid piping.

The fluid or electrical current flow through a coil can be controlled with valves or relays to regulate the temperature of the coil. As the temperature in the building space changes, controllers modulate the temperature or flow rate of the heat transfer medium. A faster flow provides greater heat transfer, which has a greater heating or cooling effect.

Alternatively, the supply air temperature can be controlled by changing the airflow volume over the coils, while the coil temperature remains constant. Faster airflow transfers more heat between the coils and the air.

Building Zones

Different areas within a building have different HVAC requirements. For simplicity, older HVAC systems attempted to provide an acceptable middle ground for the entire building. However, this can result in some areas receiving unnecessary climate control, which wastes energy, and other areas not being adequately controlled. HVAC systems with separate control over individual zones are becoming increasingly common, even in residential buildings.

A *zone* is an area within a building that shares the same HVAC requirements. A building may have many zones, divided by the use of the space, to create manageable units of HVAC conditioned spaces. **See Figure 7-10.** For example, a commercial building with a mixture of office and storage areas may divide the HVAC system functions into zones by these different areas. With many occupants, the office space climate should be carefully controlled for comfort. If this space is very large, it may be further divided into multiple zones. However, since fewer occupants are in the storage area, it may allow less stringent control over its indoor air climate, and can thus be a single separate zone. This multizone approach makes efficient use of energy while maximizing the comfort of the building occupants.

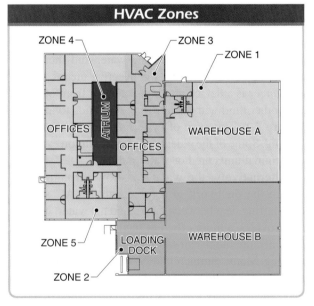

HVAC Zones

Figure 7-10. Buildings may be divided into multiple zones according to HVAC requirements, size, and location.

Single-Zone HVAC Systems. A single-zone HVAC system conditions air to only one building zone. **See Figure 7-11.** The size of the zone can vary widely but cannot be so large that the air conditions vary significantly throughout the space. Single-zone HVAC systems are identified by the heating and cooling coils that are in series and located in the central AHU. This is the simplest forced-air HVAC system configuration.

Multizone HVAC Systems. A multizone HVAC system conditions air for more than one building zone. The mixing of outside air and the filtration of the supply

air occurs at the main unit, which is located centrally. However, the devices managing the other functions, primarily heating and cooling, may be located elsewhere in the system, depending on the design. There are a few variations of multizone AHUs, which are distinguished by the locations of these devices.

The simplest versions of multizone AHUs retain all of the HVAC functions with the central unit, which outputs conditioned supply air to all of the zones. Heating and cooling coils may be located in series within the same duct or in separate parallel ducts. Either way, the supply air is fully conditioned or mixed before being distributed. Dampers at each zone control how much of this conditioned air is admitted to each zone, affecting the indoor climate of each zone differently. **See Figure 7-12.** This design is effectively a variable-air-volume system for each zone, since the zone controller can only manage the airflow into the zone. The air is conditioned to the aggregate requirements of all the zones. This system does not control every zone efficiently since the conditioned air is not customized for each zone's climate requirements.

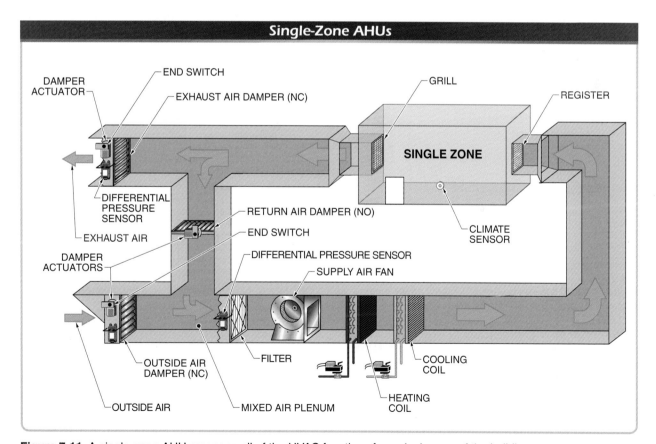

Figure 7-11. A single-zone AHU manages all of the HVAC functions for a single area of the building.

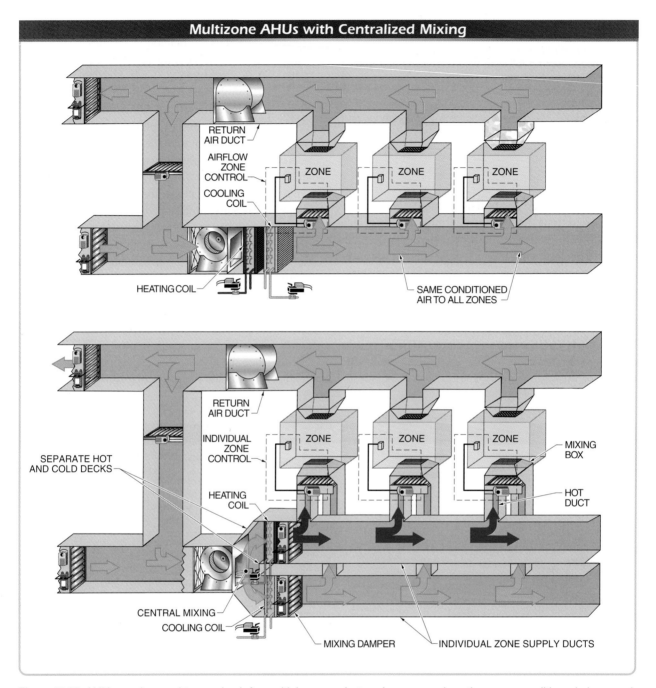

Multizone AHUs with Centralized Mixing

RETURN AIR DUCT

AIRFLOW ZONE CONTROL

COOLING COIL

ZONE ZONE ZONE

HEATING COIL

SAME CONDITIONED AIR TO ALL ZONES

RETURN AIR DUCT

INDIVIDUAL ZONE CONTROL

SEPARATE HOT AND COLD DECKS

HEATING COIL

ZONE ZONE ZONE

MIXING BOX

HOT DUCT

CENTRAL MIXING

COOLING COIL

MIXING DAMPER

INDIVIDUAL ZONE SUPPLY DUCTS

Figure 7-12. AHUs can be used to supply air for multiple zones, but each zone receives the same conditioned air, regardless of the comfort needs for the area.

There are several designs for forced-air HVAC systems that can control the supply air temperature to multiple zones individually. Some multizone AHUs can work with as many as 50 zones. The primary differences among these multizone AHU types are the locations of the heating and cooling coils and how the supply air is mixed. **See Figure 7-13.**

Dual-duct AHUs distribute hot and cold supply air separately to the building zones. The heating and cooling devices are located centrally, but in two separate ducts. The pair of ducts distributes both hot and cold air throughout the building. Dampers at or near the individual zones then mix the hot and cold air for the desired supply air temperature. This provides greater control for individual zones but is not very energy efficient since air is both heated and cooled. Also, the long lengths of ductwork can affect the supply air's outlet temperature (e.g., hot air cools and cold air warms).

Alternatively, a terminal reheat AHU places the cooling function with the central AHU and distributes supply air at a constant 55°F temperature to all building zones. Each building zone has a heating coil in the nearby ductwork that warms the supply air, as needed, to the required temperature. When cooling is required at a zone, its heating coil is shut off and the cool air flows into the zone.

COMFORT

HVAC systems are used in buildings to provide comfort to occupants. *Comfort* is the condition of a person not being able to sense a difference between themselves and the surrounding air. The five requirements for comfort are proper temperature, humidity, filtration, circulation, and ventilation. **See Figure 7-14.** Comfortable air conditions can vary among different people but generally fall within common ranges. People become uncomfortable when any of the five conditions are outside their acceptable range.

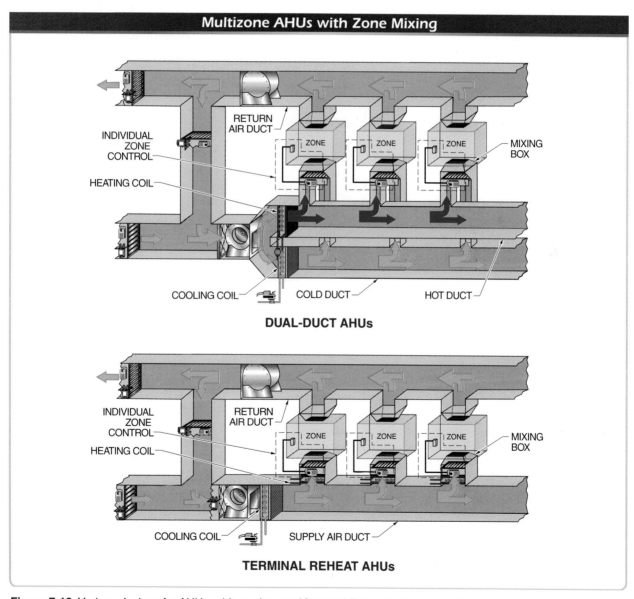

Figure 7-13. Various designs for AHUs address the need for providing customized conditioned supply air to multiple zones from a single central unit.

Comfort Requirements

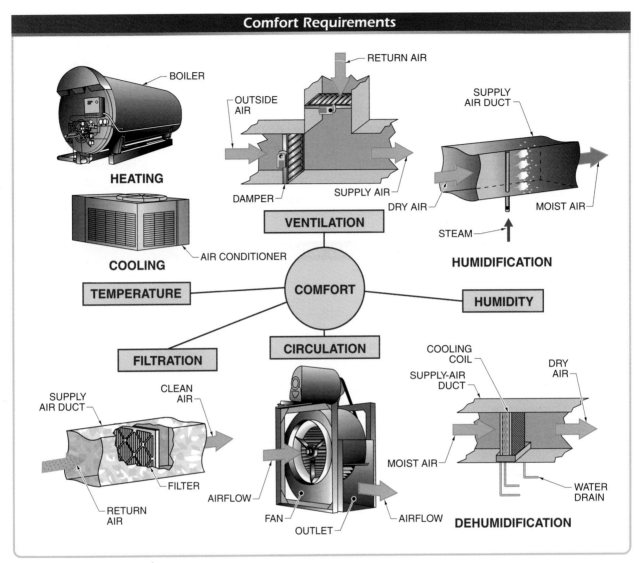

Figure 7-14. The comfort of building occupants relies on the temperature, humidity, circulation, filtration, and ventilation of the indoor climate.

Temperature

Temperature is the most important property of air that is controlled by an HVAC system. *Temperature* is the measurement of the intensity of the heat of a substance. When clothed, the human body is comfortable at an air temperature of approximately 70°F to 75°F. If the air temperature varies much above or below this range, the body begins to feel uncomfortably warm or uncomfortably cool. The HVAC system must condition indoor air to these temperatures, regardless of the outside temperature.

In the United States, the Fahrenheit scale is the most common temperature scale, though the Celsius scale may also be used. In the study of the properties of air, two different temperature measurements may be used: dry-bulb temperature and wet-bulb temperature. **See Figure 7-15.**

Dry-Bulb Temperature. The *dry-bulb temperature* is the temperature of air measured by a thermometer freely exposed to the air but shielded from radiation and moisture. Most temperature measurements are taken with dry bulb thermometers. If a temperature is not distinguished as dry-bulb or wet-bulb, it can be assumed to be a dry-bulb temperature. However, for a more detailed measurement of air conditions, both the dry-bulb temperature and wet-bulb temperature are considered.

Temperature

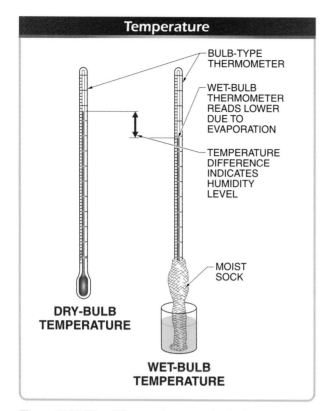

BULB-TYPE THERMOMETER

WET-BULB THERMOMETER READS LOWER DUE TO EVAPORATION

TEMPERATURE DIFFERENCE INDICATES HUMIDITY LEVEL

MOIST SOCK

DRY-BULB TEMPERATURE

WET-BULB TEMPERATURE

Figure 7-15. The difference between dry-bulb temperature and wet-bulb temperature is indicative of the amount of moisture in the air.

Wet-Bulb Temperature. The *wet-bulb temperature* is the temperature of air measured by a thermometer that has its bulb kept in contact with moisture. The bulb is wrapped in cloth that is wetted with water via wicking action. At a relative humidity below 100%, water evaporates from the cloth around the bulb, which cools the bulb to below ambient (dry-bulb) temperature. *Evaporation* is the process of a liquid changing to a vapor by absorbing heat. This lower temperature can be used to determine the relative humidity. A greater difference between the dry-bulb and wet-bulb temperatures means that water is evaporating quickly from around the wet bulb, which indicates a lower relative humidity. The precise relative humidity is determined with a psychrometric chart or by calculations.

Temperatures for automated systems are rarely measured with bulb-type thermometers. Specialized electronic sensors are used to measure some temperature and humidity properties, and the related properties are calculated from those results. However, the "bulb" names for these properties remain since they are descriptive of the relationships between temperature and humidity.

Humidity

Humidity is the amount of moisture present in the air. Some moisture is always present in air. A low humidity level indicates dry air that contains little moisture, and a high humidity level indicates damp air that contains a significant amount of moisture. *Absolute humidity* is the amount of water vapor in a particular volume of air. The most common units of absolute humidity are pounds of water per pound of dry air or grams of water per cubic meter of dry air.

Humidity ratio (W) is the ratio of the mass (weight) of the moisture in a quantity of air to the mass of the air and moisture together. Humidity ratio indicates the actual amount of moisture found in the air. Humidity ratio is expressed in grains (gr) of moisture per pound of dry air (gr/lb) or in pounds of moisture per pound of dry air (lb/lb). A *grain* is a unit of measure that equals $\frac{1}{7000}$ lb. For example, 1 lb of air contains 78 gr or 0.0111 lb of moisture $(1 \div 7000 \times 78 = 0.0111 \text{ lb})$.

Humidity represents latent heat. *Latent heat* is heat identified by a change of state and no temperature change. Therefore, latent heat cannot be measured with a thermometer. A certain quantity of heat is required to change water (liquid) into water vapor (gas). The heat required to make this change is latent heat. Because a certain amount of latent heat is required to evaporate a certain amount of water to water vapor, the amount of moisture (water vapor) in the air represents the amount of latent heat. **See Figure 7-16.** Heat must be removed from the air for the moisture to condense back to water.

Latent Heat

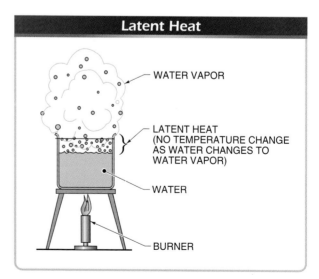

WATER VAPOR

LATENT HEAT (NO TEMPERATURE CHANGE AS WATER CHANGES TO WATER VAPOR)

WATER

BURNER

Figure 7-16. Because a certain amount of latent heat is required to evaporate a certain amount of water to water vapor, the amount of moisture (water vapor) in the air represents the amount of latent heat.

Warmer air has a greater capacity to be saturated with moisture than cooler air. Therefore, humidity is typically quantified as relative humidity, which relates the moisture content of the air within the context of current air temperature. *Relative humidity* is the ratio of the amount of water vapor in the air to the maximum moisture capacity of the air at a certain temperature. Relative humidity is represented as a percentage. A relative humidity of 50% means that the air is 50% saturated with moisture. A relative humidity of 100% represents the condition wherein the air is saturated with moisture for its temperature and dew begins to condense. *Dew point* is the air temperature below which moisture begins to condense.

Even if the amount of water in the air remains the same, the relative humidity changes as the air temperature changes. Since air has a greater moisture capacity at higher temperatures, the same amount of moisture in the air represents a smaller fraction at a higher temperature. For example, if air with 50% relative humidity is heated, the relative humidity may fall to 30% even though the amount of moisture in the air remains the same.

Humidity affects comfort because it determines the rate at which perspiration evaporates from the skin. Evaporation

of perspiration cools the body. Higher humidity slows the evaporation rate, and lower humidity increases the evaporation rate. Because of this, humidity levels can affect the perception of temperature even with no actual change in temperature. For example, at a constant temperature, an increase in humidity causes a person to feel warmer, while a decrease in humidity causes a person to feel cooler.

Psychrometrics is the scientific study of the properties of air and water vapor and the relationships between them. The properties of air found on a psychrometric chart are dry-bulb temperature, relative humidity, humidity ratio, wet-bulb temperature, enthalpy, and specific volume. Dew-point temperature is also sometimes included. When air is conditioned, one or more of the properties of air change. When one property changes, the others are affected. If any two properties of a sample of air are known, the others can be found by using a psychrometric chart.

The two properties used most often for identifying specific conditions of the air are temperature and humidity. In general, a dry-bulb temperature of 70°F to 75°F and a relative humidity of approximately 30% to 50% are considered comfortable. Therefore, comfort is defined as this area on the psychrometric chart. **See Figure 7-17.**

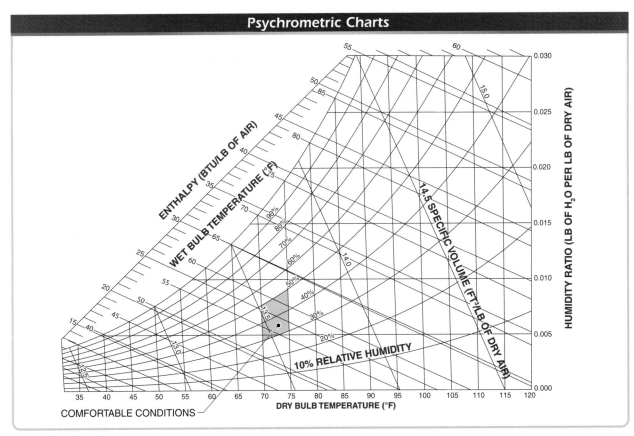

Figure 7-17. Psychrometric charts are used to show the relationships between the various properties of air at any condition.

At normal room temperatures, comfortable relative humidity levels are between about 30% and 50%. The lower end of the range is ideal for winter, while the higher end is ideal for summer. If the humidity is too low, a higher air temperature is required to feel comfortable. If the humidity is too high, a lower air temperature is required for the same feeling of comfort.

Besides controlling humidity for human comfort, it is often important to control humidity within a building for inventory or process reasons. High humidity promotes corrosion and mold growth, and low humidity promotes the buildup of electrostatic charge. Items such as sensitive electronics, food, and natural materials (e.g., paper and wood) can be damaged by excessively low or high humidity.

Air Circulation

Air in a building must be circulated continuously for maximum comfort. *Circulation* is the continuous movement of air through a building and its HVAC system. Ductwork and registers are used to move air throughout the building spaces. A *register* is a cover for the opening of ductwork into a building space. Registers include dampers to control the amount and direction of the airflow. **See Figure 7-18.** The air is blown from the supply air registers and is drawn back to the HVAC system through the return registers. When registers are sized and placed appropriately, the air flows through the entire space.

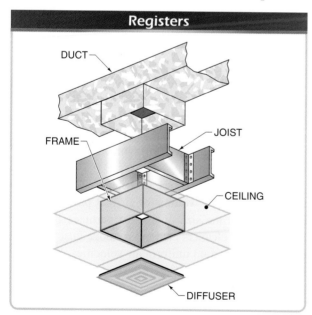

Figure 7-18. Registers are the interface between a building space and HVAC ductwork and can control the amount of airflow admitted to the space.

In a building where there is improper circulation, air rapidly becomes stale and uncomfortable. Proper air circulation also prevents temperature stratification. *Temperature stratification* is an undesirable variation of air temperature between the top and bottom of a space. **See Figure 7-19.** This is common in a building space where warm air rises to the ceiling and cool air falls to the floor. Circulating air ensures that the characteristics of indoor air are consistent throughout the space. Circulation also cycles the air through the HVAC system, which cleans and conditions the air for the desired temperature and humidity level. Stratification can also occur within AHUs, where it can cause control problems if temperature sensors are biased toward one extreme due to their location.

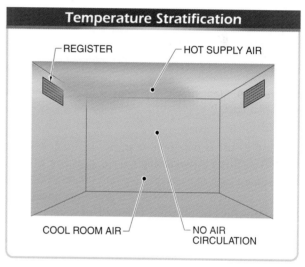

Figure 7-19. Poor air circulation can cause an uneven distribution of conditioned air (stratification) throughout a building space.

Air velocity is the speed at which air moves from one point to another. Air velocity is measured in feet per second (fps). Air circulation helps cool the body by evaporating perspiration. An increase in air velocity increases the rate of evaporation of perspiration from the skin, causing a person to feel cool. A decrease in air velocity reduces the evaporation rate, causing a person to feel warm.

Circulation is also measured in the volume of air that is cycled through the system. Air movement of 40 cfm (cubic feet per minute) per person is considered ideal. The *air changes per hour (ACH)* is a measure of the number of times the entire volume of air within a building space is circulated through the HVAC system in one hour. The requirements for air movement and ACH vary for different types of buildings and occupancy uses.

Static Pressure. *Static pressure* is the air pressure in a duct that pushes outward against the sides of the duct. It is measured at right angles to the direction of airflow. Static pressure is the pressure that has a tendency to burst a duct. *Static pressure drop* is the decrease in air pressure caused by friction between the air moving through a duct and the internal surfaces of the duct. **See Figure 7-20.**

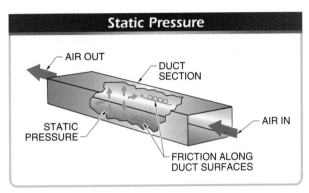

Figure 7-20. Static pressure pushes outward against the sides of a duct.

Air pressure is measured in inches of water column (in. WC). *Water column (WC)* is the pressure exerted by a square inch of a column of water. Inches of water column are used to express small pressures above and below atmospheric pressure. One inch of water column equals 0.036 psi.

Static pressure drop in a duct reduces the pressure of the air as the air moves along the duct. To measure static pressure in a duct, a small hole is drilled or punched in the side of the duct, and a flexible tube is run from the hole to one leg of a manometer. The difference between the static pressure readings at two points in a duct is the friction loss that has occurred in the duct between the two points.

Static pressure drop is the difference between the static pressure at the beginning of a duct section and the static pressure at the end of the duct section. Static pressure drop in a duct section can be calculated by dividing the length of the duct section by 100. The length of the duct section is divided by 100 because the design static pressure drop for the distribution system is per 100′ of duct.

Velocity Pressure. *Velocity pressure* is the air pressure in a duct that is measured parallel to the direction of airflow. Measuring velocity pressure without measuring static pressure is difficult. Velocity pressure is usually found by measuring total pressure and static pressure and subtracting the static pressure from the total pressure.

Dynamic pressure drop is the pressure drop in a duct fitting or transition caused by air turbulence as the air flows through the fitting or transition. Turbulence occurs wherever the airflow pattern is disturbed by a change in the airflow rate, direction of airflow, or size of the duct. Turbulence causes a drop in velocity pressure in the duct at the point where the turbulence occurs.

AIR-CLEANING METHODS

The three basic strategies used to reduce pollutant concentrations in indoor air are source control, ventilation, and air cleaning. *Source control* is a method of reducing air pollution by identifying strategies to reduce the origin of pollutants. Reducing pollutant origins may involve the use of exhaust systems at specific points where pollutants are present. The source control air-cleaning method requires the correct handling and storage of all chemicals and other pollutants found in a building. Ventilation can be used to reduce indoor air pollution by introducing various amounts of outside air in order to dilute indoor pollutants. *Air cleaning* is a method of reducing air pollution that includes electronic air cleaning, ultraviolet (UV) air cleaning, and the use of filters such as carbon filters.

Electronic Air Cleaners

An *electronic air cleaner* is a type of air cleaner in which particles in the air are positively charged in an ionizing section, and then the air moves to a negatively charged collecting area (i.e., plates) to separate the particles. **See Figure 7-21.** As the electrostatically charged particles pass through the negatively charged collecting area, the positively charged dirt particles are attracted to the negatively charged collector plates. Depending on the system, the particles clinging to the collector plates are washed off with water or blown off with compressed air.

The efficiency of electronic air cleaners is determined by the velocity and volume of air that passes through the plates and the size of the dirt particles collected. Electronic air cleaners can only remove dust particles effectively at certain uniform air velocities. Baffles or low-efficiency prefilters are used to ensure a uniform flow of air through electronic air cleaners.

Prefilters have the added benefit of eliminating larger particles that otherwise might not be captured and that contribute to plate arcing. Larger particles do not always have enough time to be properly charged and collected by the collector plates. Likewise, smaller particles do not have enough surface area to cling to the plates in the normal operation of an electronic air cleaner.

Proper care is also a major factor in electronic air-cleaner efficiency. It is important to schedule unit inspections and equipment maintenance as part of the preventive maintenance (PM) program. The manufacturer's specifications usually serve as a good baseline. Visual inspections can be used to adjust the interval of service as needed.

When cleaning electronic air cleaners, water or compressed air is used to clean all of the plate area. When not properly cleaned, dirt buildup creates a path between the negative and positive elements, which causes arcing. Arcing lowers efficiency and can take some units off-line.

The proper relative humidity should be maintained since excessive moisture can also cause arcing. As with any electronic equipment, it is important to use extreme caution when inspecting or maintaining electronic air cleaners.

Volatile organic compounds (VOCs) and any non-particulate pollutants are not removed by this method. Electronic air cleaners are used widely in residential systems but not often in large commercial systems due to their energy consumption and the difficulty of constructing plates large enough for the much higher air volumes.

A prefilter can be used to remove large particulates from air before the air passes through another filter.

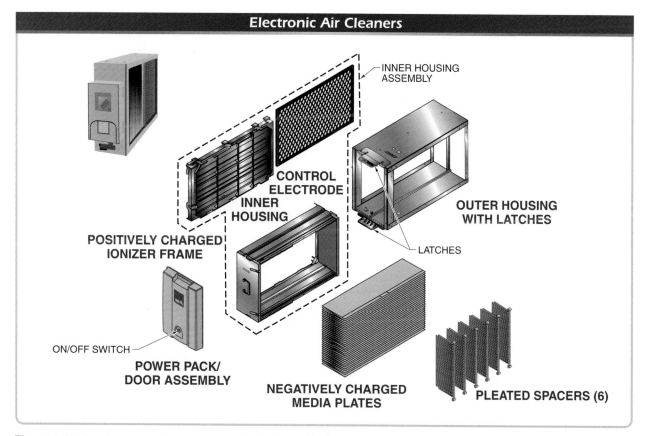

Electronic Air Cleaners

INNER HOUSING ASSEMBLY

CONTROL ELECTRODE

INNER HOUSING

POSITIVELY CHARGED IONIZER FRAME

OUTER HOUSING WITH LATCHES

LATCHES

ON/OFF SWITCH

POWER PACK/ DOOR ASSEMBLY

NEGATIVELY CHARGED MEDIA PLATES

PLEATED SPACERS (6)

Figure 7-21. An electronic air cleaner contains both positively charged and negatively charged surfaces to separate out particles in air.

Ultraviolet (UV) Air Cleaners

An *ultraviolet (UV) air cleaner* is a type of air cleaner that kills biological contaminants using a specific light wavelength. **See Figure 7-22.** Ultraviolet systems use UV light-generating equipment. Nonbiological pollutants such as particulates and VOCs are not affected by UV systems. UV systems are better suited for smaller systems or in spot use, such as in a condensate pan or condenser water system where biological growth commonly occurs.

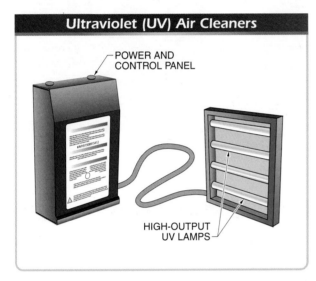

Figure 7-22. Ultraviolet systems use UV light-generating equipment to remove biological contaminants.

Carbon Filters

A *carbon filter* is a type of air cleaner in which an activated carbon medium is used to remove most odors, gases, smoke, and smog from the air by means of an adsorption process. **See Figure 7-23.** Using carbon filters, however, usually results in modifications to the filter section, higher filter costs, increased maintenance time, high-pressure drops, and increased energy costs. When an indoor-air source is the problem, the contaminants must be removed at the source using localized exhaust. By using exhaust fans, the problem can be solved without having to redesign the air-handling system.

Solving indoor air quality (IAQ) problems may require a combination of techniques. For example, when construction is taking place on a particular floor, localized exhaust prevents the contaminants from entering the main air distribution system and spreading to other floors.

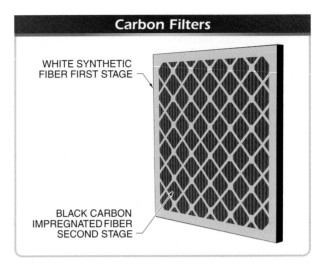

Figure 7-23. A carbon filter uses an activated carbon medium to remove most odors, gases, smoke, and smog from the air by means of an adsorption process.

Filtration

Filtration is the process of removing particulate matter from air. The most common method of filtration is forcing the contaminated air through a mechanical filter. A *filter* is a type of air cleaner in which a mechanical device is used to remove particles from the air. Many types of filters are used in commercial building applications. A filter contains tiny pores through which the molecules of air can pass but the much larger particulates cannot. **See Figure 7-24.** The larger particles become trapped in the filter. Some small particles pass through the filter. Since forced-air HVAC systems already include the equipment to effectively circulate air, filtration involves only adding a filter within the ductwork.

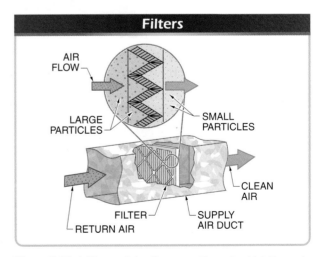

Figure 7-24. A filter contains tiny pores through which the molecules of air can pass, but the much larger particulates cannot.

Mechanical Filters. Mechanical filters contain a fiber that has two purposes. The first purpose is to trap any particulate matter, while the second purpose is for the particulates to remain attached to the fibers of the filter medium. The filter medium traps particulate matter in several different ways. The fibers may be arranged in layers or interwoven using alternating thicknesses of fiber. Sometimes adhesives are applied during the manufacture of the filter or applied as a spray product at the time of installation. Other filters use synthetic electrostatically charged material. **See Figure 7-25.**

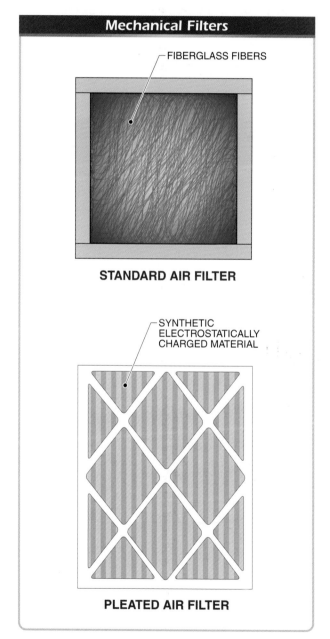

Figure 7-25. Mechanical filters are available in many forms.

Proper care of mechanical filters is important in preventing potential IAQ problems. Guidelines in caring for filters include the following:

- Ensure that filters meet the specifications of the air-handler manufacturer for size and thickness. A filter that is too thick impairs the functionality of the mechanical equipment. A filter that is too thin for the application will not stop enough particles to effectively clean the air.

- Include filter inspections and maintenance in the PM program.

- Change filters according to the manufacturer's specifications. Always replace filters immediately when damaged in any way.

- Keep filters dry by controlling water contamination and relative humidity.

- Filters should be installed so that they fit snugly with no leaks around the edges. The air-handler seams must be properly sealed. Most filters have arrows stamped on them that indicate how the filter must be placed in relation to the direction of airflow.

- Use caution when handling dirty filters. Wear gloves and a respirator to protect against particles released from the filters. Try not to disturb the dirt when moving the filters. Dispose of the filters as prescribed by maintenance procedures or local regulations.

Various grades of filters are available. Equipment manufacturer specifications must be checked for the recommended size, thickness, and flammability classification. Filters are classified as Class 1 (do not burn with negligible amounts of smoke) and Class 2 (burn moderately with some smoke). Some systems require a low flammability classification.

A thicker filter results in higher energy consumption. Stationary engineers must consult with the equipment manufacturer before changing the thickness or type of filter being used to ensure that the new filter will perform in the equipment as needed.

Antimicrobial filters are appropriate in some cases where outdoor mold and mildew are a problem. When indoor mold and mildew are a problem, filtration is not the answer. The contamination must be identified and eliminated.

When the outside-air or indoor-air source is of good quality and the filters are filtering air properly (e.g., there is no bypassing of the filter medium), the filtration system does not require modification. However, when the mixed-air source contains dust, soot, and/or other particulate matter, a high-efficiency filter may be required.

There are some drawbacks associated with replacing a filter with a high-efficiency filter. A high-efficiency filter may not fit into the filter section, the increased pressure drop caused by the high-efficiency filter may reduce airflow, and the cost may be higher than the building owner wishes to spend. When any of these problems prevent use of a high-efficiency filter, more frequent filter changes may be the only solution.

Filter Selection and Ratings. The quality of filtered air can directly affect IAQ, depending on the type and efficiency of the filtering medium. Filters are rated for efficiency, airflow resistance, and dust-holding capacity. Filter efficiency ratings, such as MERV ratings, indicate the ability of a filter to remove particulate matter from air. **See Figure 7-26.**

Filter efficiency is related to the sizes of the particulates that are caught in the filter compared to the sizes of those that pass through. The size of airborne particles is measured in microns. A *micron* is a unit of measure equal to one-millionth of a meter or 0.000039″.

Filter airflow resistance is the pressure drop across a filter at a given velocity. The *dust-holding capacity* is a filter's ability to hold dust without seriously reducing the filter's efficiency. The type and design of a filter determines its efficiency at removing particles of a given size and the amount of energy required to pull or push air through the filter. Filter media can be dry, viscous (sticky), cleanable, activated carbon, high-efficiency, or electronic.

MERV Ratings						
	ASHRAE 52.2			ASHRAE 52.1		
MERV	**3–10***	**1–3***	**0.3–1***	**Arrestance***	**Dust Spot***	**Dust Spot†**
1	<20	—	—	<65	<20	>10
2	<20	—	—	65–70	<20	>10
3	<20	—	—	70–75	<20	>10
4	<20	—	—	>75	<20	>10
5	20–35	—	—	80–85	<20	3.0–10
6	35–50	—	—	>90	<20	3.0–10
7	50–70	—	—	>90	20–25	3.0–10
8	>70	—	—	>95	25–30	3.0–10
9	>85	<50	—	>95	40–45	1.0–3.0
10	>85	50–65	—	>95	50–55	1.0–3.0
11	>85	65–80	—	>98	60–65	1.0–3.0
12	>90	>80	—	>98	70–75	1.0–3.0
13	>90	>90	<75	>98	80–90	0.3–1.0
14	>90	>90	75–85	>98	90–95	0.3–1.0
15	>90	>90	85–95	>98	>95	0.3–1.0
16	>95	>95	>95	>98	>95	0.3–1.0
17†	>99	>99	>99	—	>99	0.3–1.0
18†	>99	>99	>99	—	>99	0.3–1.0
19†	>99	>99	>99	—	>99	0.3–1.0
20	>99	>99	>99	—	>99	0.3–1.0
MERV	**Application**					
1–4	Minimum residential, minimum commercial, equipment protection					
5–8	Commercial, industrial, superior residential					
9–12	Superior commercial, superior industrial, best residential					
13–16	Medical facilities, surgical rooms, best commercial, smoke protection					
17–20	Clean rooms, hazardous materials					

* in microns and % trapped
† in microns

Figure 7-26. MERV rating indicates the ability of a filter to remove particulate matter from air.

Filters must be selected for their ability to protect both the HVAC system components and general air quality. In many buildings, the best choice is a medium-efficiency, pleated filter with a prefilter. Medium-efficiency, pleated filters have a higher level of removal efficiency than low-efficiency filters. However, medium-efficiency filters last without clogging for longer periods of time than high-efficiency filters.

Dust spot efficiency is a filter rating that measures a filter's ability to remove large particles from the air that tend to soil building interiors. Dust-holding capacity is a measure of the total amount of dust that a filter is able to hold during a dust-loading test. In the past, filters were rated by the method known as ASHRAE dust spot rating.

Facts

If a filter becomes covered with dust, dirt, or other contaminant particles, the pressure drop across the filter increases and the airflow decreases.

Filter Replacement. Filters should be checked and changed periodically to maintain clean air and system performance. A dirty filter will excessively impede airflow and the HVAC system must work harder and use more energy to continue circulating the air. Filters are usually scheduled to be replaced when the filter has performed a specific amount of work, sometimes indicated by a specific amount of time in use. To determine whether a filter must be replaced, an electronic manometer can be used to measure the increase in pressure drop across the filter and/or a visual inspection is performed to assess the amount of dirt clogging the filter. **See Figure 7-27.**

Periodic visual inspections help to establish a baseline PM program so that filters are changed at preset intervals. The intervals can be set up by month, quarter, or total hours of operation. The interval serves as a way to approximate how often filters should be changed. The intervals can be lengthened or shortened as required to conform to varying conditions of operation and outside air quality.

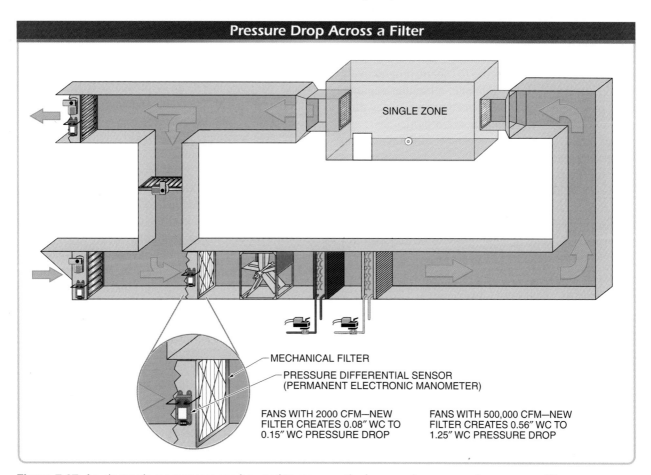

Pressure Drop Across a Filter

SINGLE ZONE

MECHANICAL FILTER

PRESSURE DIFFERENTIAL SENSOR
(PERMANENT ELECTRONIC MANOMETER)

FANS WITH 2000 CFM—NEW FILTER CREATES 0.08″ WC TO 0.15″ WC PRESSURE DROP

FANS WITH 500,000 CFM—NEW FILTER CREATES 0.56″ WC TO 1.25″ WC PRESSURE DROP

Figure 7-27. An electronic manometer can be used to measure the increase in pressure drop across a filter.

For example, a PM program may call for a filter to be changed when a manometer indicates that the pressure drop across the filter reads 0.9 in. WC. Other pressure drop values are used based on the filter design, airflow resistance, and dust-holding capacity of the filter. This may make economic sense, but other factors must be considered. In some areas, excessive pollen accumulation in the spring season will require that the filter be replaced more frequently. Other conditions may also require a filter to be changed outside of the regular time interval. For instance, water from a nearby humidifier might accidentally dampen the filter. Adjustments to the filter time interval will help prevent IAQ problems such as excessive particles in the air, reduced airflow to occupants, or microbial growth from molds or mildew.

Filters are available with different efficiencies. For example, prefilters, which are relatively inexpensive and are changed up to four times per year, have an efficiency of about 30%. This means that they capture about 30% of large particulate matter. They are changed so frequently because they are quickly clogged by the large particulates. Final filters, however, are changed every 12 to 18 months and have efficiencies of 60% to 90%. This means that they capture up to 90% of small particulates.

HVAC systems maximize the life of their filters by using two or more filters in series to capture progressively smaller particulates. For example, a prefilter captures the largest particulates, then a medium-efficiency filter captures many of the common-sized particulates, and a high-efficiency filter captures most of the remaining fine particulates.

Some filters are designed for disposal after use, while others can be cleaned and reused. Special filters are also available that remove organic and biological compounds or absorb specific types of gases and vapors from the airstream. The type of filter and size of the openings depends on the degree of filtration required.

Ventilation

Air within a building gradually becomes stale from exhaled carbon dioxide and chemical vapors, which are unhealthy for occupants. Indoor air is kept comfortable by diluting stale recirculated air with fresh outside air. **See Figure 7-28.** *Stale air* is air with high concentrations of carbon dioxide and/or other vapor pollutants. *Ventilation* is the process of introducing fresh outside air into a building. Outside air joins with return air recirculated from the building space to form mixed air, which is then conditioned as necessary and supplied back to the building space.

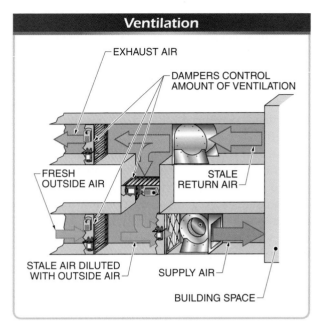

Figure 7-28. The positions of damper blades control how much fresh outside air is added to the return air to ventilate building spaces.

Stale air is quantified by measuring the CO_2 level in the return air. CO_2 is measured in parts per million (ppm) by volume. The amount of CO_2 in the atmosphere is about 390 ppm. Levels up to about 1000 ppm in indoor air are considered acceptable in many buildings.

Ventilation requirements are based on the building's occupancy and type of indoor activity, such as office work, classes, or manufacturing. For example, the ventilation requirement for a typical office building space is 20 cfm per person. Commonly, 5% to 30% of the air circulation for commercial buildings is composed of outside air. The position of the blades of AHU dampers controls the amount of ventilation.

Some industrial, research, and commercial buildings cannot recirculate any return air because of potentially dangerous air contaminants and must use 100% outside air for ventilation. If the indoor and outdoor climates are very different, such as during winter in northern locations, the amount of ventilation greatly affects the operation of the HVAC devices and the energy used.

Economizers

An *air-handler economizer* is an HVAC unit that uses outside air for free cooling. The outside air is used instead of mechanical cooling from chilled water or direct expansion air conditioning coils. A *direct expansion (DX) system* is a system in which the refrigerant expands directly inside a coil in the main airstream itself to affect the cooling of the air. Buildings with high internal heat gain require cooling even when it is cold outside. An economizer system is a control arrangement that allows cool outside air to be used to provide free cooling for buildings with high internal heat gain. For economizer systems to operate effectively, special control schemes must be used that allow economizer operation only when outside air is at a proper temperature. Building economizer systems fall into two groups: dry-bulb economizers and enthalpy economizers.

Dry-Bulb Economizers. A *dry-bulb economizer* is a type of economizer that operates strictly in proportion to the outside-air temperature, with no reference to humidity values. As long as the outside-air temperature is below a specific value, the air can be used for economizer cooling. **See Figure 7-29.** Usually, the temperature of the outside air must be below 65°F. The exact setting is dependent upon the prevailing climate in specific geographical regions. When the outside-air temperature rises and is too warm for free economizer cooling, the economizer cycle is ended. Ending the economizer cycle forces the outside-air damper to a minimum value, which allows enough outside airflow for ventilation purposes only.

Enthalpy Economizers. An *enthalpy economizer* is a type of economizer that uses temperature and humidity levels of the outside air to control operation. *Enthalpy* is the total heat content of a substance. For enthalpy

control to work, a combination temperature and humidity sensor can be mounted in the outside air. However, due to the problems associated with humidity sensors in the outdoor environment, humidity readings can be taken in the mixed-air airstream using new algorithms for the controller. **See Figure 7-30.** A controller uses the temperature and humidity values to calculate the current enthalpy of the outside air in British thermal units per pound (Btu/lb). When the enthalpy of the outside air is low enough, the economizer cycle of the HVAC system is used.

A pressure transmitter can send a signal to a controller indicating that an air filter needs to be replaced.

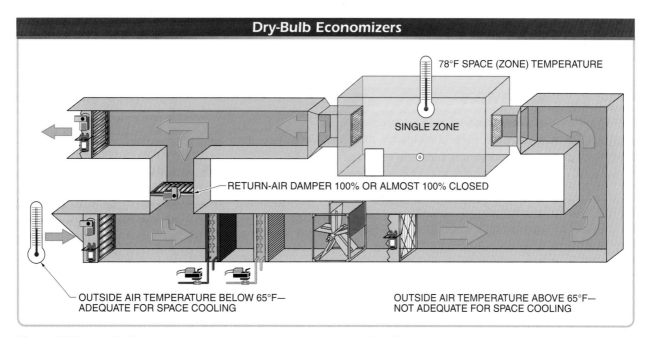

Figure 7-29. A dry-bulb economizer uses outside-air temperature only, without regard to humidity, to control its operation.

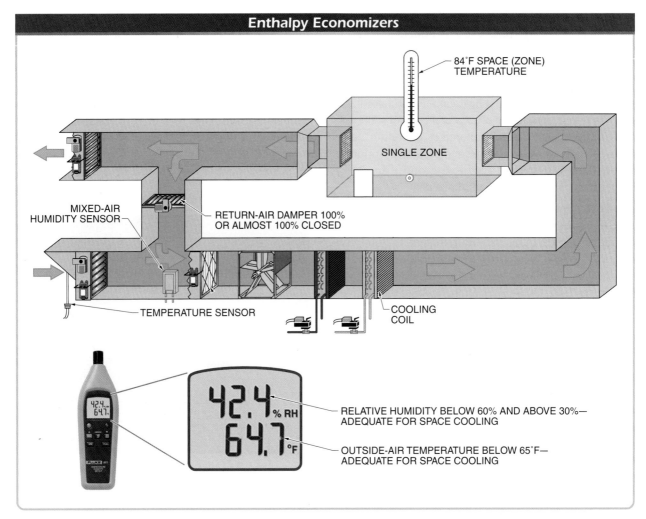

Figure 7-30. An enthalpy economizer uses temperature and humidity levels of the outside air to control its operation.

HVAC CONTROL DEVICES

HVAC systems are composed of many devices and subsystems used together in some combination to accomplish the types of conditioning functions expected for the local climate. For example, an HVAC system in a warm climate includes devices for cooling but may not include those for heating.

These devices are all involved in measuring or controlling the characteristics of the indoor climate that affect comfort. HVAC control devices include sensors to measure the current climate conditions and the presence of occupants within a zone, devices to control the temperature and humidity of the supply air, and devices to control the distribution of air, water, or steam throughout the HVAC system.

HVAC Schematic Representations

HVAC system documents, particularly those specifying the control devices and sequences, typically include schematic representations of the system and its components. **See Figure 7-31.** These are not meant to be exact depictions of the physical size and shape of the equipment, but rather a representation of the basic operation of the system and the relationships between the components. Similar schematics may also appear on human-machine interfaces (HMIs) for monitoring system operation.

HVAC system schematics use symbols to represent many of the control devices involved in the system. These symbols are not standardized across the HVAC controls industry, but most are similar and recognizable as correlating to a certain physical component. The symbols may also include information about what type of data they share or receive from controllers or other devices, such as digital inputs (DIs), analog inputs (AIs), digital outputs (DOs), and analog outputs (AOs).

Some devices with similar functions may be either analog or digital, depending on the needed information. For example, the valve controlling a heating coil may accept a digital or an analog signal, corresponding to a completely open/closed position or any position in between, depending on the type of valve or heating system.

It is important to note that inputs and outputs are designated from the point of view of the HVAC system controller, such as an AHU controller. For example, a temperature sensor outputs a signal, but that signal is received as an input to the controller. Therefore, the temperature sensor is considered an input device.

Likewise, the controller outputs a signal to a damper actuator, which receives the signal as an input.

Some devices interface with more than one signal, or even more than one type of signal. For example, dampers receive an analog output signal to move to a different position, but they may also include an end switch that provides a digital input to the controller if the damper closes completely.

Facts

Electronic actuators provide precise control of HVAC systems. Many electronic actuators can operate on a variety of input signals from a controller. The inputs to the actuator can be specified by setting DIP switches on the actuator. In addition, the direction and length of stroke can be changed.

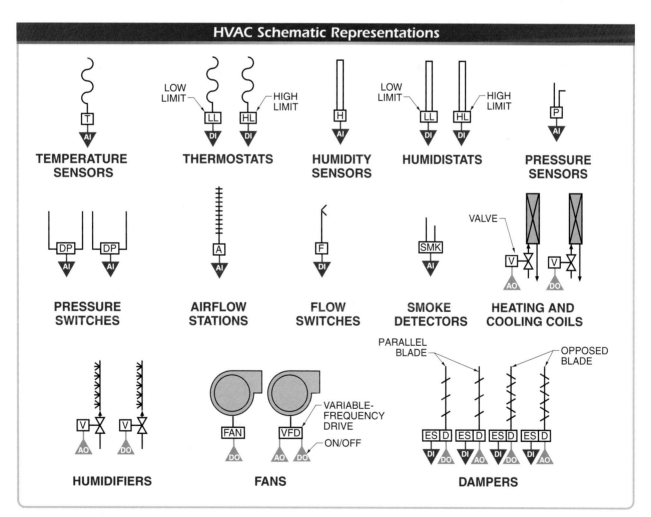

Figure 7-31. HVAC system documents, particularly those specifying the control devices and sequences, typically include schematic representations of the system and its components.

Climate Sensors

Climate sensors measure the comfort-related properties of indoor air: temperature, humidity, circulation, filtration, and ventilation. There may be many climate sensors in a building, or even in an individual zone. Climate sensors are also installed within air-handling systems to monitor the mixing and conditioning of the supply air.

These electronic sensors communicate the values of these properties to local or central controllers, which then make decisions on HVAC system operation based on this information. The goal is to continually adjust the system functions until the sensors read values within an optimal range. This is an example of a closed-loop control system.

Temperature Sensors. A *temperature sensor* is a device that measures a property related to temperature. Temperature sensors are the most common inputs used in building automation systems and are used to measure temperatures in ducts, pipes, and rooms. Temperature sensors can be designed for measuring either air temperature or water temperature. **See Figure 7-32.** The technology for both types is the same, but the device packaging usually varies for different types of mounting or for protecting the sensor from the different environments.

Air-sensing temperature sensors can be mounted on walls or inside ducts. Wall-mount temperature sensors measure air temperatures within building spaces. Duct-mount temperature sensors are used to measure air temperatures inside ductwork. Since airflow through large ducts can actually have slightly different temperatures within the duct cross-section, small temperature sensors mounted in one spot can be inaccurate. Instead, an averaging temperature sensor is used. Averaging temperature sensors have a long, coiled sensing element that is mounted within a large duct to measure an overall air temperature. Temperature sensors for outside-air temperatures are housed in weather-resistant enclosures.

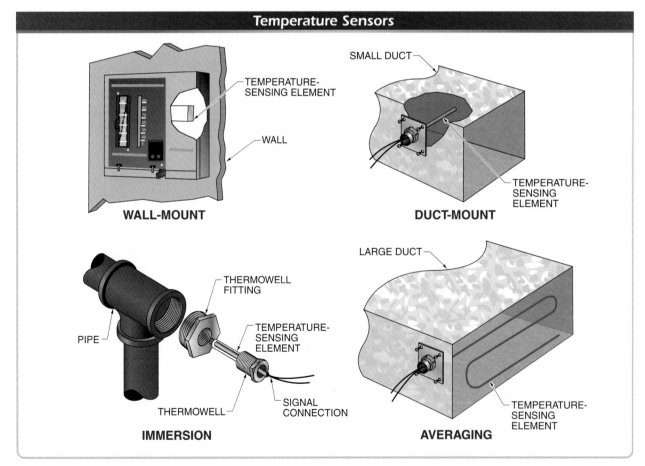

Temperature Sensors

WALL-MOUNT

DUCT-MOUNT

IMMERSION

AVERAGING

Figure 7-32. Temperature sensors are available in many types of packages for installation in different parts of an HVAC system.

Temperature sensors used for liquid measurements are known as immersion temperature sensors. The sensing element of an immersion temperature sensor is encased within a small, watertight, thermally conductive casing, known as a thermowell, that is mounted inside the pipe, vessel, or fixture containing the water to be measured. Special fittings may be necessary to allow the connection of leak-proof conduit for the conductors to pass outside the water-containing vessel.

There are several types of electric temperature sensors that can measure and communicate temperature. These sensors are simple in design and relatively inexpensive but offer significant capabilities when coupled with sensor electronics. Each of these can be combined with electronics to convert their output to standard analog signals or structured digital network message information that corresponds to the temperature. **See Figure 7-33.** The primary types of electric temperature-sensing elements are thermocouples, resistance temperature detectors (RTDs), and thermistors.

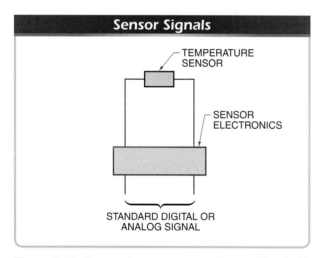

Figure 7-33. Temperature sensors can be combined with electronics to convert their output to standard analog signals or structured digital network message information.

Thermocouples. A *thermocouple* is a temperature-sensing element consisting of two dissimilar metal wires joined at the sensing end that generate a small voltage that varies in proportion to the temperature at the hot junction. **See Figure 7-34.** The sensing end is usually called the hot junction even though it may be used to measure cold objects. The cold junction is located at the other end of the wires at the voltmeter and is normally kept at 32°F. Modern meters use a separate temperature measurement instead of the cold junction and use software to compensate for the fact that the cold junction is not at the standard temperature. With a sensitive voltmeter, the temperature can be accurately measured over a wide range. Many different combinations of wires can be used, but a common combination is copper and constantan.

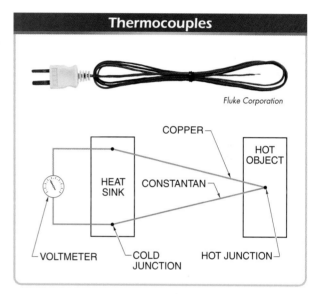

Figure 7-34. A thermocouple consists of two dissimilar metal wires joined at the sensing end that generate a small voltage related to the temperature.

Facts

Many measurement errors are caused by unintended thermocouple junctions. Any time wires of two different metals are connected together, a thermocouple junction is formed. Care must be taken to ensure that the proper extension wires are used.

Resistance Temperature Detectors. A *resistance temperature detector (RTD)* is a temperature-sensing element made from a metal or alloy conductor with an electrical resistance that changes with temperature. RTDs are usually made from platinum, so they are relatively expensive. **See Figure 7-35.** Since RTDs do not generate a voltage of their own, they must have a power source and a measuring circuit. All RTDs of similar design are normally manufactured to have a fixed resistance at a certain temperature. This allows them to be interchangeable.

Resistance Temperature Detectors (RTDs)

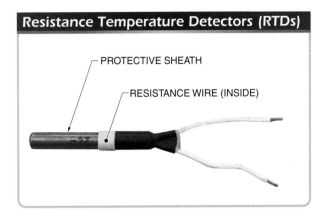

Figure 7-35. An RTD is made from a metal or alloy conductor with an electrical resistance that changes with temperature.

Thermistors. A *thermistor* is a temperature-sensing element made from a semiconductor with an electrical resistance that changes with temperature. Thermistors are available in several shapes such as rods, disks, beads, washers, and flakes. **See Figure 7-36.** An external voltage source is connected to the thermistor and the voltage drop across its resistance is measured with a voltmeter. If the voltage source remains constant, the measured voltage drop changes with the thermistor's resistance and is proportional to the temperature.

Thermistors are similar to RTDs in operating principle but are made from different materials and therefore have different electrical/temperature characteristics. They typically have a larger resistance than RTDs, giving them a narrower possible temperature range. They also have a faster response to temperature changes.

Since the measured thermistor voltage can be much greater than thermocouple voltages for comparable temperatures, thermistor measurements are usually more precise. However, thermistors are susceptible to inaccuracies when the thermistor and the sensor electronics are connected with long conductors. Long conductors add resistance to a circuit, which affects the voltage measurements. If this is unavoidable, the sensor or system software may allow offset or calibration features to compensate for these inaccuracies.

Thermistors can typically be used over a temperature measurement range of –22°F to 212°F. However, prolonged exposure to elevated temperatures, even below the maximum limit specified by the manufacturer, can cause permanent damage. A typical application uses a thermistor for only a fraction of that range. RTDs and thermistors have different advantages and disadvantages. **See Figure 7-37.** The choice of the proper sensor depends on the application.

Thermistors

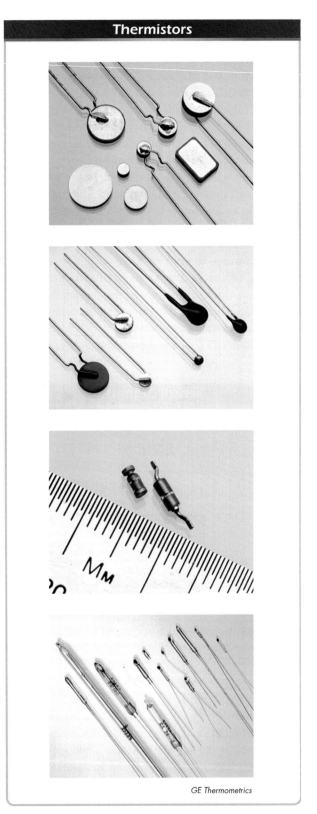

GE Thermometrics

Figure 7-36. A thermistor is made from a semiconductor material with an electrical resistance that changes in response to changing temperatures.

Comparison of Thermistors and RTDs		
Type	**Advantages**	**Disadvantages**
Thermistor	• Fast response • Small size • High resistances eliminate most lead resistance problems • Rugged, not affected by shock or vibration • Inexpensive	• Nonlinear • Narrow span for any singular input • Interchangeability limited unless matched pairs are used
RTD	• Linear over wide range • Wide temperature range • High temperature range • Interchangeable over wide range • Better stability at high temperatures	• Low sensitivity • More expensive • No point sensing • Affected by shock and vibration • Requires 3- or 4-wire operation • Can be affected by contact resistance

Figure 7-37. RTDs and thermistors have different advantages and disadvantages.

Thermostats. Most temperature inputs for automated HVAC systems require analog temperature values. However, there may be certain circumstances that require only a comparison of a temperature to a setpoint. Thermostats can be used in these applications. A *thermostat* is a switch that activates at temperatures either above or below a certain setpoint. For example, a thermostat may be used to close a set of contacts when a low-temperature limit is reached, shutting down cooling processes to avoid ice formation in the HVAC ducts. When they are applicable, thermostats may be more desirable than thermometers because they are simpler to configure and may be less expensive.

Thermostats provide only contact closure outputs, which are normally open and/or normally closed. The temperature sensing may be done with either an electronic or electromechanical device. Electromechanical temperature-sensing devices include an element that moves in proportion to the ambient temperature. For example, a bimetallic strip is a temperature-sensing element composed of two dissimilar metals fused together, each with a different coefficient of thermal expansion. **See Figure 7-38.** As one material expands more than the other, the bonded strip bends toward one side. This movement can be used to open or close contacts at a certain temperature.

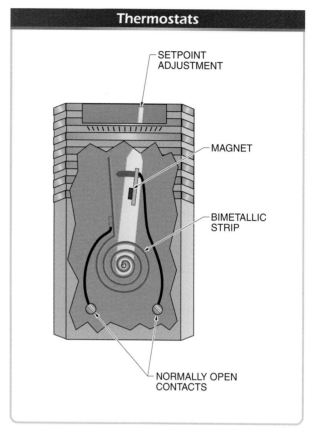

Thermostats

SETPOINT ADJUSTMENT

MAGNET

BIMETALLIC STRIP

NORMALLY OPEN CONTACTS

Figure 7-38. A thermostat is a switch that activates at temperatures either above or below a certain setpoint.

The setpoint temperature is typically set manually with an adjustment screw in the device. In the simplest systems, the contact closure is used to switch another device, such as a fan, on or off through a relay. With electronic controllers, the contact closure is read as a digital input into the control program.

One of the most common applications of thermostats is as a low-temperature limit alarm. These thermostats include a manual reset that, once tripped, forces building personnel to physically investigate the reason for the alarm. Low-limit thermostats have multiple connections to allow a direct connection to the fan safety circuit and monitoring by controllers.

Hygrometers. A *hygrometer*, also known as a humidity sensor, is a device that measures the amount of moisture in the air. **See Figure 7-39.** The most common hygrometers measure percent relative humidity (% rh), while others measure dew point or absolute humidity. Humidity can also be calculated from the dry-bulb and wet-bulb temperature measurements.

Hygrometers

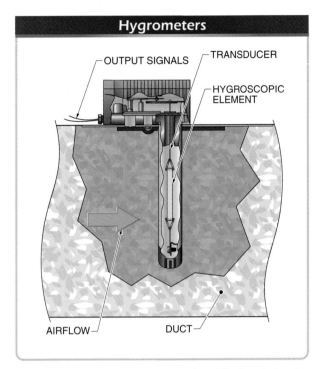

Figure 7-39. Hygrometers sense humidity levels by detecting very small changes in the length of a hygroscopic element that is sensitive to moisture.

Accurate humidity sensing is more difficult than temperature sensing. Since the sensing element must be directly exposed to the air, it is susceptible to contamination, which can affect readings. The accuracy of hygrometer measurements can drift over time, commonly 1% per year. Hygrometers should be checked annually for accuracy. Any instrument used to check the accuracy of hygrometers must be regularly calibrated against a known standard.

Most hygrometers incorporate a hygroscopic element that absorbs moisture from the air. A *hygroscopic element* is a material that changes its physical or electrical characteristics as the humidity changes. Electronic hygrometers apply a voltage to the hygroscopic element, measure its electrical characteristics to determine its moisture content, and then calculate the associated air humidity. Changes in either the capacitance or conductivity of the element indicate moisture content. Onboard electronics output the measurement as a standard analog signal or structured network message.

Since hygrometers and temperature sensors are often needed in similar locations, these two sensors are often combined in a single unit. In fact, some hygrometers require temperature information to convert absolute humidity values to relative humidity percentages.

Humidistats. Some humidity-sensing applications can be satisfied with the simple humidistat. A *humidistat* is a switch that activates at humidity levels either above or below a setpoint. Humidistats include normally open and/or normally closed contacts, which are used to directly operate other devices, such as a humidifier, or as a digital signal to an electronic controller. Humidistats are normally less expensive than analog hygrometers. A common application of a humidistat is measuring the amount of humidity in ductwork and taking action as needed to prevent condensation. **See Figure 7-40.**

Humidistats

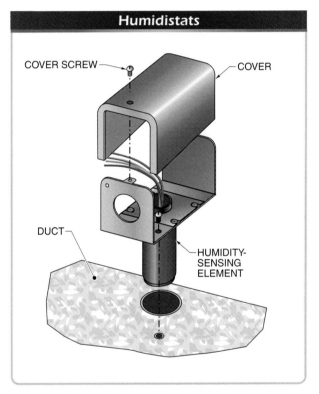

Figure 7-40. A humidistat is a switch that activates at humidity levels either above or below a setpoint.

Pressure Sensors. A *pressure sensor* is a sensor that measures the pressure exerted by a fluid, such as air or water. In HVAC systems, measured pressure is typically a differential pressure, rather than an absolute pressure. *Differential pressure* is the difference between two pressures. HVAC system differential pressure measurements are usually quantified in units of inches of water column (in. WC). Differential pressure sensors are often an integral part of air temperature, liquid flow, and steam flow measurements.

Differential pressure is used to compare the air pressure between two points within a duct, which is used to monitor fan operation or filter condition. Differential pressure sensors are connected to the two measuring points with a pair of tubes. **See Figure 7-41.** For example, a low differential pressure between the upstream and downstream sides of a filter means that the air is flowing freely, indicating a clean filter. However, a dirty filter impedes the airflow significantly, which is indicated by a relatively large differential pressure. Therefore, differential air pressure sensors can be used to monitor the dirtiness of filters, indicating the level of air contamination within the building.

pressure applied to it. One of the most common types of elastic deformation pressure elements is a diaphragm. One side of the flexible diaphragm is open to a pressure to be measured and the other is open to either another pressure (for differential pressure) or the atmosphere. The diaphragm moves into the space on the side with the lower pressure in response to the force of the higher-pressure side. The amount of displacement is proportional to the pressure.

Motion of the pressure element is transferred to a transducer, where it is converted into an electrical signal. A *transducer* is a device that converts one form of energy into another form of energy. There are many different types of transducers that can produce electrical signals from some other type of signal, each taking advantage of a different electromechanical or material property. For example, piezoelectric elements are pressure-sensitive crystals that produce small voltages when squeezed. A pressure sensor may couple a diaphragm with a piezoelectric element. When the diaphragm moves to one side, it puts pressure on the piezoelectric element, which outputs a tiny voltage in proportion to the pressure. Additional electronics convert the transducer outputs into standard sensor output signals, such as 4 mA to 20 mA. A pressure transmitter then sends a signal to a controller. **See Figure 7-42.**

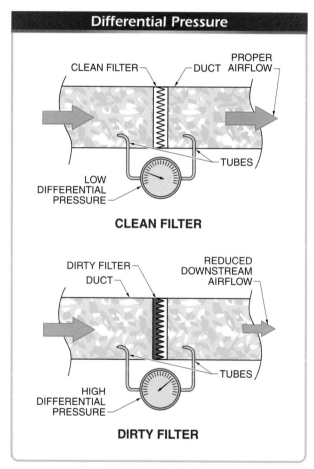

Figure 7-41. The differential pressure between the upstream and downstream sides of a filter can be used to indicate the condition of the filter.

Pressure sensors include some type of elastic deformation pressure element, which is a piece of material that flexes, expands, or contracts in proportion to the

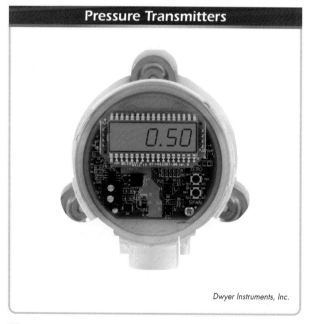

Dwyer Instruments, Inc.

Figure 7-42. Pressure sensors use electronics to convert the small movement of a flexible diaphragm into a signal proportional to the pressure.

Differential Pressure Switches. A *differential pressure switch* is a switch that activates at a differential pressure either above or below a certain value. Differential pressure switches are often used across a fan or filter and indicate conditions that are impeding airflow. Like other HVAC sensor switches, differential pressure switches include a setpoint adjustment and normally open and/or normally closed contacts.

Airflow Stations. An *airflow station* is a sensor that measures the velocity of the air in a duct system. The velocity multiplied by the duct cross-sectional area yields a volume expressed in cubic feet per minute (cfm). Airflow stations are installed either in a fan inlet or in ductwork. Airflow stations are very sensitive, so they must be located some distance downstream of anything causing turbulence, which can produce erroneous readings.

There are two types of airflow stations. The most common type is a pitot tube station, which is a differential pressure sensor comparing the airflow's total pressure (velocity pressure plus static pressure) and the static pressure. **See Figure 7-43.** The difference is proportional to the air velocity. The second type of airflow station is a hot wire anemometer. In this type, a temperature sensor is placed on a wire and current is applied to the wire to keep it at a constant temperature as air moves across the device. As more air moves across the wire, more current is needed to keep the constant temperature. This current draw is proportional to the air velocity. In both types of airflow stations, the air velocity reading is converted into an airflow rate and output as a standard analog signal.

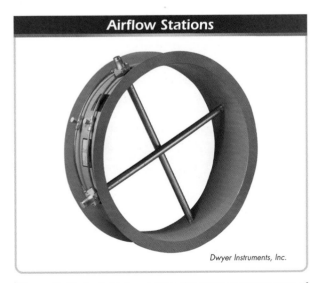

Airflow Stations

Dwyer Instruments, Inc.

Figure 7-43. A pitot tube station, the most common type of airflow station, compares the total pressure of the airflow and the static pressure within a duct.

Flow Switches. A *flow switch* is a switch with a vane that moves from the force exerted by the water or air flowing within a duct or pipe. Flow switches are installed so that the vane is inside the duct or pipe and the rest of the device is outside. **See Figure 7-44.** The pressure of the air or water flow pushing on the vane causes it to move downstream slightly, which activates the switch. Switch connections include normally open and/or normally closed contacts.

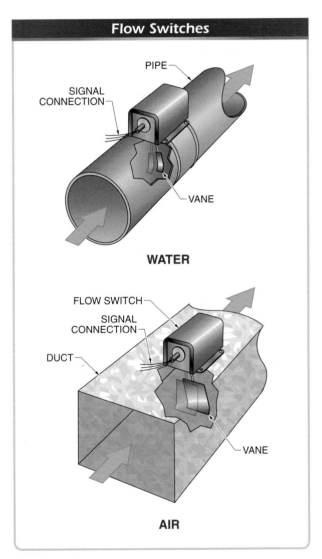

Flow Switches

PIPE

SIGNAL CONNECTION

VANE

WATER

FLOW SWITCH

SIGNAL CONNECTION

DUCT

VANE

AIR

Figure 7-44. Both water flow switches and airflow switches use a vane deflected by the fluid stream to detect the movement of the fluid.

When the flow switch is activated, some of the force of the fluid flow must be removed before the switch returns to its previous state. This results in a deadband in the switch's response to changing flow. The vane deflection is a function of the vane area, the pipe or duct size, and the characteristics of the flowing air or water. Setpoint adjustments determine how much vane deflection, and therefore flow rate, is required to actuate and deactivate the switch.

Like differential pressure switches, flow switches are used in HVAC systems to indicate the flow status of a pump or fan. However, due to fan or pump flow surges, flow switches are susceptible to false activations. For this reason, the use of differential pressure switches is more common.

Carbon Dioxide Sensors. The primary feature of stale air is an excess concentration of carbon dioxide. Ventilation incorporates outside air into the building space, reducing the carbon dioxide level. However, the ventilation system may not run continuously, especially if there are no heating or cooling requirements at the time. Therefore, carbon dioxide sensors are used to control ventilation systems so that the system runs only when needed.

A *carbon dioxide (CO$_2$) sensor* is a sensor that detects the concentration of CO$_2$ in air. The reading is output as a standard analog signal. Controllers use measured CO$_2$ levels to activate and deactivate the ventilation system. They may also increase the percentage of outside air mixed into the supply air when the CO$_2$ level rises rapidly, such as when many people occupy a space, or decrease the percentage when spaces are unoccupied.

This ventilation system control, however, can be overridden by other HVAC functions. For example, in economizer mode, the system will incorporate a large amount of cool outside air into the building for cooling, regardless of the indoor CO$_2$ level.

Occupancy Sensors

An *occupancy sensor* is a sensor that detects whether an area is occupied by people. Occupancy sensors are currently more commonly used with lighting and security control systems than with HVAC systems, though they are becoming more popular for the control of ever-smaller HVAC zones, such as individual rooms and workstations. They are used to save energy by deactivating certain HVAC functions to a zone when no one is in the room or area.

Occupancy sensors are mounted on the ceiling or high on a wall to maximize their field-of-view. **See Figure 7-45.** The sensors provide an ON/OFF signal that is used to control a relay or send a digital signal to a controller.

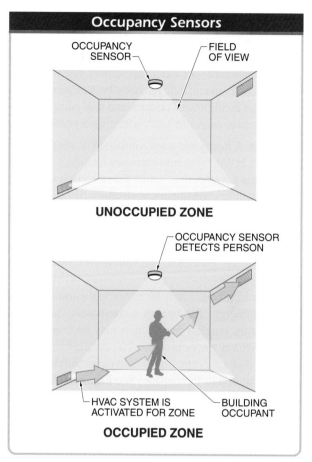

Figure 7-45. Occupancy sensors can save energy from the operation of HVAC systems by turning off some HVAC functions when no occupants are within a certain building zone.

Different technologies can be used to detect occupants. A *passive infrared (PIR) sensor* is a sensor that activates when it senses the heat energy of people within its field-of-view. Infrared occupancy sensors are line-of-sight devices and require the entire detection area to be within view. An *ultrasonic sensor* is a sensor that activates when it senses changes in the reflected sound waves caused by people moving within its detection area. The sensor emits low-power, high-frequency sound waves and listens for a phase shift in the reflected sound returning to the sensor, which indicates a moving target. This even works to some degree around corners, so the detection area is larger than the line-of-sight field-of-view. A *microwave sensor* is a sensor that activates when it senses changes in the reflected microwave energy caused by people moving within its field-of-view. Dual-technology occupancy sensors include a combination of two types of occupancy sensing, such as a passive infrared sensor with an ultrasonic sensor.

Temperature Control Devices

The main aspect of controlling building occupant comfort is regulating indoor air temperature. Temperature control devices maintain a specified temperature range within the building zones and include furnaces, boilers, electric heat sources, air conditioners, chillers, and heat pumps. These devices deliver or absorb heat via heating or cooling coils.

Furnaces. A *furnace* is a self-contained heating unit for forced-air HVAC systems. Furnaces include a fan, heat source, and controls that handle the operation of all the valves, switches, regulators, and safety devices within the furnace. Typically, the only outside input needed to operate a furnace is an ON/OFF signal when the building space requires heating. The on-board controls manage the startup, operation, and shutdown of the furnace automatically.

Furnaces produce heat by either fuel combustion or electrical energy. The heat in a combustion furnace is produced by burning fuel, which may be coal, wood, fuel oil, natural gas, or propane. The combustion and the resulting gases are contained within pipes and ducts until they are vented to the outside. **See Figure 7-46.** These pipes absorb heat from the combustion, forming a heat exchanger. As cooler air flows past the heat exchanger, it gains heat energy, which is then added to the air within the heated building space.

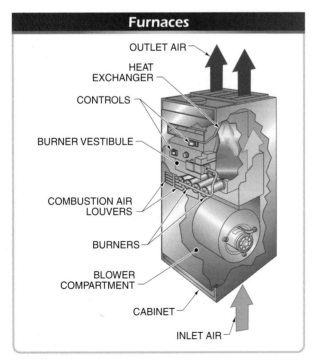

Figure 7-46. Furnaces are incorporated into forced-air HVAC systems to heat air quickly and efficiently.

Electric furnaces work in much the same way, except that there are no combustion gases to segregate or exhaust. Electricity flows through heating elements that become very hot due to their high resistance. Air is heated as it flows directly by the hot electric elements.

Boilers. A *boiler* is a closed metal container that heats water to produce steam or hot water. The heat is produced primarily by fuel combustion, though small electric units are becoming more common for limited on-demand applications, such as radiant floor heating for a single zone. Similar to the operation of a furnace, the hot gases from combustion heat water in a heat exchanger. **See Figure 7-47.**

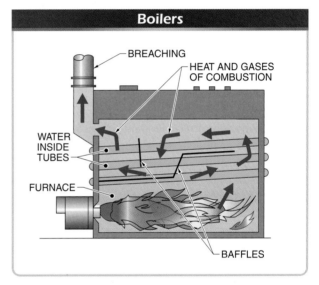

Figure 7-47. Boilers heat water to provide hot water or steam to heating coils throughout the building.

The resulting steam or hot water is used to carry heat through piping systems to other parts of the building, where other heat exchangers transfer the heat to the building space. The high amount of heat energy contained in steam causes the air temperature in building spaces to rise quickly.

In some urban areas, it is possible for building owners to purchase steam or chilled water from a very large centralized heating and cooling plant. For building owners, this eliminates the expense of purchasing, operating, and maintaining the boiler or chiller equipment themselves. These district heating and cooling plants distribute steam and chilled water to many separately owned buildings via a piping network and charge according to the consumption.

The water or steam flows through the system in a continuous loop, propelled by water pumps or steam pressure. As the steam gives up its heat, some of it condenses back into water. Steam heating systems include steam traps attached to the discharge sides of heating units to separate the condensate and return it to the boiler. *Condensation* is the formation of liquid (condensate) as moisture or other vapor cools below its dew point.

Like furnaces, boilers include electronic controls to manage the operation of the boiler and its associated equipment. However, instead of simple ON/OFF commands from outside controllers, the boiler controls may receive analog signals corresponding to specific temperatures, within an acceptable range. For example, as the outside air temperature drops from 55°F to –10°F, an HVAC controller may change the boiler's hot water setpoint from 100°F to 190°F. The boiler controller modulates the heating function and a three-way mixing valve (incorporating cool water) in order to obtain the correct water temperature. Also, the supply of hot water or steam to the heat exchangers conditioning the building spaces can be modulated with valves. For example, the steam supply to a heating coil in an AHU can be increased or decreased as needed by changing the position of the valve.

Unlike furnaces, which are relatively simple to turn on and off and produce heated air quickly, boilers require outlet water temperature control. In fact, since large boilers require much more complicated start-up and shut-down procedures, it is common to leave them operating, but at a lower water output temperature, when the heating demand is low.

Electric Heat Sources. Electricity can be used to create heat with electric resistance heating elements. Resistance to the flow of electric current causes the heating elements to become hot. **See Figure 7-48.** The electrical power supply to the unit is simple to control for both ON/OFF and staged heating. Since they require only an electrical connection and no piping for water or steam distribution, electric heating units are relatively simple and inexpensive to install. The disadvantages of electric heating systems include the high cost of electricity and the precautions that must be taken to ensure that the heating elements and building are not damaged due to excessive heat.

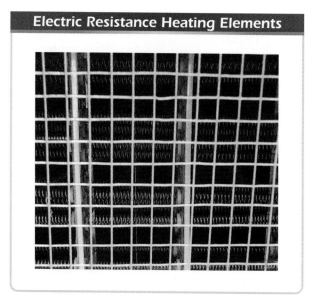

Electric Resistance Heating Elements

Figure 7-48. Electric heat sources create heat from electrical current flowing through elements with high resistances.

Electric heating elements can be used as a heating coil within an AHU. This type of heating is most common in systems with many VAV terminal boxes, since the electrical supply is relatively simple to distribute to many heating coils, as opposed to additional piping for hot water or steam supplies. These units include fans to direct the airflow over the heating elements, thus warming the air.

A *silicon-controlled rectifier (SCR)* is a solid-state power controller that provides proportional current to a heating element in response to an analog control signal. An SCR is able to use a low-power controller signal to regulate current as high as 1200 A. An SCR conducts only in one direction, so it is ideal for controlling DC loads. An SCR continues to conduct as long as the current in the load circuit is higher than the holding current value of the SCR. The current must be broken by an external switch in order to deenergize an SCR. A related device called a triac conducts current in two directions. A typical application of a triac is controlling the amount of heat generated by a heating element by quickly turning the power on and off. **See Figure 7-49.** SCRs and triacs have no moving parts and provide long operating life.

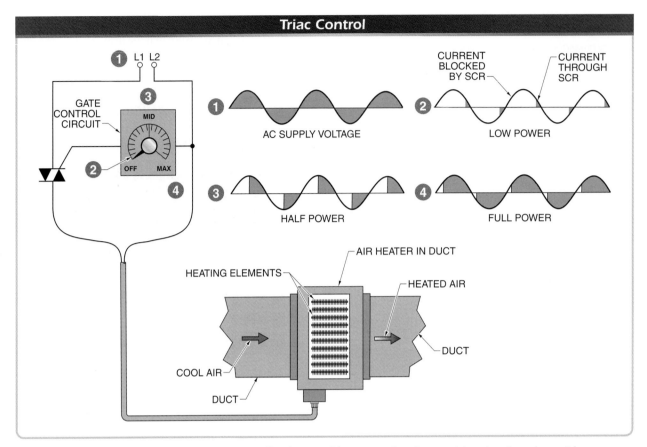

Triac Control

Figure 7-49. A triac is a solid-state power controller that provides proportional current to a heating element in response to an analog control signal.

Air Conditioners. Furnaces and boilers create heat energy from other types of energy, such as chemical fuel energy. However, heat energy cannot be destroyed to produce a cooling effect. Cooling appliances such as air conditioners can only move heat energy to where the heat energy can be used or where its effect is negligible, such as outdoors. An *air conditioner* is a self-contained cooling unit for forced-air HVAC systems.

Air conditioners rely on a mechanical compression refrigeration cycle to move heat. This process uses the phase-change properties of a refrigerant to absorb heat in one part of the system and release it in another. A *refrigerant* is a fluid that is used for transferring heat.

Air conditioners are divided into two parts. The evaporator is located inside the building, where the refrigerant cools the air by absorbing heat. The condenser is located outside the building, where the refrigerant rejects heat from the indoor air into the outside air. The refrigerant flows in one direction through a loop of piping that connects the two parts. **See Figure 7-50.**

Liquid refrigerant is metered into the evaporator by an expansion device, which lowers the pressure and temperature of the refrigerant. An *evaporator* is a heat exchanger that adds heat to low-pressure refrigerant gas. Warm return air from the building is blown over the evaporator coils, transferring heat to the refrigerant and causing it to heat up. This cools the air. Refrigerant leaves the evaporator as a vapor and flows to the compressor.

The compressor increases the pressure of refrigerant vapor and circulates the refrigerant through the system. Because the refrigerant absorbs heat from the compression process, the refrigerant that leaves the compressor is hotter than the refrigerant in the rest of the refrigeration system. The hot refrigerant vapor discharged from the compressor flows to the condenser.

A *condenser* is a heat exchanger that removes heat from high-pressure refrigerant vapor. A fan in the condenser blows warm outside air over the condenser coils. Heat transfers from the refrigerant to the air, lowering the temperature of the vapor until it condenses into a liquid. This liquid returns to the expansion device to begin the cycle again.

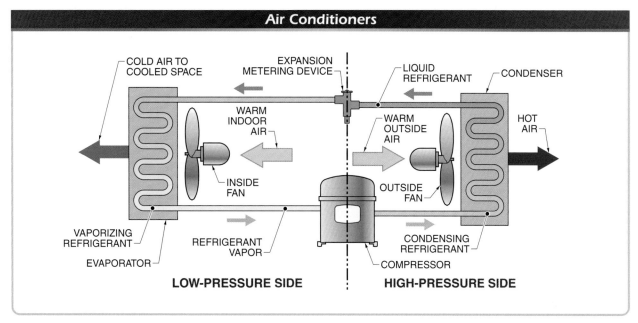

Figure 7-50. Air conditioners use the evaporating and condensing properties of a refrigerant to transfer heat from one area to another.

Air conditioners are simply switched on or off to start the refrigeration cycle. However, the amount of refrigerant metered into the evaporator coils may be modulated by valves controlled by analog signals.

Chillers. Chillers are very similar to air conditioners. A *chiller* is a refrigeration system that cools water. Chillers use the same mechanical compression refrigeration cycle as air conditioners, but instead of the refrigerant absorbing or rejecting heat to the air in the heat exchangers, chillers exchange heat with loops of water, on both sides, which then exchange heat with air. Chiller compressors circulate the water and refrigerant throughout their closed-loop systems. Other types of chillers use an expansion device that vaporizes some of the refrigerant, which then absorbs heat from nearby liquid refrigerant.

Commercial buildings with large cooling loads often use chillers. Water is cooled to about 45°F in the chiller evaporator and pumped throughout a building for cooling purposes. **See Figure 7-51.** Heat from building spaces is absorbed by the water within terminal devices located in the building spaces or cooling coils located in an AHU.

Warmer return water (approximately 55°F) is pumped into the chiller and its heat is used to vaporize the refrigerant in the evaporator. Liquid refrigerant surrounds the coils. The resulting cooled water is sent back to the building to absorb heat and cool the building air. The heat taken in by the evaporator is rejected in the condenser.

The condenser is cooled by water that has been cooled in a cooling tower located outside. A *cooling tower* is a device that uses evaporation and airflow to cool water. A fan forces air upward through the tower as warm water is sprayed downward and drips through rows of tiles. Some of the water is evaporated and carried away by the upward airflow, removing heat from the water left behind. The cool water falls to the bottom of the tower for reuse in the condenser. Makeup water is added to replace the water lost to evaporation. The cooling tower range is the temperature difference between the water entering the tower and the water leaving the tower. Cooling tower efficiency is limited by the temperature of the water entering the tower and the difference between the wet-bulb air temperature and the temperature of the water leaving the tower.

Like boilers, chillers are not turned on or off for short periods of time, since it takes some time during startup for the unit to establish normal operating temperatures. Rather, the chiller remains operating most of the time and analog-controlled water valves at the terminal devices or cooling coils provide temperature control.

Facts

Older chiller systems often contained refrigerants that were targeted by the Montreal Protocol in the 1990s and have been replaced or retrofitted for new refrigerants.

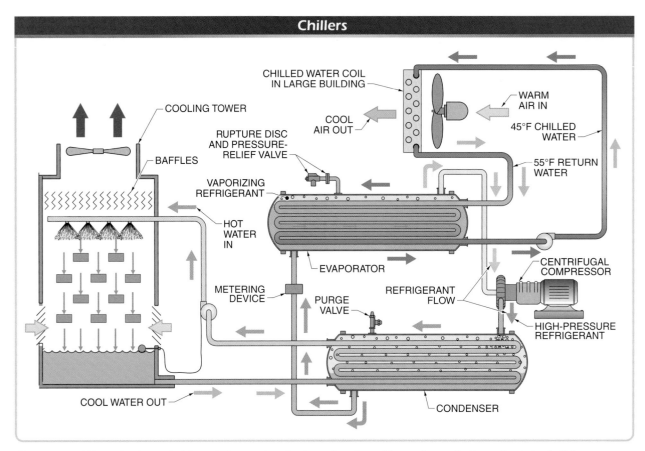

Chillers

CHILLED WATER COIL
IN LARGE BUILDING

COOLING TOWER

COOL
AIR OUT

WARM
AIR IN

45°F CHILLED
WATER

BAFFLES

RUPTURE DISC
AND PRESSURE-
RELIEF VALVE

55°F RETURN
WATER

VAPORIZING
REFRIGERANT

HOT
WATER
IN

EVAPORATOR

METERING
DEVICE

PURGE
VALVE

REFRIGERANT
FLOW

CENTRIFUGAL
COMPRESSOR

HIGH-PRESSURE
REFRIGERANT

COOL WATER OUT

CONDENSER

Figure 7-51. Water that is cooled in a chiller can be distributed and used in cooling coils throughout the building.

Heat Pumps. A *heat pump* is a mechanical compression refrigeration system that moves heat from one area to another area. Heat pumps are nearly identical to air conditioning systems, except that heat pumps can reverse the flow of refrigerant to switch between heating and cooling modes. **See Figure 7-52.** Heat pumps are turned on or off as needed to provide heating or cooling to the indoor supply air.

When a heat pump is in the cooling mode, it moves heat from inside the building to outside the building, just like an air conditioner. The indoor unit is the evaporator and the outdoor unit is the condenser. However, when a heat pump is in the heating mode, the flow of refrigerant is reversed, which moves heat from outside the building to inside the building. In heating mode, the indoor unit is the condenser and the outdoor unit is the evaporator.

The heat is moved between the building air and either the outside air, water, or the ground. Air-to-air heat pumps use the outside air as the heat source (for heating) and heat sink (for cooling). Commercial heat pumps are commonly water-to-air heat pumps, which use water from lakes, streams, wells, or retention ponds

as the heat source and heat sink. Water-to-air heat pumps have a coil heat exchanger with water as the heat transfer medium. A *geothermal heat pump,* or *geoexchange heat pump,* is a heat pump that uses the ground below the frost line as the heat source and heat sink.

Cooling towers cool water by allowing some of the water to evaporate and transferring heat to the atmosphere.

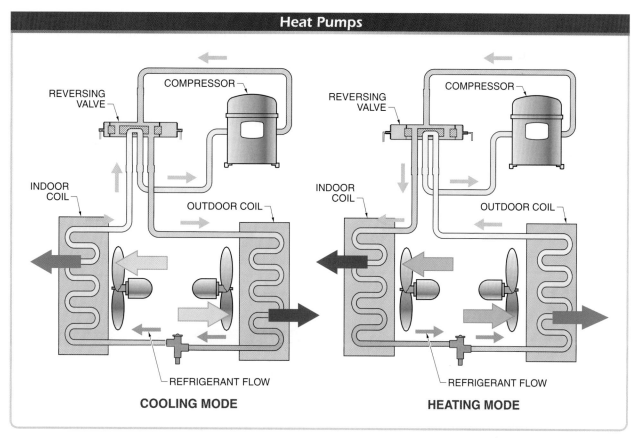

Heat Pumps

REVERSING VALVE
COMPRESSOR
INDOOR COIL
OUTDOOR COIL
REFRIGERANT FLOW
COOLING MODE

REVERSING VALVE
COMPRESSOR
INDOOR COIL
OUTDOOR COIL
REFRIGERANT FLOW
HEATING MODE

Figure 7-52. Heat pumps are nearly identical to air conditioners in operation, except that heat pumps can reverse the flow of refrigerant for heating or cooling needs.

Humidification Control Devices

Humidification control devices are needed to control the amount of moisture present in the air. These devices rely on forced-air HVAC systems to affect the humidity of the supply air and distribute it throughout the building space. Humidification is controlled by humidifiers and dehumidifiers. **See Figure 7-53.**

Humidifiers. Humidification requirements depend primarily on the local climate. Buildings in warm climates may not require humidification if temperatures are rarely low, though there are some regions with extremely dry and warm weather. Buildings in northern climates, however, require humidification during cold weather to increase the relative humidity, which falls during the heating process.

A *humidifier* is a device that adds moisture to the air by causing water to evaporate into the air. Humidifiers are typically installed as central units in an AHU after the heating coil (if there is one). This ensures that the air can absorb as much water as possible, since warm air can hold more moisture than cool air. Humidifiers can also be installed downstream of air terminal units for special applications.

Methods of humidification in commercial buildings include steam jets, flow-through metal filters, and drum-type humidifiers. Steam jets are the most common type and simply inject steam into the air flowing through the supply air duct. The humidification is easily modulated by controlling the steam flow with valves operated by analog signals. The latter two types of humidification rely on air being blown through a filter-like material that holds water, and the air picks up moisture by evaporating some of the water. Drum types of humidifiers are increasingly uncommon due to IAQ concerns.

Facts

Commercial and industrial buildings often require the use of humidifiers and dehumidifiers to maintain humidity within the comfort zone.

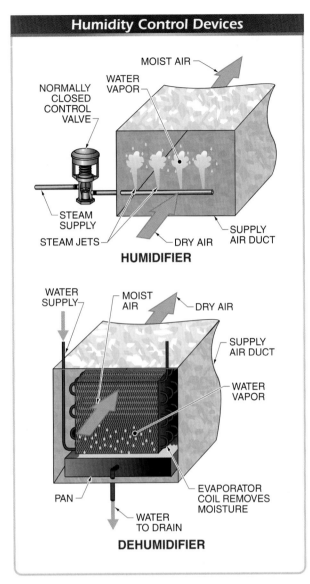

Figure 7-53. Humidifiers and/or dehumidifiers are installed within the supply duct of an AHU to add moisture to or remove moisture from the conditioned air.

Dehumidifiers. Dehumidification is separate from humidification, and the two respective devices are not always included together, depending on the building's HVAC system requirements. A *dehumidifier* is a device that removes moisture from the air by causing the moisture to condense. It does this by cooling the air until it reaches its dew point, or 100% relative humidity. Further cooling causes the moisture to condense into water. The water is either collected in a container to be emptied later or, more commonly, drains directly into the wastewater plumbing system. Air conditioners typically dehumidify air.

Dehumidifiers incorporated into building HVAC systems are very similar to air conditioners. The moist air is blown over a cooling coil in the AHU that is chilled with low-pressure refrigerant (though chilled water could also be used). The resulting air is both cooler and drier. If cooling the air is not desired, the air then flows over the warm condensing coils of the refrigeration system. This extra step does not affect the moisture content of the air, but the relative humidity falls again by warming the air temperature.

Dehumidifiers are not easily modulated to different levels of dehumidification, so they are generally only turned on or off as needed. However, since temperature and relative humidity are closely related, with changes in one property often affecting the other, heating and cooling functions also have some control over humidification and dehumidification.

> **Facts**
>
> *Wet bulb and dew point temperatures are used in various HVAC applications. For example, cooling towers release heat to the ambient air through evaporation. The ability of a cooling tower to reject heat is directly related to the ambient wet bulb temperature. Cooling towers are often rated by the amount of heat they can reject at specific wet bulb temperatures.*

Airflow Distribution Control Devices

Airflow is essential to a forced-air system, which relies on the movement of air over temperature and humidity control devices to affect its properties and to distribute the conditioned air throughout the building space. Airflow is also important to the efficiency of the HVAC system. Conditioned air should flow to the areas where it is needed and not be wasted in areas where it is not needed. The design of a building and the installation of its HVAC system greatly influence airflow. However, fans and dampers within the system can be used to further adjust the way air flows through the ducts and the building spaces.

Fans. Fans are primary components in forced-air HVAC systems. Separate supply air and return air fans are used to move air within different parts of the duct system. Fans are also included with some unit devices, such as furnaces. A *fan* is a mechanical device with spinning blades that move air. Fans are used within air ducts and in open environments such as air conditioners and heat pumps to force outside air over the coils located outside of the building.

Fan designs fall within two main categories: centrifugal fans and axial-flow fans. **See Figure 7-54.** Air enters a centrifugal fan through the center of the impeller. An *impeller* is a bladed, spinning hub of a fan or pump that forces fluid to the perimeter of a housing. As the impeller rotates, the blades move rapidly through the air, forcing the air to move outward. The housing of the fan, known as the scroll, directs this airflow to the fan outlet. Because of this design, centrifugal fans change the direction of the airflow by 90°. Axial-flow fans, however, do not change the direction of airflow. Fan hubs for either type may be mounted directly on the shaft of a motor or may be turned by a motor with a pulley and belt arrangement.

Centrifugal fans are often used to move air in ductwork.

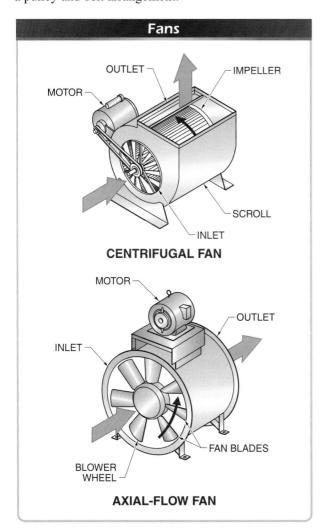

Figure 7-54. Centrifugal fans change the direction of the airflow by 90°, while axial-flow fans have no effect on the airflow direction.

Fan airflow output is modulated in variable-air-volume AHUs in a variety of ways. The velocity of the airflow is affected by the pitch (angle) of the blades, the restriction of the fan inlet (vortex dampers), and the fan speed. **See Figure 7-55.**

Some fans can be adjusted for different blade pitches, within an allowable range. A small, geared motor in the hub rotates the attachment points of each blade, changing its angle relative to the direction of airflow. Within limits, steeper pitches increase the air velocity and shallower pitches decrease the air velocity. This allows the air velocity to be controlled without changing the motor speed.

Alternatively, vortex dampers can be used to control the air volume output of a fan by restricting the air flowing into the fan. A *vortex damper* is a pie-shaped damper at the inlet of a centrifugal fan that reduces the ability of the fan to grip and move air. Like other dampers, vortex dampers can be set at various positions to allow a percentage of air to pass through.

Most commonly, though, airflow output is controlled by adjusting the speed of the electric motor with a variable-frequency drive. A *variable-frequency drive (VFD)* is a motor controller that is used to change the speed of an AC motor by changing the frequency of the supply voltage. This drive changes the characteristics of the electrical power to the motor so that the motor operates at different speeds. The drive can even change the direction of rotation, causing the air to flow in the opposite direction. All three methods of airflow control are very effective, but variable-frequency electric motor drives are the most energy efficient and the most common in new installations.

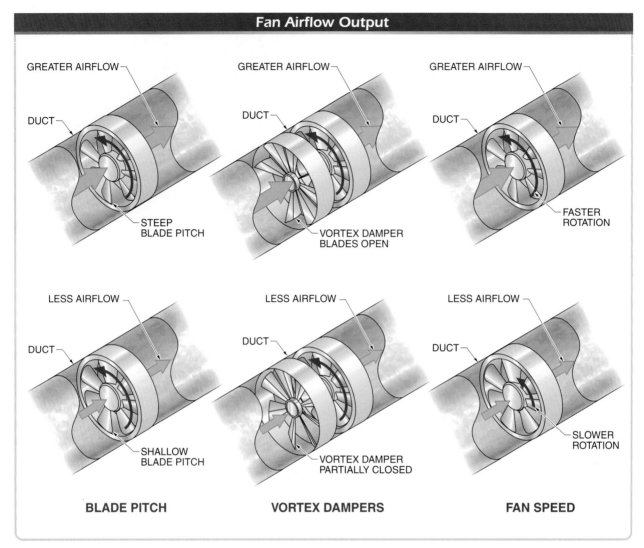

Fan Airflow Output

GREATER AIRFLOW

DUCT

STEEP BLADE PITCH

GREATER AIRFLOW

DUCT

VORTEX DAMPER BLADES OPEN

GREATER AIRFLOW

DUCT

FASTER ROTATION

LESS AIRFLOW

DUCT

SHALLOW BLADE PITCH

LESS AIRFLOW

DUCT

VORTEX DAMPER PARTIALLY CLOSED

LESS AIRFLOW

DUCT

SLOWER ROTATION

BLADE PITCH   **VORTEX DAMPERS**   **FAN SPEED**

Figure 7-55. The volume and velocity of airflow can be controlled by the blade pitch and vortex dampers but are most commonly controlled by changing the fan speed.

Fan airflow output control requires an analog input to produce the desired airflow. Depending on the fan, inputs may be either a standard analog signal, such as 0 VDC to 10 VDC, or a structured network message that includes the desired settings.

Dampers. A *damper* is a set of adjustable metal blades used to control the amount of airflow between two spaces. Typical damper designs include parallel-blade, opposed-blade, and round-blade dampers. **See Figure 7-56.** Dampers can be used to control the flow of air in HVAC systems. For example, dampers are used throughout the AHUs and ductwork to control outside air, return air, and exhaust airflow. Dampers are also used at the ends of ductwork, in air terminal units, where the air flows into the building spaces. Depending on the type and application of the AHU, dampers may also be used to control airflow across heating or cooling coils. There may also be manually operated dampers used to balance the HVAC system, though these are not involved in automation.

Damper positions range from fully open, which has minimal effect on airflow through the damper, to fully closed, which almost completely seals a section or space off from the airflow on the other side. Dampers can also be set to any position in between to modulate the volume of airflow through the damper. **See Figure 7-57.**

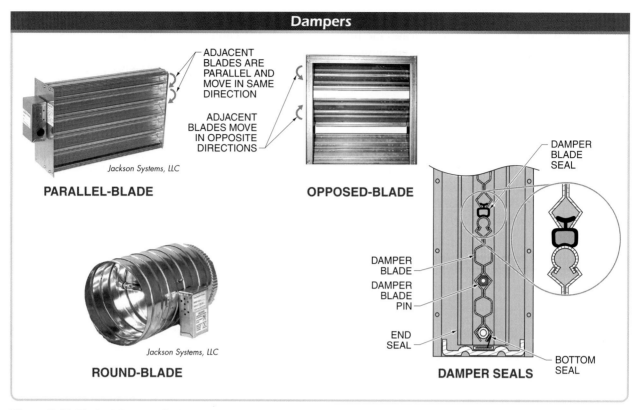

Figure 7-56. Typical damper designs include parallel-blade, opposed-blade, and round-blade dampers.

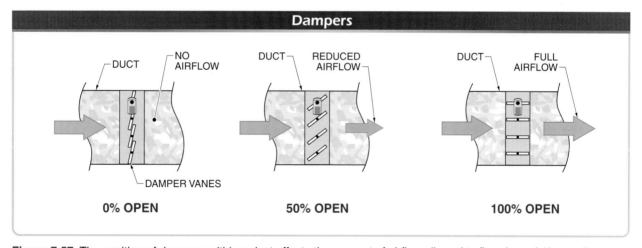

Figure 7-57. The position of dampers within a duct affects the amount of airflow allowed to flow through the section.

The angle of the damper blades is controlled by an actuator containing a small, geared electric motor. The blades are connected together so that they all move simultaneously. The actuator responds to signals from a controller to set the damper position. Digital (ON/OFF) signals correspond to completely open or completely closed.

Tristate actuators use two digital signals to drive the actuator open or closed. When a signal stops, the actuator stalls in its current position. Alternatively, many actuators accept standard analog signals, which correspond to any position between 0% open (100% closed) and 100% open, allowing the damper opening to be positioned precisely.

End switches are commonly installed on many dampers. An *end switch* is a switch that indicates the fully actuated damper positions. End switches are similar to limit switches. Its contacts are actuated by the position of the damper, providing a digital (ON/OFF) output signal. Some end switches are installed in the damper actuator, proving that the actuator moved to a full position in response to a control signal. Other end switches are installed near one of the damper blades, proving that the damper blades moved to a full position. **See Figure 7-58.** This distinction may be important if an actuator linkage breaks; in this case, the actuator may move, but the blades may not.

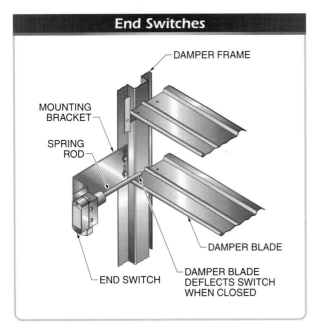

End Switches

DAMPER FRAME

MOUNTING BRACKET

SPRING ROD

DAMPER BLADE

END SWITCH

DAMPER BLADE DEFLECTS SWITCH WHEN CLOSED

Figure 7-58. End switches are installed on dampers to verify damper position.

Dampers vary in their response to a control system problem or power failure. Spring return dampers return to a default position, either fully open or fully closed, without the actuator's influence. Others stall in their current position. The choice between the two damper design defaults depends on the damper's function and placement within the HVAC system. For example, exhaust-air and outside-air dampers are normally closed dampers, which means that they automatically close if the damper motor control loses power. The return-air damper, however, is normally open.

Some dampers are used exclusively during a building fire by the fire alarm system. These dampers help control the spread of fires and smoke or purge smoke outside the building. However, because they provide life safety functions, the control of smoke dampers is heavily regulated. Engineered smoke control system applications may be controlled only by the fire protection system, which is the only building system UL®-listed for reliability.

Dampers can also be used to improve HVAC system control by avoiding temperature stratification within an AHU. The outside-air and return-air damper blades can be arranged so their airstreams collide, thoroughly mixing the air. Also, a two-section outside-air damper, with one stage opening before the other, keeps the outside air velocity fast in order to promote mixing. The first stage of the damper is called the minimum outside-air damper, while the second stage is an economizer damper.

Water Distribution Control Devices

Systems that use either hot or chilled water require pumps and valves to circulate and control the water through the pipes and heat exchangers. A pump and several valves are usually integral to the water-conditioning device, such as a boiler or chiller, and are controlled by that device's on-board controller. Additional pumps and valves, particularly those involved in the supply of the water to terminal devices, may be controlled separately by an HVAC system controller.

Pumps. A *pump* is a device that moves water through a piping system. A pump moves water from a lower pressure to a higher pressure, overcoming this pressure difference by adding energy to the water. Water pumps are normally driven by electric motors. The motor for smaller pumps is typically integrated into the pump housing. For pumps with a pumping rate above 100 gpm (gallons per minute), the motor is normally connected by a coupling between the shaft and motor.

The motors can be turned on or off with relays or motor starters. With a VFD, the speed of the electric motor can be controlled to achieve the desired pumping rate.

Centrifugal pumps are the most widely used type of pump in HVAC systems. **See Figure 7-59.** A *centrifugal pump* is a pump with a rotating impeller that uses centrifugal force to move water. Centrifugal pumps are very similar to centrifugal fans, even though they move different fluids. Water enters the pump through the center of the impeller. The impeller rotates at a relatively high speed, imparting a centrifugal force to the water, which moves it outward. The shape of the casing directs the water to the outlet.

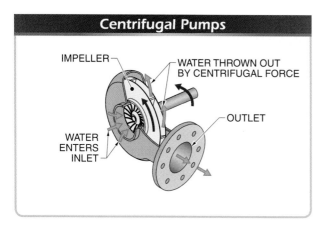

Figure 7-59. Pumps, such as centrifugal pumps, circulate water from a water heating or chilling device, such as a boiler, throughout a hydronic system.

Valves. Indoor-air temperature control in a hydronic system may be maintained with a valve, which controls the flow of chilled water, heated water, or steam to a heat exchanger. The amount of flow has a direct impact on the rate of heat transfer in the exchanger. A *valve* is a fitting that regulates the flow of water within a piping system.

A *full-way valve,* also known as a shutoff valve, is a valve designed to operate in only the fully open or fully closed positions. When open, the internal design permits a straight and unrestricted fluid flow through the valve, resulting in very little pressure loss due to friction.

A *globe valve* is a valve designed to control water flow by partially opening or closing to allow for throttling, mixing, and diverting of the water. Due to the internal configuration, water flowing through the valve changes direction several times, resulting in flow resistance and a pressure drop. This makes them ideal for reducing flow and pressure. **See Figure 7-60.**

Valves used in automation systems are essentially the same as manually actuated valves except that they include electronically controlled actuators, which are usually small geared electric motors mounted to the rotating valve stems. **See Figure 7-61.** However, these motors require on-board controls to allow for operation in both directions (for both opening and closing) and any position in between (for throttling). Since globe valves can be used for any position between fully open and fully closed, the actuators for these valves accept analog inputs.

The electronic valve packages include the necessary electronics to operate the motor and accept input signals that indicate the desired valve position. For example, full-way valves may require only digital inputs, which correspond to completely open and completely closed. Throttling valves require analog values to determine intermediate positions. Either type may be available for use with structured network message systems.

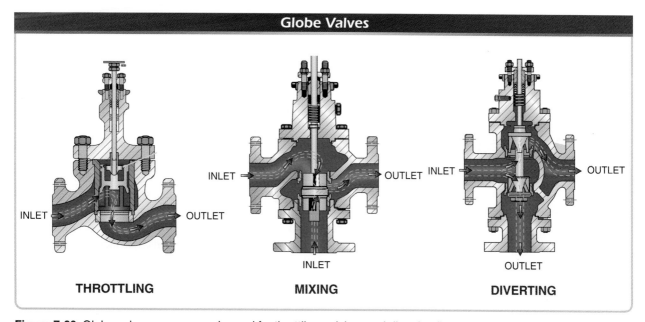

Figure 7-60. Globe valves are commonly used for throttling, mixing, and diverting flows.

Figure 7-61. Many types of water valves are used to control the flow of water for an HVAC system. The actuators attached to the valves receive signals from controllers to open or close the valve.

STRATEGIES AND ENERGY CONSIDERATIONS

Many building automation system and energy control strategies are generally compatible with IAQ strategies, provided that the strategies are instituted with certain IAQ protections. Energy retrofits and control strategies must include provisions to protect IAQ and provide additional outside air to meet the ventilation requirements of ANSI/ASHRAE Standard 62.1-2010, *Ventilation for Acceptable Indoor Air Quality*. Implementing IAQ strategies usually results in energy savings. **See Figure 7-62.**

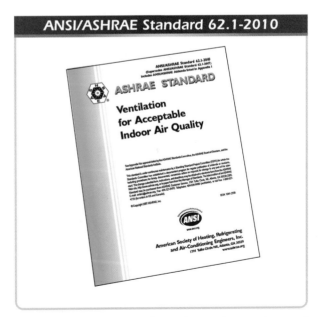

Figure 7-62. Energy retrofits and control strategies must include provisions to meet the requirements of ANSI/ASHRAE Standard 62.1-2010, *Ventilation for Acceptable Indoor Air Quality.*

Energy and operational strategies that are compatible with IAQ include:
- improving the energy efficiency of the building shell
- reducing internal loads (such as lighting and office equipment upgrades)
- upgrading fans, motors, and drive systems
- upgrading chillers and boilers
- using energy recovery ventilation (ERV) systems
- prudent equipment downsizing
- PM of entire HVAC system
- using an air-side economizer
- night precooling
- reducing demand charges
- using supply-air temperature reset
- reducing light usage during unoccupied hours

Attempts in the past to save energy have needlessly compromised IAQ and resulted in occupant complaints and/or serious illness. Energy and operational strategies that are not compatible with IAQ include:
- reducing outside air ventilation below standards
- reducing HVAC operating hours
- relaxing temperature and/or humidity setpoints below standards

Strategies for Air-Handling Systems

Air-handling systems are classified as either constant volume (CV) or variable-air-volume (VAV). The strategies used to control both types differ. In a CV air-handling system, variations in the temperature requirements of a space are satisfied by varying the temperature of the constant volume of air being delivered to the space. A constant percentage of outside air (ventilation air) means that a constant volume of outside air is delivered to occupied spaces. The volume of outside air must be set to satisfy applicable ventilation standards. While CV systems are less energy efficient than VAV systems, controlling the amount of outside air to a space is easier to do than varying the temperature of the air.

In a VAV air-handling system, temperature requirements of a space are satisfied by varying the volume of air that is delivered to the space at a constant temperature. VAV systems reduce HVAC energy costs by 10% to 20% in comparison to CV systems, but they complicate the delivery of outside air.

An inadequate outside air percentage, combined with an inadequate VAV box minimum setting, may result in inadequate outside airflow (ventilation) to occupant spaces. This occurs during partial-load conditions.

VAV systems also complicate pressure relationships in a building and make testing, adjusting, and balancing more difficult. Most of the year, the volume of outside air is reduced to about a third of the outside air volume at design load. This could result in IAQ problems.

Having separate controls to ensure adequate outside airflow (ventilation) to occupant spaces year-round does not increase energy costs. Some new VAV systems incorporate ventilation controls. Important air-handling strategies include CO_2-controlled ventilation and air-side economizer usage.

CO_2-Controlled Ventilation

CO_2-controlled ventilation modulates the supply of outside air in response to CO_2 levels in building spaces or zones, which is used as an indicator of occupancy. CO_2 controls may be useful for reducing energy use for places where occupancy is highly variable and irregular such as general meeting rooms, studios, theaters, and educational facilities. **See Figure 7-63.** CO_2 levels are typically measured in parts per million (ppm). This is often called demand-control ventilation.

A typical HVAC system increases ventilation when CO_2 levels increase to the range of 600 ppm to 800 ppm so that levels do not exceed 1000 ppm. **See Figure 7-64.** The system must incorporate a minimum outside-air setting to dilute building-related contaminants during low-occupancy periods. CO_2 sensors must be calibrated periodically and setpoints adjusted based on outside CO_2 levels around the building.

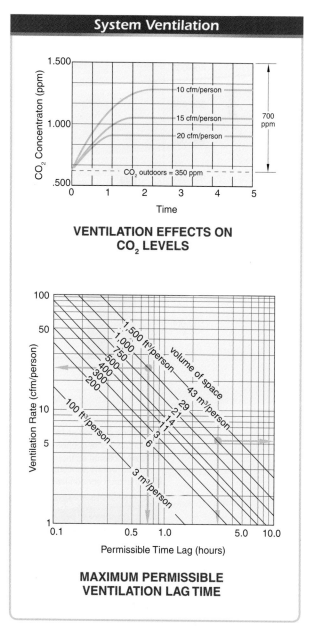

VENTILATION EFFECTS ON CO_2 LEVELS

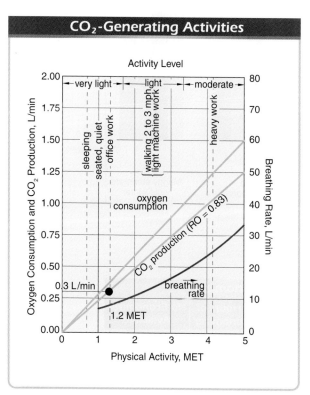

Figure 7-63. A significant source of CO_2 is human occupancy, making CO_2 controls useful in places where occupancy is highly variable.

Figure 7-64. A typical HVAC system increases ventilation when CO_2 levels increase.

Air-Side Economizers

Air-side economizers use outside air to provide free cooling. The operation of these economizers potentially improves IAQ by helping to ensure that the outside-air ventilation rate meets IAQ requirements. **See Figure 7-65.** With the exception of use in dry climates, moisture control must also be incorporated. Air-side economizers are not practical or advisable in hot-humid climates. An air-side economizer is disengaged when a problem occurs involving outside air pollution. These economizers usually reduce annual HVAC energy costs when used in cold or temperate climates.

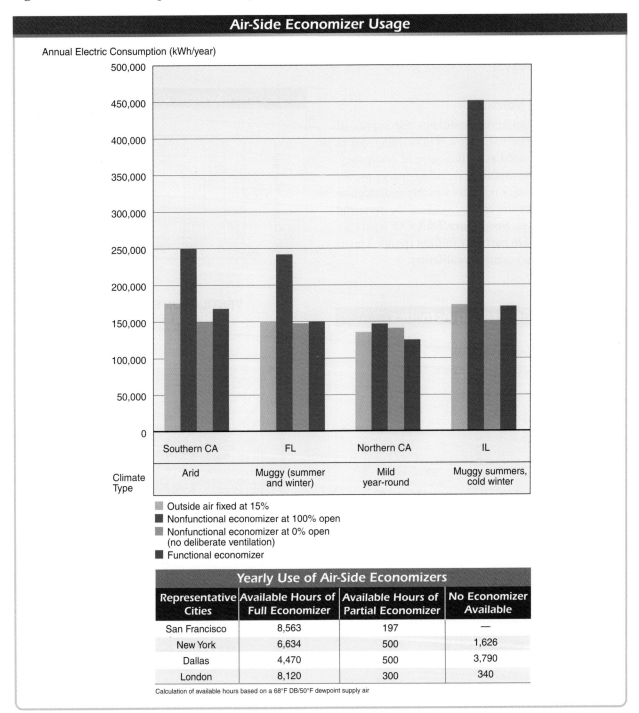

Figure 7-65. Air-side economizers use outside air to provide free cooling.

REVIEW QUESTIONS

1. What is the primary difference between a forced-air system and a hydronic system?

2. How can separate HVAC zones improve energy efficiency and occupant comfort?

3. Why do relative humidity measurements depend on the air temperature?

4. Why is ventilation important?

5. Why do some HVAC systems use a prefilter in addition to a standard filter in air ducts?

6. How can differential pressure sensors be used to indicate filter condition?

7. What is the primary difference between a temperature sensor and a thermostat?

8. How can occupancy sensors be used to save energy?

9. What is the difference between a boiler and a furnace?

10. How does a heat pump change between functioning as a heating device and as a cooling device?

Chapter 7 Review and Resources

HVAC System Applications

HVAC system control is the most common application of building automation, but it can be quite complex. Multiple zones with different requirements, constantly changing variables, and the interrelationships between conditioned air characteristics make HVAC system control seem particularly complicated. However, the key to understanding HVAC system controls is to study each application individually. Most applications can be broken down into a simple sequence of inputs, decisions, and outputs. Additional considerations of optional controls, safeties, alarms, and energy-saving measures only build further upon these basic sequences.

HVAC SYSTEM APPLICATIONS

Each HVAC system sequence specifically addresses a single application with a desired outcome. Many control sequences occur simultaneously within an HVAC system to manage different aspects of comfort, such as temperature, humidity, and airflow. The sequences are managed locally by controllers that collect input signals from HVAC sensors and other building automation devices, make decisions on how to operate the unit to maintain the desired setpoints, and generate output signals to the necessary devices to change their operation in accordance with these decisions. The controllers also share HVAC operating information with other devices on the building automation network.

However, some sequences also affect each other. **See Figure 8-1.** The output from one sequence may become the input of another, or the change of a setpoint may affect the calculations of another. Also, many sequences involve some of the same devices. Some devices send out control signals that are used in more than one sequence, and some devices receive control signals from more than one sequence. Priority information may be necessary to ensure that a device receiving multiple control inputs behaves appropriately.

All of these interactions can make HVAC control applications especially tricky to balance. The consulting-specifying engineer and the controls contractor must pay careful attention to sequences and controller programming in order to appropriately manage the system as a whole.

Control applications in HVAC systems may also include special considerations for safety functions. While some failures in an HVAC control application cause relatively minor discomfort for the building occupants, other failures or adverse conditions may cause equipment damage and even life safety issues. Therefore, some applications require special safety functions. The most vital safeties are hardwired as separate circuits that are independent of the automation system, which ensures reliable operation regardless of the status of the building automation system. **See Figure 8-2.**

The opening of the circuit by any safety control device on the circuit immediately shuts down part of the system by deenergizing the variable-frequency drives (VFDs) or motor starters, regardless of the current mode of operation. Safety circuits also typically provide an input to controllers to generate alarms. Other safeties may be present only as a feature that is programmed into the controller.

Since an HVAC system is one of the largest consumers of electrical and fuel energy in a building, some control applications allow variations in the sequence in order to reduce energy consumption. Depending on the application, energy-saving measures may involve a small sacrifice in comfort. The decision to implement these measures requires input from the building owner to balance the energy savings with less-than-optimal occupant comfort.

201

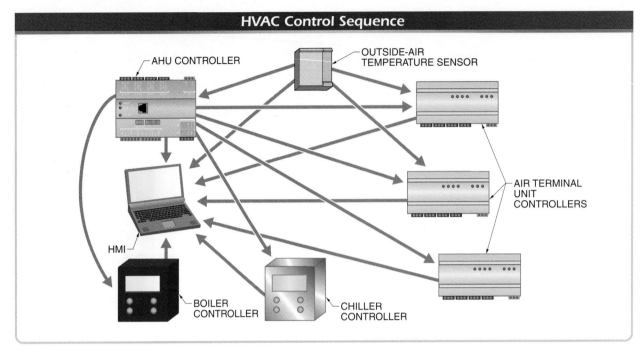

Figure 8-1. Many sensor inputs and controller outputs are shared between some of the common HVAC control sequences.

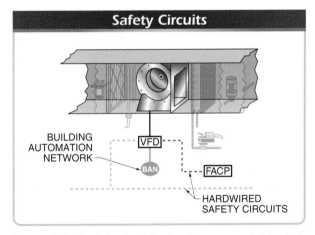

Figure 8-2. Hardwired safety circuits ensure reliable shutdown of the system for a major problem, regardless of the status of the automation system.

HVAC control applications can be divided by the primary unit or controller involved in executing the sequence. Air-handling units are the most common central HVAC unit used in commercial buildings. Their controllers manage all the control devices found in the unit in order to properly condition the air supplied to building spaces. Located at the building spaces, air terminal units are typically managed by a separate controller to further condition the supply air for the precise requirements of an individual building zone. Other HVAC control applications involve more specific heating and cooling control with hydronic and refrigeration systems.

AIR-HANDLING UNIT CONTROL APPLICATIONS

The hub of a forced-air HVAC system is an air-handling unit. This device combines many HVAC control devices to fully condition the air in a central location before it is distributed throughout the building. An *air-handling unit (AHU)* is a forced-air HVAC system device consisting of some combination of fans, ductwork, filters, dampers, heating coils, cooling coils, humidifiers, dehumidifiers, sensors, and controls to condition and distribute supply air. Depending on the application, AHUs may vary in size and configuration, but most are arranged similarly.

Supply-Air Fan Control

The supply-air fan moves conditioned air from the AHU to the building spaces via ductwork. There must be enough supply airflow to adequately distribute air through the ducts. Control of the supply-air fan is managed by the AHU controller and is based primarily on the static air pressure within the supply-air duct, which should be maintained at a setpoint. **See Figure 8-3.**

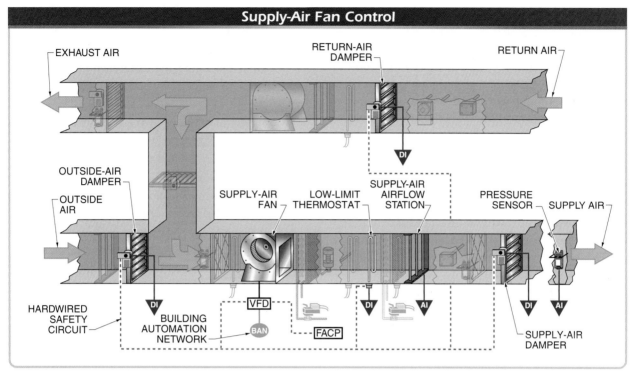

Figure 8-3. Supply-air fan control maintains an adequate static air pressure in the supply-air duct.

The air pressure is raised or lowered as necessary by changing the speed of the VFD operating the supply-air fan (or energizing a motor starter if the AHU is a constant-volume type).

Control Sequence. A typical sequence for controlling the supply-air fan in an AHU includes the following:

1. The AHU controller monitors inputs from the building automation system.
   - The AHU controller receives a digital input indicating desired occupied/unoccupied fan operation based on a preprogrammed schedule.
   - The AHU controller receives digital input indicating an override of a scheduled unoccupied period in order to prepare it in anticipation of occupancy.

2. The AHU controller monitors inputs from physical sensors.
   - A static air pressure sensor provides an analog input indicating the supply-air duct static air pressure. If there are multiple supply-air duct static pressure sensors, such as one on every floor in a multistory building, the sensor with the lowest static pressure (greatest differential from setpoint) is used for control.

   - A control device provides a digital input indicating an override of the occupancy schedule, such as an occupancy sensor that indicates unscheduled use of the building space.

3. If the digital inputs indicate that the supply-air fan should be on, the controller compares the measured static air pressure to the setpoint.
   - If the measured static air pressure is lower than the setpoint, the supply-air fan speed must be increased.
   - If the measured static air pressure is higher than the setpoint, the supply-air fan speed must be decreased.

4. The AHU controller generates an analog signal to the supply-air fan VFD to modulate the speed of the supply-air fan. If a larger AHU has multiple fans, each receives individual but identical signals during normal operation.

Facts

Compared to mechanical controls, VFDs are less expensive to maintain and provide more accurate control with fewer adjustments.

Return-Air Fan Control

The return-air fan assists the supply-air fan in moving conditioned air into the building spaces by drawing air out of the spaces. There must be enough return airflow to adequately circulate air through the building space. Control of the return-air fan in the first sequence is based on the static air pressure within the building, similar to the supply-air fan sequence, and involves many of the same AHU components. **See Figure 8-4.** The air pressure is raised or lowered as necessary by changing the speed of the VFD operating the return-air fan (or energizing a motor starter if the AHU is a constant-volume type).

Control Sequence. A typical sequence for controlling the return-air fan in an AHU includes the following:

1. The AHU controller monitors inputs from the building automation system.
   - The AHU controller receives a digital input indicating desired occupied/unoccupied fan operation based on a preprogrammed schedule.
   - The AHU controller receives a digital input indicating an override of a scheduled unoccupied period in order to prepare it in anticipation of occupancy.

2. The AHU controller monitors inputs from physical sensors.
   - A differential pressure sensor provides an analog input indicating the difference in air pressure between the building space and outside.
   - A control device provides a digital input indicating an override of the occupied/unoccupied schedule, such as an occupancy sensor that indicates unscheduled use of the building space.

3. If the digital inputs indicate that the return-air fan should be in occupied mode, the controller compares the measured differential air pressure to the setpoint.
   - If the measured differential pressure is below the setpoint (outside pressure is greater than inside pressure), the return-air fan speed must be decreased.
   - If the measured differential pressure is above the setpoint (outside pressure is less than inside pressure), the return-air fan speed must be increased.

4. The AHU controller generates an analog signal to the return-air fan VFD to modulate the speed of the return-air fan.

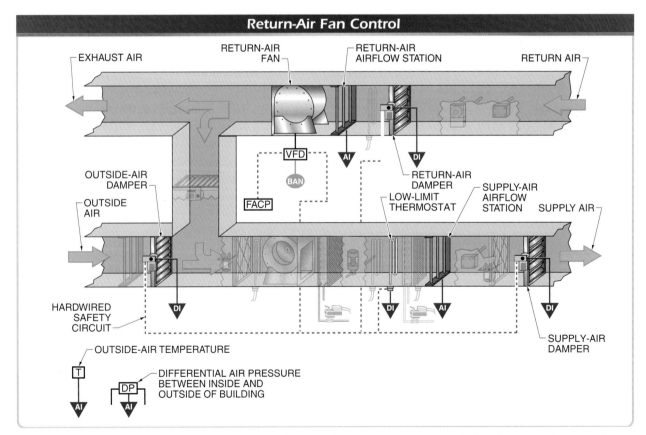

Figure 8-4. Return-air fan control can be based on differential building pressure or a calculated return airflow setpoint.

While this sequence is simple to install, it may be unstable. Building pressure is influenced by wind gusts, wind direction, and turbulence around the building itself. Instead, an alternate sequence relies on airflow measurements. These measurements are then used to calculate a return airflow setpoint with the following formula:

$$RA = SA - EA - p$$

where

RA = return airflow (in cfm)

SA = supply airflow (in cfm)

EA = exhaust airflow (in cfm)

p = pressurization factor (in cfm)

The exhaust airflow is a fixed value from the fan schedule and should include only the fans that are turned on. The pressurization factor is an airflow that represents 2% to 3% (adjustable) of the supply airflow.

Alternative Control Sequence. An alternative sequence for controlling the return-air fan in an AHU includes the following:

1. The AHU controller monitors inputs from the building automation system.
   - The AHU controller receives a digital input indicating desired occupied/unoccupied fan operation based on a preprogrammed schedule.
   - The AHU controller receives a digital input indicating an override of a scheduled unoccupied period in order to prepare it in anticipation of occupancy.

2. The AHU controller monitors inputs from physical sensors.
   - A differential pressure sensor provides an analog input indicating the difference in air pressure between the building space and outside.
   - An airflow station provides an analog input indicating the supply airflow.
   - An airflow station provides an analog input indicating the return airflow.
   - A control device provides a digital input indicating an override of the occupancy schedule, such as an occupancy sensor that indicates unscheduled use of the building space.

3. If the digital inputs indicate that the return-air fan should be in occupied mode, the controller compares the measured differential air pressure to the calculated setpoint.
   - If the differential pressure is below the setpoint (outside pressure is greater than inside pressure), the pressurization factor must be increased in order to reduce return airflow.

- If the differential pressure is above the setpoint (outside pressure is less than inside pressure), the pressurization factor must be decreased in order to increase return airflow.

4. The AHU controller generates an analog signal to the return-air fan VFD to modulate the speed of the return-air fan to maintain the new return airflow setpoint.

Outside-Air and Return-Air Damper Control

The outside-air and return-air dampers control the relative proportions of the two components of supply air: outside air and return air from the building spaces. Together, these dampers serve two purposes. The first is to satisfy the ventilation requirements, thus maintaining good indoor air quality. The second is to take advantage of the outside air conditions for reducing building energy use. This function opens the outside-air damper even more than the minimum outside air setting required for adequate ventilation.

A building space may require cooling even when the outdoor weather is cool. This is because building occupants and electrical equipment, such as lighting, can significantly warm the air within a space. In this situation, the building space can be cooled very efficiently by ventilating with the cool outside air. *Economizing* is a cooling strategy that adds cool outside air to the supply air. This is a very energy-efficient method of cooling but can only be used when the outside-air temperature falls below the supply-air temperature. Economizers will work as the outside-air temperature continues to fall until the economizer airflow is equal to the minimum ventilation airflow. When the minimum ventilation airflow exceeds the economizer airflow, the AHU activates the preheat coil.

Control of the outside-air and return-air dampers for economizing is based primarily on outside temperature and airflow. **See Figure 8-5.** The mixed-air temperature is raised or lowered as necessary by changing the positions of the outside-air and return-air dampers. The two dampers maintain equal but opposite positions.

The return airflow setpoint is calculated using the following formula:

$$OA = EA + p$$

where

OA = outside airflow (in cfm)

EA = exhaust airflow (in cfm)

p = pressurization factor (in cfm)

The exhaust airflow is a fixed value from the fan schedule and should include only the fans that are on. The pressurization factor is an airflow that represents 2% to 3% (adjustable) of the supply airflow.

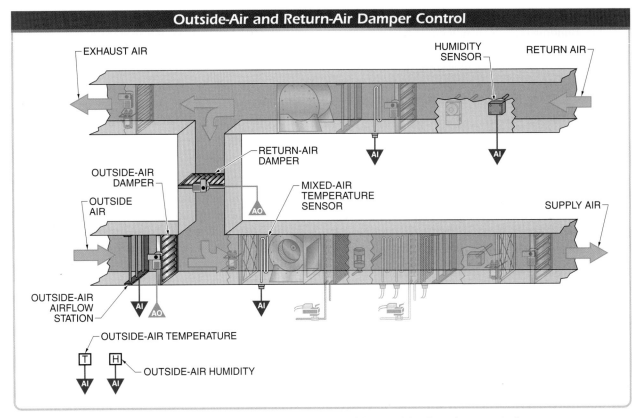

Outside-Air and Return-Air Damper Control

Figure 8-5. Economizing involves control of the relative positions of the outside-air and return-air dampers.

Control Sequence. A typical sequence for controlling the outside-air damper and the return-air damper in an AHU includes the following:

1. The AHU controller monitors inputs from the building automation system.
   - The AHU controller receives a digital input indicating desired occupied/unoccupied fan operation based on a preprogrammed schedule.

2. The AHU controller monitors inputs from physical sensors.
   - A temperature sensor provides an analog input indicating the outside-air temperature.
   - A temperature sensor provides an analog input indicating the mixed-air temperature.
   - An airflow station provides an analog input indicating the outside airflow.

3. The controller compares the measured outside airflow to the calculated outside airflow setpoint.
   - If the outside airflow is below the setpoint, the outside-air damper must be opened.
   - If the outside airflow is above the setpoint, the outside-air damper must be closed.

4. If the outside-air temperature is below the supply-air discharge temperature, the economizer is enabled. The controller compares the mixed-air temperature measurement to the setpoint.
   - If the mixed-air temperature is above the setpoint, the outside-air damper must be opened.
   - If the mixed-air temperature is below the setpoint, the outside-air damper must be closed.

5. The AHU controller generates analog signals to the two damper actuators. The outside-air damper is slowly opened and the return-air damper is slowly closed. However, any direction to close the damper can be overridden to maintain the minimum ventilation requirements. The dampers are stopped in their current position when the mixed-air temperature reaches the setpoint.

Facts

Pneumatic actuators continue to be used because they are easy to troubleshoot and repair in the field, are relatively inexpensive, and create a large force using compressed air.

Exhaust-Air Damper Control

There are many possible sequences for controlling building pressure (relative to outside air pressure), though modulation of the exhaust-air damper is one of the more common and stable methods. While it is possible to control the exhaust-air dampers with the mixed-air damper signal (to the outside-air and return-air dampers), controlling it independently is more stable. This sequence works in conjunction with the return airflow setpoint calculation to control building pressure. The exhaust-air damper only opens if the return-air fan discharge plenum is positive, which prevents outside air from entering through the exhaust-air damper.

Control of the exhaust dampers is managed by the AHU controller and is based primarily on differential air pressure across the exhaust-air damper. **See Figure 8-6.** The damper position is modulated as necessary to maintain the differential air pressure setpoint.

Control Sequence. A typical sequence for controlling the exhaust-air damper in an AHU includes the following:

1. The AHU controller monitors inputs from the building automation system.
   - The AHU controller receives a digital input indicating desired occupied/unoccupied fan operation based on a preprogrammed schedule.

2. The AHU controller monitors an input from a physical sensor.
   - A differential pressure sensor provides an analog input indicating the differential air pressure across the exhaust-air damper.

3. The controller compares the measured differential air pressure to the setpoint.
   - If the differential pressure is above the setpoint, the exhaust-air damper must be opened.
   - If the differential pressure is below the setpoint, the exhaust-air damper must be closed.

4. The AHU controller generates a signal to modulate the position of the exhaust-air damper actuator.

Cooling Coil Control

The AHU cooling coil both chills (as measured in dry-bulb temperature) and dehumidifies the supply air. Both of these actions are accomplished by controlling the cooling coil to a temperature setpoint. Typically, dehumidification is not controlled via humidity sensors. Instead, temperature sensors, which are less expensive and more accurate, can control humidity by taking advantage of the relationships between temperature and humidity.

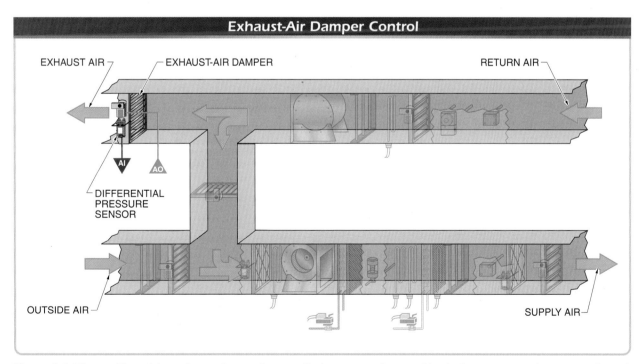

Figure 8-6. The position of the exhaust damper is controlled to ensure that outside air does not enter the system through the exhaust-air duct.

Some AHU configurations place the supply-air fan upstream of the cooling coil (blow-through AHUs) and some place the supply-air fan downstream of the cooling coil (draw-through AHUs). In either case, the supply-air temperature sensor must be downstream of both the cooling coil and the supply-air fan. The fan motor generates heat that can raise the temperature of the supply air, which must be accounted for in the temperature of the conditioned air supplied to the building space.

The control of the cooling coil is managed by the AHU controller based on the supply-air temperature. **See Figure 8-7.** The cooling function is raised or lowered as necessary by opening or closing the valve supplying the coil with chilled water. If a cooling coil is on a chilled water system with variable-speed pumping, the chilled water valve is a two-way control valve. If it is a constant-speed pumping system, the valve is a three-way valve, which bypasses excess chilled water back to the chiller.

Control Sequence. A typical sequence for controlling the cooling coil in an AHU includes the following:
1. The AHU controller monitors inputs from the building automation system.
   • The AHU controller receives a digital input indicating desired occupied/unoccupied fan operation based on a preprogrammed schedule.

2. The AHU controller monitors inputs from physical sensors.
   • A temperature sensor provides an analog input indicating the supply-air temperature.
   • A humidity sensor provides an analog input indicating the return-air humidity.
3. The controller compares the measured supply-air temperature to the setpoint.
   • If the supply-air temperature is above the setpoint, the chilled water valve must be opened.
   • If the supply-air temperature is below the setpoint, the chilled water valve must be closed.
4. The AHU controller generates an analog signal to modulate the actuator controlling the chilled water valve.

Safeties. There is a significant difference of opinion as to whether a cooling coil should have a spring return on the valve actuator upon loss of power or control signal. One opinion contends that the cooling coil should close in order to stabilize the operation of the chiller. Another opinion states that cooling coils should be open in case the owner does not drain the cooling coil in winter. Yet another theory claims that the valve should fail in its last known position, thereby leaving at least some cooling capacity. The ultimate decision is left to the specifying engineer.

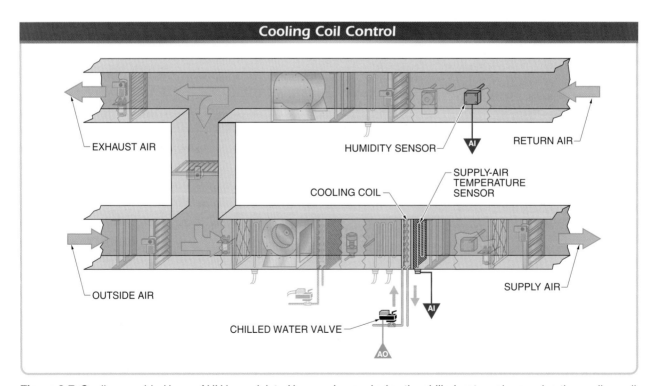

Figure 8-7. Cooling provided by an AHU is modulated by opening or closing the chilled water valve serving the cooling coil.

Alarms. The controller transmits an alarm for the following conditions:

- The supply-air temperature falls below or rises above the setpoint by a predetermined amount.
- The return-air humidity rises above a setpoint.

Heating Coil Control

The heating coil heats (as measured in dry-bulb temperature) the supply air by controlling the heating coil to a temperature setpoint. When the heating coil is upstream of the supply fan, the heating coil discharge temperature sensor setpoint may be reset by the supply-air temperature sensor downstream of the fan. The fan motor generates heat that can raise the temperature of the supply air, which must be accounted for in the temperature of the conditioned air supplied to the building space.

If the heating coil is supplied by a glycol heating system with variable-speed pumping, the valve controlling fluid to the coil is a two-way control valve. If it is a constant-speed pumping system, the valve is a three-way valve that bypasses the excess hot water back to the heat exchanger.

Control of the heating coil is managed by the AHU controller based primarily on temperature measurements. A blow-through AHU uses the supply-air temperature sensor. **See Figure 8-8.**

> **Facts**
>
> *Heating boilers use air as part of the combustion process. Preheating the incoming combustion air helps maximize boiler system efficiency. Also, waste heat from another nearby process can be used as a source of low-cost or free heat.*

A draw-through AHU requires an additional temperature sensor at the discharge of the heating coil. **See Figure 8-9.** The heating coil is then modulated by controlling the heating coil valve actuator.

Control Sequence. A typical sequence for controlling the glycol heating coil in an AHU includes the following:

1. The AHU controller monitors inputs from the building automation system.
   - The AHU controller receives a digital input indicating desired occupied/unoccupied fan operation based on a preprogrammed schedule.
2. The AHU controller monitors inputs from physical sensors.
   - A temperature sensor provides an analog input indicating the supply-air temperature.
   - A temperature sensor (for the draw-through AHU) provides an analog input indicating the heating coil discharge temperature.

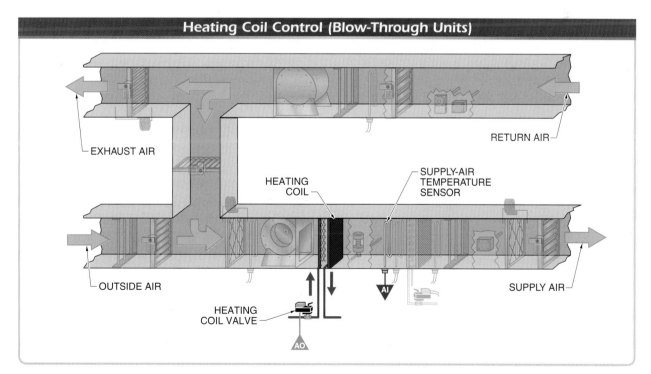

Heating Coil Control (Blow-Through Units)

EXHAUST AIR

RETURN AIR

HEATING COIL

SUPPLY-AIR TEMPERATURE SENSOR

OUTSIDE AIR

SUPPLY AIR

HEATING COIL VALVE

AI

AO

Figure 8-8. In a blow-through AHU, heating from glycol heating coils is modulated by opening or closing the fluid valve.

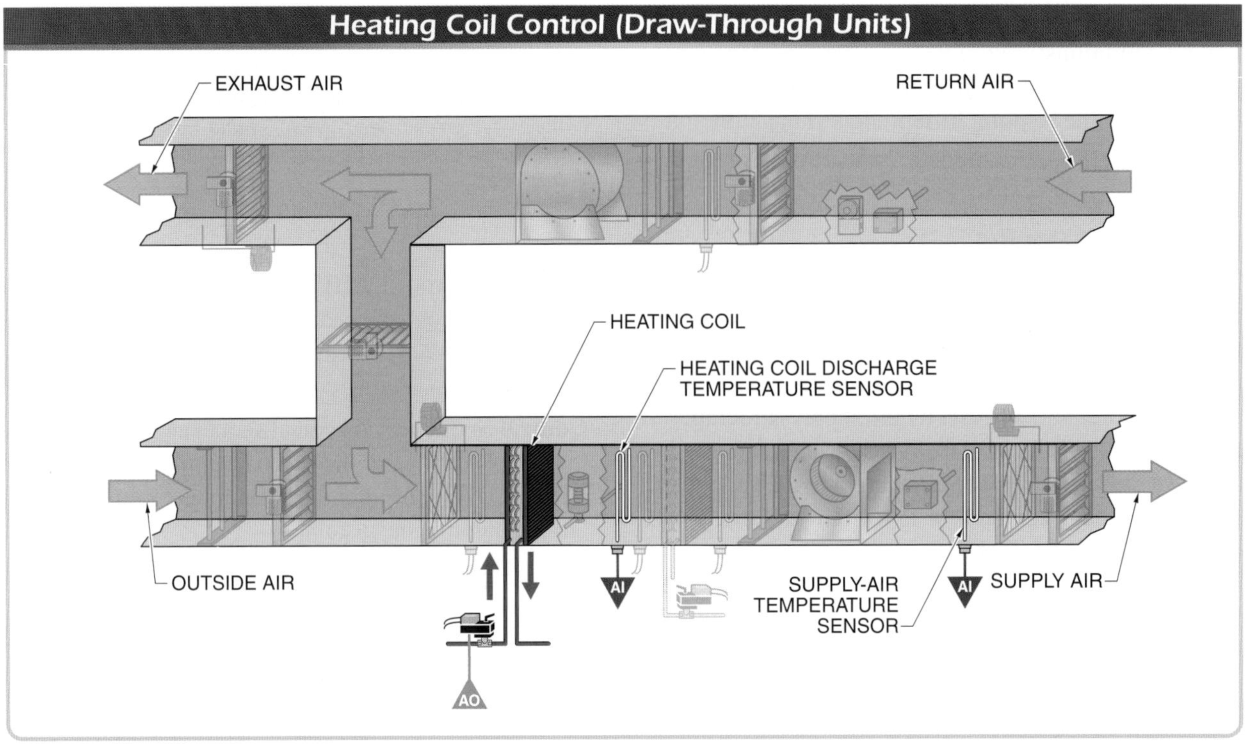

Figure 8-9. Heating with a draw-through AHU requires an additional temperature measurement from a sensor at the discharge of the heating coil.

3. In a draw-through AHU, the controller compares the measured supply-air temperature to the supply-air temperature setpoint.

- If the supply-air temperature is above the setpoint, the heating coil discharge temperature setpoint is decreased.
- If the supply-air temperature is below the setpoint, the heating coil discharge temperature setpoint is increased.

4. The controller compares the measured heating coil discharge temperature to the setpoint.

- If the discharge temperature is above the setpoint, the heating coil valve must be closed.
- If the discharge temperature is below the setpoint, the heating coil valve must be opened.

5. The AHU controller generates an analog signal to the actuator controlling the glycol heating water valve.

If a heating coil is supplied by steam, there is an airflow control damper in addition to the fluid (steam) control valve. This is known as a face-and-bypass heating coil. In this configuration, the face (cross-sectional area) of the AHU is divided into two components.

The first is the coil component and the second is the by-pass component that allows air to move around the heating coil without picking up heat. A bypass damper diverts air from one component to another. **See Figure 8-10.**

Control Sequence. A typical sequence for controlling the steam heating coil in an AHU includes the following:

1. The AHU controller monitors inputs from the building automation system.

- The AHU controller receives a digital input indicating desired occupied/unoccupied fan operation based on a preprogrammed schedule.

2. The AHU controller monitors inputs from physical sensors.

- Temperature sensors provide analog inputs indicating the supply-air temperature and the outside-air temperature.

3. The AHU controller compares the measured outside-air inputs from the physical sensors to the setpoints.

- If the outside-air temperature is above freezing, the bypass damper closes to move air over the coil, and the steam valve modulates to maintain the supply-air setpoint.

- If the outside-air temperature is below freezing, the steam valve is set to 100% open, and the bypass damper modulates to push air over the coil or through the bypass.
4. The AHU controller generates appropriate signals to the actuators controlling the bypass damper and the steam valve.

When the outside-air temperature is below freezing, the AHU controller opens the steam valve to 100% to prevent any condensate from freezing in the return lines. Full-pressure steam inside the heating coil ensures that there is enough pressure to push condensate through the steam trap serving the coil.

Facts

Steam traps are designed to remove air and condensate from steam systems. Failed steam traps must be identified and repaired quickly to prevent problems from developing.

Humidification Control

While dehumidification is often accomplished with cooling coils, humidification requires a separate device. The most common method of humidification is the addition of steam into the supply air. The steam is piped to the AHU from a boiler system.

Control of the humidifier is managed by the AHU controller based primarily on the humidity sensor. **See Figure 8-11.** The amount of moisture added to the air can be modulated by controlling the position of a valve in the humidifier.

Control Sequence. A typical sequence for controlling humidification in an AHU includes the following:
1. The AHU controller monitors inputs from the building automation system.
   - The AHU controller receives a digital input indicating desired occupied/unoccupied fan operation based on a preprogrammed schedule.
2. The AHU controller monitors an input from a physical sensor.
   - A humidity sensor provides an analog input indicating return-air humidity.
3. The controller compares the measured return-air humidity to the setpoint.
   - If the return-air humidity is above the setpoint, the humidifier valve must be closed.
   - If the return-air humidity is below the setpoint, the humidifier valve must be opened.
4. The AHU controller generates an analog signal to the actuator controlling the humidifier steam valve.

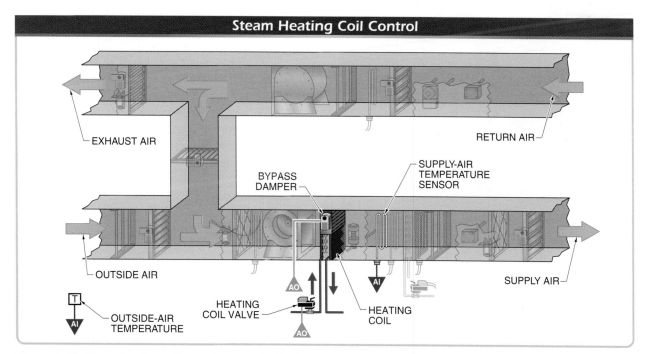

Steam Heating Coil Control

EXHAUST AIR

RETURN AIR

BYPASS DAMPER

SUPPLY-AIR TEMPERATURE SENSOR

OUTSIDE AIR

HEATING COIL VALVE

HEATING COIL

SUPPLY AIR

OUTSIDE-AIR TEMPERATURE

Figure 8-10. Steam heating coils require control of both the steam valve and a bypass damper that directs air away from the coil when the outside-air temperature is below freezing.

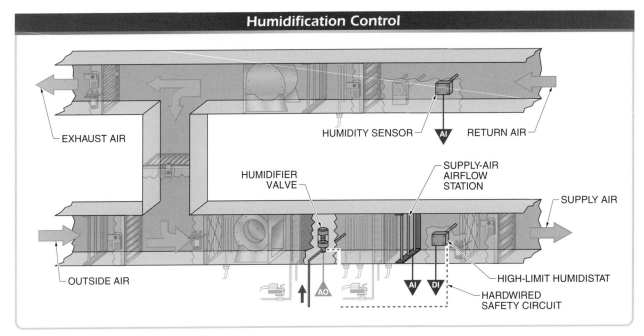

Humidification Control

EXHAUST AIR

HUMIDITY SENSOR — AI — RETURN AIR

HUMIDIFIER VALVE

SUPPLY-AIR AIRFLOW STATION

SUPPLY AIR

OUTSIDE AIR

AO

AI — DI

HIGH-LIMIT HUMIDISTAT

HARDWIRED SAFETY CIRCUIT

Figure 8-11. Humidification is modulated by opening or closing the steam valve serving the humidifier.

Fluke Corporation

Humidity can be measured with a portable meter.

Filter Pressure Drop Monitoring

The monitoring of the filter pressure drop is managed by the AHU controller based on differential pressure across the filter. **See Figure 8-12.** Dirty filters cause a greater pressure drop and can have a significant effect on fan energy consumption. Filters cannot be controlled by the automation system, so this sequence only provides the information to the operator workstation.

The output must be the highest recorded value over the previous three days. The filter pressure drop will be higher on a summer afternoon than a summer morning because more air is being forced through the filter on a VAV system when it is hot. The sliding window to cancel out the effect that unoccupied weekend days may have on the readings is three days.

A VAV system that monitors pressure drop across filter banks may use a differential pressure switch. The switch may require adjustment to prevent the building management system or rooftop unit from shutting the unit down with a false fan-failure signal or from generating a nonlatching diagnostic code for dirty filters.

Control Sequence. A typical sequence for monitoring the filter bank's pressure drop includes the following:
1. The AHU controller monitors an input from a physical sensor.
   • A differential pressure sensor provides an analog input indicating pressure drop across a filter.
2. The AHU controller generates an analog signal to the operator workstation.

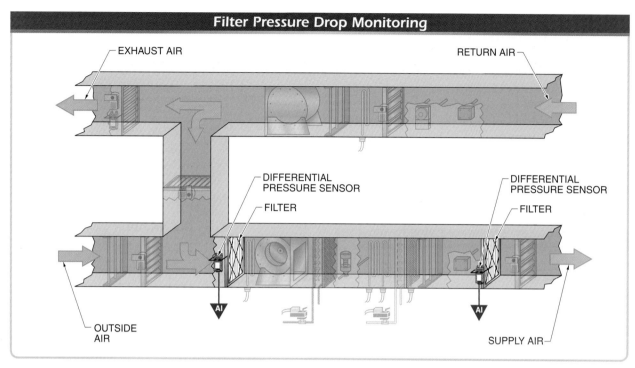

Figure 8-12. The condition of the filter is monitored by a differential pressure sensor and communicated to an operator workstation.

SMOKE CONTROL APPLICATIONS

There are many regulations governing the installation and control of smoke control systems in buildings. NFPA standards, other applicable fire protection codes, and even an owner's fire insurer all affect the operation of smoke control sequences in AHUs. Smoke control sequences must be tightly regulated because failures can easily cause smoke to be spread throughout the building, endangering all occupants.

Smoke purge sequences are controlled to a default fire response position by the fire protection system. For example, the fire alarm control panel (FACP) automatically closes smoke dampers to isolate areas and contain the smoke. Then, a manual command by responding fire personnel may override this action and control the fans and dampers to instead purge the smoke, but only if a subsequent failure returns them to the default fire response position. Therefore, if there is a failure in the purge sequence, the smoke is at least contained again.

Alternatively, an engineered smoke control sequence is executed entirely automatically, but may only be controlled by the building's fire protection system, which is the only building system that must be UL®-listed for reliability. For example, an engineered smoke control

system, via the fire protection system, opens the exhaust-smoke dampers and turns on the fans automatically.

Smoke Detectors

A *smoke detector* is an initiating device that is activated by the presence of smoke particles. In addition to gases, smoke is composed of solid particles and liquid droplets suspended in the air, often invisible to the unaided eye. Smoke detectors sense these smoke particles as the particles enter a small sensing chamber within the device. Smoke detectors should be protected from dust during construction to prevent possible damage to the sensing elements of the smoke detector. The primary difference among smoke detectors is the technology used to detect the particles. The two basic types are photoelectric and ionization smoke detectors.

Facts

Mold spores are much larger than one micron in size and can be removed by applying air filtration according to ASHRAE. A MERV 6 pleated-panel filter has an efficiency rating of over 80% for most molds.

Photoelectric smoke detectors are the most common type of smoke detector. This type senses light scattered by smoke particles within the sensing chamber. An infrared LED light source and sensing photocell are both aimed into the sensing chamber, but not at each other. **See Figure 8-13.** When no smoke is present, the light from the source does not reach the sensing photocell. However, smoke particles will scatter some light onto the photocell, which results in an alarm condition. The device resets once smoke concentrations fall below threshold levels.

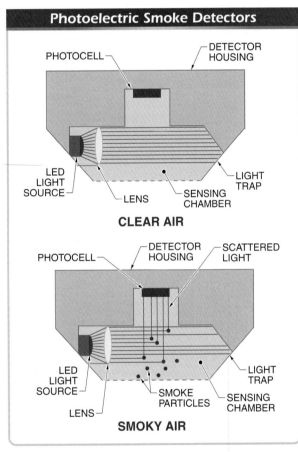

Figure 8-13. Photoelectric smoke detectors sense the presence of smoke particles by watching for the scattering of light they cause.

Ionization smoke detectors use a small radioactive source to ionize (electrically charge) the air in the sensing chamber. **See Figure 8-14.** This causes the air to conduct a small current across the chamber. However, smoke particles entering the sensing chamber will interfere with the normal current flow. The reduction in current is detected by the device, which initiates an alarm.

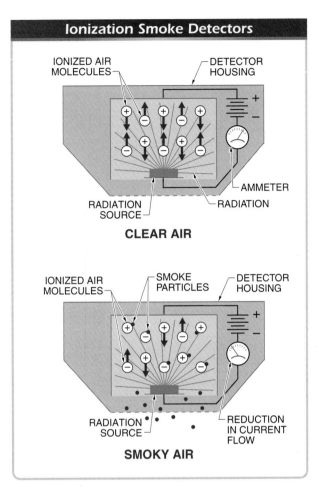

Figure 8-14. Ionization smoke detectors sense the presence of smoke particles by the change they induce in the current flow across a chamber of ionized air.

Smoke Control Modes

In a smoke control system, the signals controlling smoke dampers can be put into one of three different modes. The three modes are the smoke evacuation, smoke purge, and combination evacuation/purge modes. **See Figure 8-15.**

The smoke evacuation mode creates a negative pressure in the smoke area by running only the return-air fan as an exhaust-air fan. This mode is common in buildings without much perimeter wall area in comparison to floor area, and when there is a minimal risk of contaminating adjacent zones.

The smoke purge mode creates a positive pressure in the smoke area by running only the supply-air fan. This carries a significant risk of contaminating adjacent zones, but it is the best option in certain circumstances. For example, in a chemistry lab, a negative smoke control sequence may reverse the flow in a safety fume hood, pulling chemical fumes back into a fire. Instead, a positive pressure helps push contaminated air out of the area, though there must be adequate outside exposure so that the contaminated air can escape.

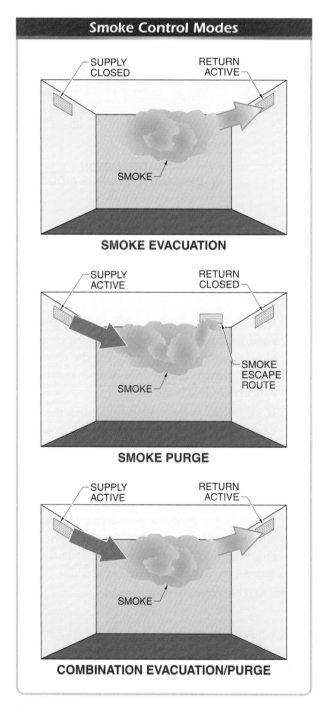

Figure 8-15. Control signals for smoke dampers can be put into one of three different modes for smoke control.

The combination evacuation/purge mode runs both the supply-air fan and the return-air fan at the same time. The supply-air fan brings in 100% outside air and the return-air fan exhausts 100% of the return airflow to outside. This completely ventilates the area with fresh, outside air. However, this function requires an oversized mechanical system to ensure the AHU can deliver 100% outside air even in the middle of winter.

Smoke Control with AHU Dampers

AHU smoke-isolation dampers prevent the AHU from recirculating air contaminated with smoke back into the building spaces. The smoke detector in the return-air duct closes the return-air isolation damper and the smoke detector in the supply-air duct closes the supply-air isolation damper. **See Figure 8-16.**

Control Sequence. A typical sequence for controlling smoke with the AHU includes the following:

1. Upon detection of smoke within the AHU, the FACP automatically closes the smoke dampers within the unit.

2. Upon a manual command, the AHU controller initiates one of the three smoke control modes. It sends analog and/or digital output signals to the AHU fans and dampers according to the operating mode.

   • If the smoke evacuation mode is indicated, the AHU control dampers are positioned so that the outside-air damper is closed, the return-air damper is open, and the exhaust-air damper is open. The AHU smoke-isolation dampers are positioned so that the supply-air damper is closed and the return-air damper is open. The return-air fan operates at a preset fixed speed and the exhaust-air fan turns on.

   • If the smoke purge mode is indicated, the AHU control dampers are positioned so that the outside-air damper is open, the return-air damper is closed, and the exhaust-air damper is closed. The AHU smoke-isolation dampers are positioned so that the supply-air damper is open and the return-air damper is closed. An independent exhaust-air fan turns on and pulls the smoky air out through a separate smoke escape route, such as up a safety fume hood in a chemistry lab. The supply-air fan operates to maintain an air balance using standard control based on supply-air-duct static pressure. If this signal is not available, the fan operates at a preset fixed speed.

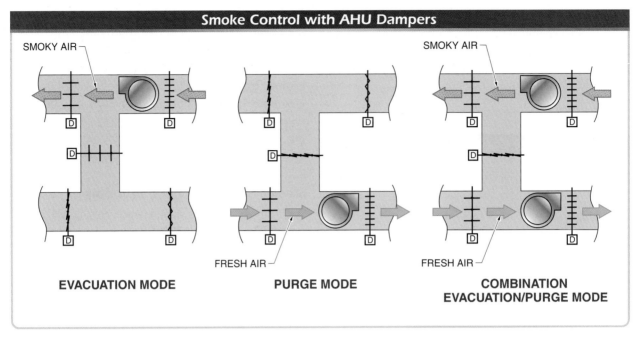

Figure 8-16. Isolation dampers prevent an AHU from circulating smoke-contaminated air throughout the building.

• If the combination evacuation/purge mode is indicated, the AHU control dampers are positioned so that the outside-air damper is open, the return-air damper is closed, and the exhaust-air damper is open. The AHU smoke-isolation dampers are positioned so that both the supply-air and return-air dampers are open. The supply-air fan operates by standard control. The return-air fan operates with fan-tracking control. The exhaust-air fan turns on.

Smoke Control with Zone Dampers

Zone smoke dampers can also be used to isolate specific floors or partial floors of a building. The smoke detector located in the building space or in the air-distribution ducts controls the zone smoke dampers. **See Figure 8-17.**

Control Sequence. A typical sequence for controlling smoke in a certain zone served by an AHU includes the following:

1. Upon detection of smoke within a certain zone, the FACP automatically closes the smoke dampers for that zone.

2. Upon a manual command, the AHU controller initiates one of the three smoke control modes. It sends analog and/or digital output signals to the AHU fans and dampers according to the operating mode.

   • If the smoke evacuation mode is indicated, the zone damper serving the supply-air duct remains closed, the zone damper serving the return-air duct is opened, and the zone damper serving the exhaust-air duct is opened.

   • If the smoke purge mode is indicated, the zone damper serving the supply-air duct is opened, the zone damper serving the return-air duct is closed, and the zone damper serving the exhaust-air duct is opened.

   • If the combination evacuation/purge mode is indicated, the zone damper serving the supply-air duct is opened, the zone damper serving the return-air duct is opened, and the zone damper serving the exhaust-air duct is opened.

3. The AHU operates in parallel mode.

Stairwell Pressurization

Another major smoke control sequence is to control the pressurization of the escape stairwells. This sequence is not controlled nor overridden by the building automation system, but it is important to understand this element of smoke control.

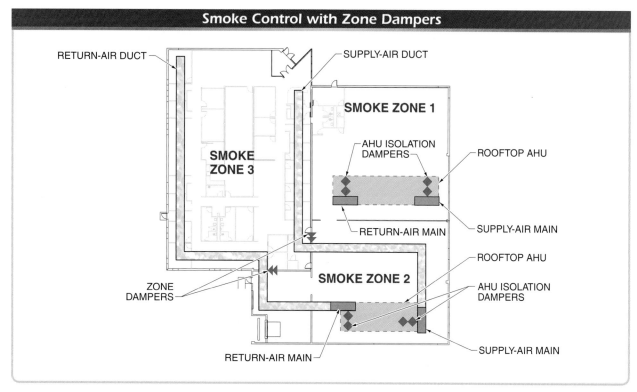

Smoke Control with Zone Dampers

RETURN-AIR DUCT

SUPPLY-AIR DUCT

SMOKE ZONE 1

SMOKE ZONE 3

AHU ISOLATION DAMPERS

ROOFTOP AHU

RETURN-AIR MAIN

SUPPLY-AIR MAIN

ROOFTOP AHU

ZONE DAMPERS

SMOKE ZONE 2

AHU ISOLATION DAMPERS

SUPPLY-AIR MAIN

RETURN-AIR MAIN

Figure 8-17. Zone smoke dampers are used to isolate certain areas of the building to prevent the spread of smoke-contaminated air.

A fan brings in outside air to pressurize the stairwell so that if someone exits a zone filled with smoke via the stairwell, the smoke is not introduced into the escape route. **See Figure 8-18.** Control of the pressurization fan is hardwired directly into two circuits. The differential pressure sensor measuring the stairwell pressure is tied directly into the fan's VFD, bypassing the building automation system. The other circuit provides an input to the building automation system to generate alarms.

Control Sequence. A typical sequence for controlling the stair pressurization fan includes the following:

1. Upon activation by the fire alarm system, the pressurization fan VFD initiates the stair pressurization mode.

2. The VFD monitors an input from a physical sensor.

   • A differential air pressure sensor provides an analog input indicating the stairwell static air pressure.

3. The VFD compares the static air pressure measurement to the setpoint.

   • If the static air pressure is below the setpoint, the pressurization fan speed must be increased.

   • If the static air pressure is above the setpoint, the pressurization fan speed must be decreased.

4. The VFD modulates the speed of the fan to maintain the static air pressure.

Facts

Flow control valves are commonly used in pneumatic systems to decrease, rather than increase, actuator speed by restricting the airflow exiting the actuator. The most common flow control valves used in pneumatic systems are needle valves. Needle valves allow the possible best control of airflow.

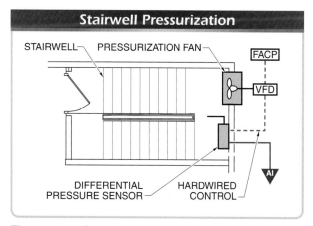

Stairwell Pressurization

STAIRWELL   PRESSURIZATION FAN

FACP

VFD

DIFFERENTIAL PRESSURE SENSOR

HARDWIRED CONTROL

AI

Figure 8-18. Controlling stairwells to a pressure higher than the rest of the building keeps smoke from entering the escape route.

AIR TERMINAL UNIT CONTROL APPLICATIONS

After being distributed from the AHU, air enters the building space through a terminal unit. A terminal unit is the end point in an HVAC distribution system where the conditioned medium (air, water, or steam) is added to or directly influences the environment of the conditioned building space. Air terminal units (ATUs) include dampers to modulate the amount of conditioned supply air into the space and may include other devices to further condition the supply air. A single AHU can serve multiple ATUs.

ATU Control

The two common designs of ATUs are single-duct units and dual-duct units. Single-duct units are provided with cool air from the AHU, which the ATUs then heat as needed. **See Figure 8-19.** In older buildings, it is common to have ATUs with reheat coils on the exterior zones and ATUs without reheat coils on the interior zones. However, with the evolution of the ventilation code, it is often necessary to include reheat coils on every ATU to maintain both comfort and indoor air quality.

The ATU controller uses temperature to determine the airflow setpoint. The change in airflow rate into the room causes the temperature to change. The temperature is also increased or decreased as necessary by modulating the heating function.

Single-Duct Control Sequence. A typical sequence for controlling conditioned air to a certain zone with a single-duct ATU includes the following:

1. The ATU controller monitors inputs from the building automation system.
   - The ATU controller receives a digital input indicating desired occupied/unoccupied fan operation based on a preprogrammed schedule. This schedule also contains minimum and maximum airflow setpoints as well as temperature setpoints for various situations.
2. The ATU controller monitors inputs from physical sensors.
   - A temperature sensor provides an analog input indicating zone temperature.
   - An airflow station provides an analog input indicating airflow at the inlet of the ATU.
3. The controller compares the zone temperature measurement to the zone cooling setpoint.
   - If the zone temperature is above the setpoint, the airflow must be increased.
   - If the zone temperature is below the setpoint, the airflow must be decreased.
4. The controller generates a signal to modulate the damper to maintain the new airflow setpoint.
5. The controller compares the zone temperature measurement to the zone heating setpoint.
   - If the zone temperature is above the setpoint, the heating function must be decreased.
   - If the zone temperature is below the setpoint, the heating function must be increased.
6. The controller generates an analog signal to modulate the heating function to maintain the temperature setpoint. This involves either opening or closing a hot water valve or turning electric heating elements on or off.

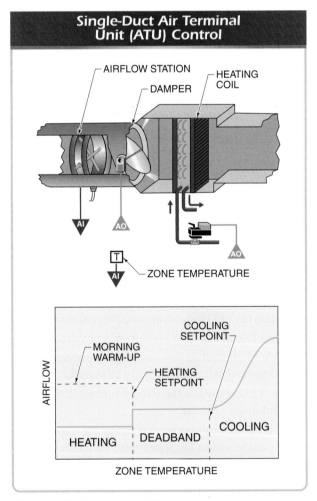

Figure 8-19. Single-duct ATUs control cooling via airflow and heating via a reheat device.

A dual-duct ATU is essentially two single-duct ATUs operating in parallel with the airstreams mixing before going to the diffuser. Dual-duct systems were popular 25 years ago because of their stability in constant-volume systems. However, with the advent of more sophisticated and more accurate control systems, implementation in new buildings has decreased. However, many existing buildings continue to use these systems.

Control of the dual-duct ATU is managed by the ATU controller and based primarily on zone temperature. **See Figure 8-20.** The temperature is also increased or decreased as necessary by modulating the relative contribution of airflow from each duct.

Dual-Duct Control Sequence. A typical sequence for controlling conditioned air to a certain zone with a dual-duct ATU includes the following:

1. The ATU controller monitors inputs from the building automation system.
   - The ATU controller receives a digital input indicating desired occupied/unoccupied fan operation based on a preprogrammed schedule. This schedule also contains minimum and maximum airflow setpoints as well as temperature setpoints for various situations.
2. The ATU controller monitors inputs from physical sensors.
   - A temperature sensor provides an analog input indicating zone temperature.
   - An airflow station provides an analog input indicating airflow in the hot duct.
   - An airflow station provides an analog input indicating airflow in the cold duct.
3. The controller compares the zone temperature measurement to the zone cooling setpoint.
   - If the zone temperature is above the setpoint, the cold-duct airflow must be increased.
   - If the zone temperature is below the setpoint, the hot-duct airflow must be increased.
4. The controller generates two signals to modulate the positions of the duct dampers: one for the cold-duct damper and one for the hot-duct damper.

Many ATUs can control airflow only by modulating their integral dampers, but some include a circulation fan that operates in parallel with the airflow from the central AHU to assist with heating. During cooling, the fan is off and cool supply air is served from the central AHU. A backdraft damper prevents reverse flow through the fan. During heating, though, the ATU may turn on its circulating fan to take advantage of the warm air that has risen to near the ceiling. Recirculating some of the warm air reduces the heating load.

Facts

A minimum amount of outdoor air must be introduced into most AHUs to ensure good IAQ and to pressurize a building. The flow rate must match or exceed the amount of exhaust air taken from the building unless the building or area needs to operate at a negative pressure.

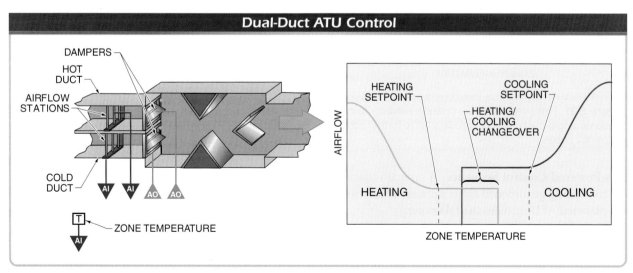

Figure 8-20. Dual-duct ATUs control heating and cooling by modulating the relative position of dampers in the hot and cold ducts, respectively.

The control of the fan-powered ATU is managed by the ATU controller and is based primarily on zone temperature. **See Figure 8-21.** The necessary output signals to change the operation of the ATU are sent to the heating function actuators.

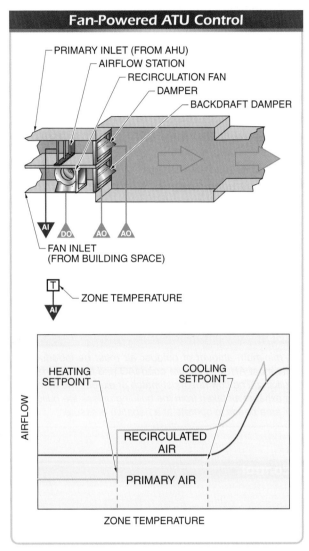

Figure 8-21. Fan-powered ATUs operate similarly to conventional single-duct units but include a fan to further control airflow for cooling.

Fan-Powered Control Sequence. A typical sequence for controlling conditioned air to a certain zone with a fan-powered ATU includes the following:

1. The ATU controller monitors inputs from the building automation system.
   - The ATU controller receives a digital input indicating desired occupied/unoccupied fan operation based on a preprogrammed schedule.

This schedule also contains minimum and maximum airflow setpoints as well as temperature setpoints for various situations.

2. The ATU controller monitors inputs from physical sensors.
   - A temperature sensor provides an analog input indicating zone temperature.
   - An airflow station provides an analog input indicating airflow in the primary inlet of the ATU.

3. The controller compares the zone temperature measurement to the zone cooling setpoint.
   - If the zone temperature is above the setpoint, the airflow must be increased.
   - If the zone temperature is below the setpoint, the airflow must be decreased.

4. The controller generates an analog signal to the duct damper to modulate the airflow.

5. The controller compares the zone temperature measurement to the zone heating setpoint.
   - If the zone temperature is above the setpoint, the heating function must be decreased.
   - If the zone temperature is below the setpoint, the heating function must be increased.

6. The controller generates an analog signal to modulate the heating function to maintain the temperature setpoint. This involves either opening or closing a hot water valve or turning electric heating elements on or off.

7. If the zone is unoccupied and requires heating, the controller generates a digital signal to turn on the circulation fan. Heat is picked up as the recirculated air is drawn from near the ceiling.

Perimeter Heating System Control

The perimeter heating system counteracts the loss of heat through the building envelope. Perimeter heating systems generally include either radiant heating panels, which heat with infrared (IR) energy, or fintube radiator units, which use hot water or steam to heat via convection airflow. **See Figure 8-22.** Each type has advantages and disadvantages. For example, fintube radiators prevent occupants from placing furniture against an outside wall because it would block airflow, but radiant heating panels complicate the mounting of nearby window coverings. However, both are controlled in the same manner.

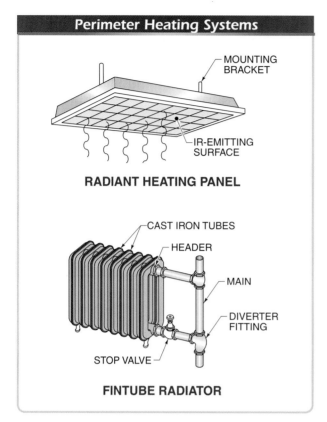

Figure 8-22. Perimeter heating systems generally include either radiant heating panels or fintube radiator units.

Perimeter heating systems are generally used in cold climates in facilities with large amounts of exterior glass or relatively high humidity levels throughout the year, such as healthcare facilities. Perimeter heating systems are usually controlled by extra connection points on an ATU controller. This is a complementary relationship because the perimeter heat must be coordinated with the ATU reheat system.

The perimeter heating system is activated in relationship to when the heating device inside the ATU is activated. The two systems can operate at the same time, but the perimeter heating system can also be programmed to activate either before or after the overall zone temperature measurements call for heating. All three options are valid, and any one option may be implemented by the specifying engineer.

Control of the perimeter heating system is managed by the ATU controller based on temperature. The temperature is raised or lowered as necessary by modulating the heating function. Depending on the application, the control of perimeter heating systems may be influenced

by the fact that radiant heat and convective air loops have very long response times. This also means that perimeter heat must never be controlled by ON/OFF or triac control techniques.

Control Sequence. A typical sequence for controlling perimeter heating systems includes the following:

1. The ATU controller monitors an input from a physical sensor.
   - A temperature sensor provides an analog input indicating overall zone temperature. (This is the same sensor used by the ATU for other sequences.)
2. The controller compares the zone temperature measurement to the zone heating setpoint.
   - If the zone temperature is above the setpoint, the heating function must be decreased.
   - If the zone temperature is below the setpoint, the heating function must be increased.
3. The controller generates an analog signal to modulate the hot water valve or electric heating element to maintain the temperature setpoint.

Unit Heater Control

A unit heater is a relatively small device used for spot heating, meaning that there is no air distribution system. Unit heaters typically serve utility spaces such as penthouses, loading docks, exterior stairwells, and entrance vestibules. Unit heaters are often controlled with stand-alone controls, but it is possible to integrate them into a building automation system. Unit heaters may use either hot water coils or electric heating elements.

If the hydronic unit is supplied by a heating system with variable-speed pumping, the valve controlling the heating water to the coil is a two-way control valve. If it is a constant-speed pumping system, the valve is a three-way valve that bypasses the excess hot water back to the heat exchanger.

On a hydronic unit heater, when the outside-air temperature is above freezing and heating is required, the fan cycles on and the valve modulates to maintain the setpoint. When the outside-air temperature is below freezing, the heating valve is 100% open and the fan cycles to maintain setpoint. Constant circulation prevents the water from being exposed to freezing temperatures, even though it requires constant pumping.

Control of the unit heater is managed by the unit heater controller based on temperature. **See Figure 8-23.** Ideally, the zone temperature sensor should be mounted far from the unit heater so that convective air loops do not influence the sensor when the fan is turned off.

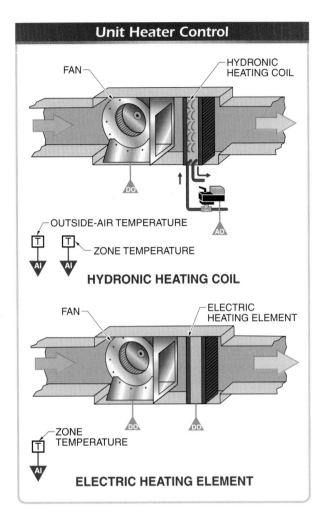

Figure 8-23. Unit heaters are small, individual units with their own controls.

Control Sequence. A typical sequence for controlling hydronic unit heaters includes the following:

1. The unitary controller monitors inputs from physical sensors.
   - A temperature sensor provides an analog input indicating the zone temperature.
   - A temperature sensor provides an analog input indicating the outside-air temperature.
2. The controller compares the measured outside-air temperature to the programmed setpoint (typically 38°F).
   - If the outside-air temperature is below the setpoint, the heating coil valve must be opened.
   - If the outside-air temperature is above the setpoint, the heating coil valve must be closed.
3. The controller generates an analog signal to modulate the actuator controlling the hot water valve.

4. The controller compares the measured zone temperature to the setpoint.
   - If the zone temperature is below the setpoint, the fan must be turned on.
   - If the zone temperature is above the setpoint, the fan must be turned off.
5. The controller outputs a digital signal to turn the fan on or off.

Fan Coil Control

Fan coils are similar to unit heaters. They are installed for similar applications and the heating function is controlled in the same way. The difference is that a fan coil also includes a cooling function. **See Figure 8-24.** Again, the zone temperature sensor should be mounted far from the unit heater so convective air loops do not influence the sensor when the fan is turned off.

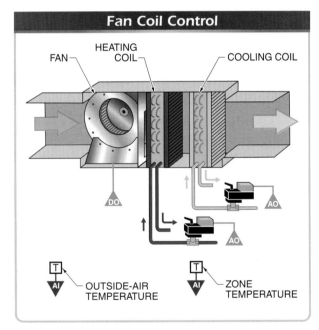

Figure 8-24. Fan coil units are similar to unit heaters, but they also include a cooling function.

Facts

Ventilation requirements can be very strict for residential and light commercial buildings. Since buildings that do not have enough ventilation air can suffer from poor IAQ, following current standards and using the proper equipment can help ensure that the ventilation requirements are met.

Control Sequence. A typical sequence for controlling fan coils includes the following:
1. The unit controller monitors inputs from physical sensors.
   - A temperature sensor provides an analog input indicating the zone temperature.
   - A temperature sensor provides an analog input indicating the outside-air temperature.
2. The controller compares the measured zone temperature to the zone cooling setpoint.
   - If the zone temperature is below the setpoint, the cooling coil valve must be closed.
   - If the zone temperature is above the setpoint, the cooling coil valve must be opened.
3. The controller generates an analog signal to modulate the cooling coil valve to maintain the temperature setpoint. However, if the heating function is operating, the cooling valve must be closed.
4. The controller compares the measured zone temperature to the zone heating setpoint.
   - If the zone temperature is above the setpoint, the fan must be turned on.
   - If the zone temperature is below the setpoint, the fan must be turned off.
5. The controller generates a digital signal to turn the fan on or off.

Exhaust-Air Fan Control

An exhaust-air fan removes contaminated air from the building. **See Figure 8-25.** Exhaust requirements can range from relatively safe to highly toxic, which affects the physical configuration of exhaust system components, as well as the operating sequence. For example, toilet exhaust in an office building can be turned off at night, but fume hood exhaust must be turned on for 24 hours per day, 7 days a week.

Control Sequence. A typical sequence for controlling the exhaust-air fan includes the following:
1. The controller monitors inputs from the building automation system.
   - The AHU controller receives a digital input indicating desired occupied/unoccupied fan operation based on a preprogrammed schedule.
2. The controller monitors an input from a physical sensor.
   - A digital input from the fan indicates its ON/OFF status.

3. The controller generates a digital signal to turn the exhaust fan on or off based on the schedule.

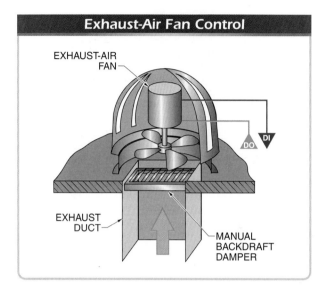

Figure 8-25. Exhaust-air fans operate according to a pre-programmed schedule.

HYDRONIC AND STEAM HEATING CONTROL APPLICATIONS

A hydronic HVAC system distributes water or steam throughout a building as the heat-transfer medium for heating and cooling systems. Hydronic HVAC systems use pipes, valves, and pumps to distribute water throughout the building. Hot water or steam is generated by boilers and is distributed in a piping loop for heating use. Water systems require a pump to circulate the hot water throughout the building, but steam heating systems do not require circulating pumps because steam flows easily due to the difference in pressures between the boiler and the terminal units.

The water or steam is piped into a heat exchanger that transfers heat between the water or steam and the indoor air. A *heat exchanger* is a device that transfers heat from one fluid to another fluid without allowing the fluids to mix.

Package Boiler System Control

A package boiler system produces hot water or steam for hydronic heating. Control within package boiler units is managed by the factory-installed controller. A package boiler is only enabled or disabled as needed by the building automation system. **See Figure 8-26.**

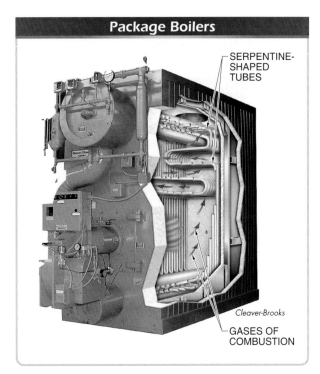

Figure 8-26. A package boiler system produces hot water or steam for hydronic heating.

Heat Exchanger Control

A heat exchanger transfers heat between two fluids. A common application of heat exchangers is the transfer of heat energy from a steam system to a hydronic (hot water) heating system. **See Figure 8-27.** A hydronic heating system then serves hydronic reheat coils in ATUs and perimeter heating systems. Substituting a glycol mixture in the tube side of the heat exchanger creates the glycol heating system used in AHU preheat coils.

Heat exchangers can experience extreme load swings, from a 100% load to a load less than 5%, which complicates the selection of a steam valve for all circumstances. Valves sized for full load can become unstable when turned down. One solution is to use two valves on the steam inlet. The first valve is sized at approximately one-third full load and the second is sized at approximately two-thirds full load. Both valve actuators receive the same controller output signal, but the smaller valve is set to open first, thereby giving good control at both ends of the spectrum.

Control Sequence. A typical sequence for controlling a heat exchanger includes the following:
1. The controller monitors inputs from physical sensors.
   - A temperature sensor provides an analog input indicating the water inlet temperature.
   - A temperature sensor provides an analog input indicating the water outlet temperature.
2. The controller compares the measured outlet temperature to the heat exchanger setpoint.
   - If the outlet temperature is below the setpoint, the heat exchanger valve must be opened.
   - If the outlet temperature is above the setpoint, the heat exchanger valve must be closed.
3. The controller generates an analog signal to modulate the heat exchanger valves to maintain the setpoint temperature.

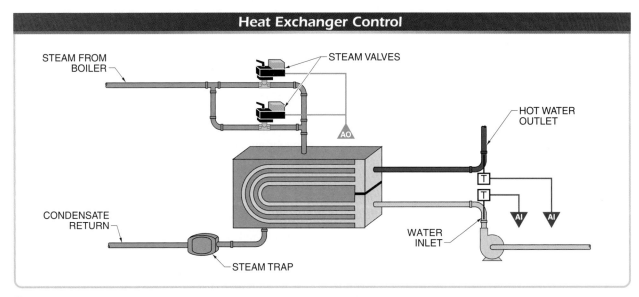

Figure 8-27. Heat exchangers control the transfer of heat between fluids by modulating valves.

Hot Water Pumping Control

A hydronic heating system distributes hot water to various terminal devices such as hydronic heating coils in ATUs, unit heaters, and perimeter heating systems. A pump circulates hot water throughout the hot water distribution system and must provide adequate flow. A differential pressure sensor needs to be installed between the hot water supply pipe and the hot water return pipe. The differential water pressure between the hot water supply pipe and the hot water return pipe indicates if there is enough energy to push water through the heating coils. **See Figure 8-28.**

Control Sequence. A typical sequence for controlling hot water pumping includes the following:

1. The controller monitors an input from a physical sensor.
   - A differential water pressure sensor provides an analog input indicating the differential water pressure between the hot water supply and hot water return lines.
2. The controller compares the measured differential pressure to the pressure setpoint.
   - If the differential pressure is below the setpoint, the pump speed must be increased.
   - If the differential pressure is above the setpoint, the pump speed must be decreased.
3. The controller generates an analog signal to the pump motor VFD to modulate the speed of the pump.

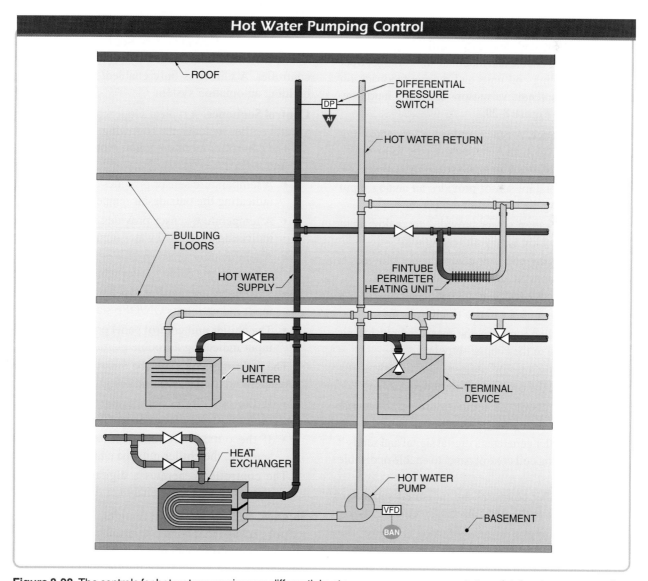

Figure 8-28. The controls for hot water pumping use differential water pressure measurements to maintain adequate water flow.

COOLING CONTROL APPLICATIONS

HVAC systems use refrigerant or chilled water for cooling air in an AHU. Refrigerant is controlled with a direct expansion (DX) system and chilled water is controlled with a water chiller system. Either cold fluid flows through a cooling coil within the ductwork of a forced-air HVAC system, transferring heat from the air to the fluid. Moving air provides a more efficient method of heat transfer.

DX Cooling Coil Control

DX cooling coils use a refrigeration cooling cycle. These units are also known as air conditioners, though this term is more commonly applied to residential and light commercial cooling installations. These are typically installed as package units that manage their own internal controls and devices for efficient operation. External signals enable and modulate the cooling effect of the DX cooling coil units and receive feedback on their status. A solenoid valve stops and starts fluid flow as needed. **See Figure 8-29.**

Control Sequence. A typical sequence for controlling a DX cooling coil in an AHU includes the following:
1. The controller monitors inputs from physical sensors.
   - A temperature sensor provides an analog input indicating the outside-air temperature.
   - A temperature sensor provides an analog input indicating the supply-air temperature.
   - The cooling coil unit control panel provides a digital input indicating its ON/OFF status.
   - The cooling coil unit control panel provides a digital input indicating a trouble alarm.
2. The controller compares the measured outside-air temperature to the cooling setpoint.
   - If the outside-air temperature is below the setpoint, the cooling coil must be disabled.
   - If the outside-air temperature is above the setpoint, the cooling coil must be enabled.
3. The controller generates a digital signal and sends it to the cooling coil control panel to enable or disable the unit.
4. The controller compares the measured supply-air temperature to the AHU discharge setpoint.
   - If the supply-air temperature is below the setpoint, the cooling function must be decreased.
   - If the supply-air temperature is above the setpoint, the cooling function must be increased.

Refrigerant lines can be insulated to save energy.

Water Chiller Control

Chilled water is distributed to cooling coils in various terminal devices such as AHUs, fan coil units, and computer room air conditioners. **See Figure 8-30.** Control of the chiller is managed by the factory-installed controller. A chiller is only enabled/disabled by the building automation system.

Control Sequence. A typical sequence for controlling a water chiller includes the following:
1. The controller monitors inputs from physical sensors.
   - A temperature sensor provides an analog input indicating the outside-air temperature.
   - A temperature sensor provides an analog input indicating the chilled water supply temperature.
   - A temperature sensor provides an analog input indicating the chilled water return temperature.
   - The chiller unit control panel provides a digital input indicating its ON/OFF status.
   - The chiller unit control panel provides a digital input indicating a trouble alarm.
2. The controller compares the measured outside-air temperature to the setpoint.
   - If the outside-air temperature is below the setpoint, the chiller must be disabled.
   - If the outside-air temperature is above the setpoint, the chiller must be enabled.
3. The controller generates a digital signal to the chiller control panel to enable or disable the unit.

Safeties. All chiller safety features are part of the control package that comes shipped from the factory. The chiller control panel monitors all the system control devices and generates a digital alarm signal for operating problems.

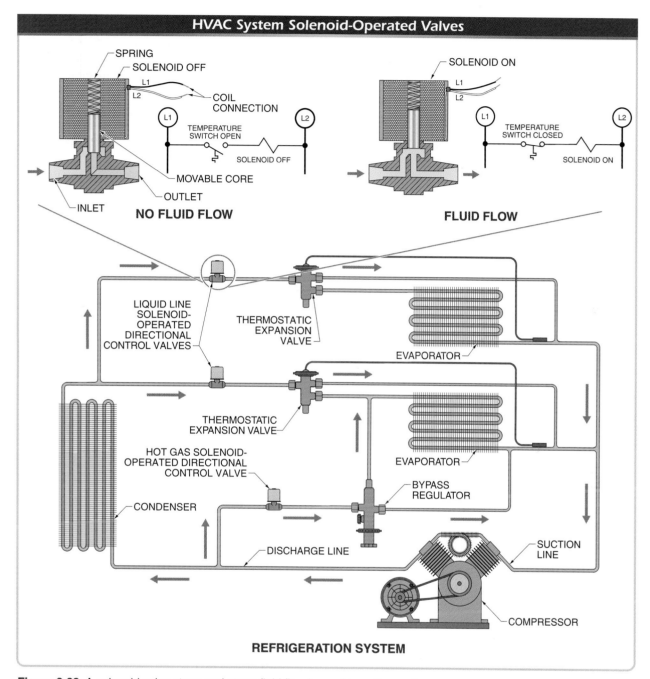

HVAC System Solenoid-Operated Valves

SPRING
SOLENOID OFF
L1
L2
COIL CONNECTION
L1   L2
TEMPERATURE SWITCH OPEN
SOLENOID OFF
MOVABLE CORE
OUTLET
INLET
NO FLUID FLOW

SOLENOID ON
L1
L2
L1   L2
TEMPERATURE SWITCH CLOSED
SOLENOID ON
FLUID FLOW

LIQUID LINE SOLENOID-OPERATED DIRECTIONAL CONTROL VALVES
THERMOSTATIC EXPANSION VALVE
EVAPORATOR
THERMOSTATIC EXPANSION VALVE
HOT GAS SOLENOID-OPERATED DIRECTIONAL CONTROL VALVE
EVAPORATOR
BYPASS REGULATOR
CONDENSER
SUCTION LINE
DISCHARGE LINE
COMPRESSOR
REFRIGERATION SYSTEM

Figure 8-29. A solenoid valve stops and starts fluid flow in a refrigeration system.

Alarms. An alarm is triggered for any of the following conditions:

- The chiller is enabled, the temperature sensor in the chilled water header is indicating a high temperature, and the chiller is turned off.
- The chilled water temperature sensor temperature rises above a preset setpoint.
- The chiller activates its trouble alarm.

Facts

HVAC units operate most efficiently with the correct amount of airflow across the heating and cooling coils. Any significant variation in airflow causes a unit to lose efficiency. Large commercial systems may have air-velocity sensors permanently mounted in the ductwork. An air-velocity sensor enables an operator to check the air velocity on a building automation system computer.

Chilled Water Systems

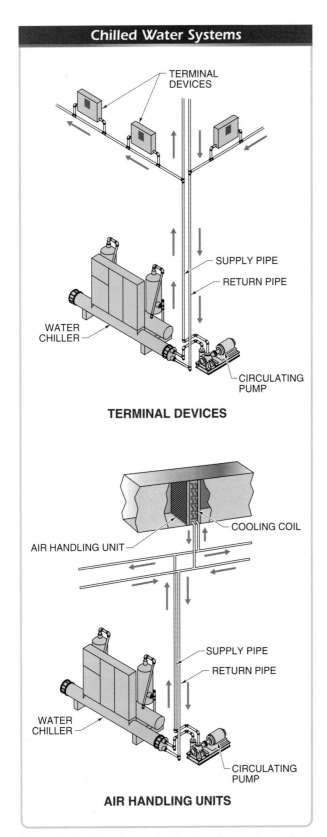

TERMINAL DEVICES

TERMINAL DEVICES

SUPPLY PIPE

RETURN PIPE

WATER CHILLER

CIRCULATING PUMP

COOLING COIL

AIR HANDLING UNIT

SUPPLY PIPE

RETURN PIPE

WATER CHILLER

CIRCULATING PUMP

AIR HANDLING UNITS

Figure 8-30. Chilled water is distributed to cooling coils in various terminal devices as needed.

Cooling Tower Control

A cooling tower is a large structure where air moves upward to mix with falling water, resulting in evaporative cooling of the water. Cooling towers are more efficient than DX units because the condenser fluid is exposed to the environment, thereby allowing it to take advantage of evaporative cooling. The condenser temperature can fall 10°F to 15°F below ambient dry-bulb temperatures, thereby saving compressor energy.

Control of the cooling tower is managed by the controller based primarily on water outlet temperature. **See Figure 8-31.** The temperature is increased or decreased as necessary by changing the speed of the VFD operating the cooling tower fan.

Cooling Tower Control

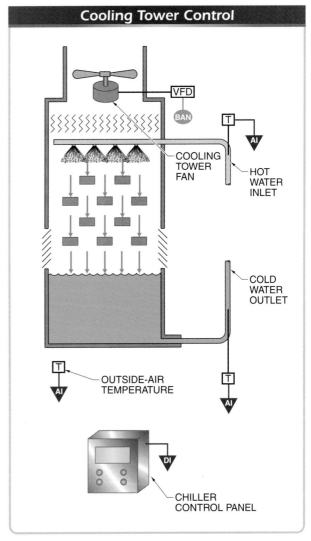

VFD

BAN

COOLING TOWER FAN

HOT WATER INLET

COLD WATER OUTLET

OUTSIDE-AIR TEMPERATURE

CHILLER CONTROL PANEL

Figure 8-31. Cooling towers discharge the heat absorbed by a chilled water system by evaporating some of the water.

Control Sequence. A typical sequence for controlling a cooling tower includes the following:

1. The controller monitors inputs from physical sensors.
   - A temperature sensor provides an analog input indicating the outside-air temperature.
   - A temperature sensor provides an analog input indicating the cooling tower water inlet temperature.
   - A temperature sensor provides an analog input indicating the cooling tower water outlet temperature.
   - The chiller unit control panel provides a digital input indicating its ON/OFF status.
2. The controller compares the measured water outlet temperature to the setpoint.
   - If the water outlet temperature is below the setpoint, the cooling tower fan's speed must be decreased.
   - If the water outlet temperature is above the setpoint, the cooling tower fan's speed must be increased.
3. The controller generates an analog signal to the cooling tower fan VFD to modulate the fan motor speed.

Chilled Water Pumping Control

Chilled water pumping requires more control than hot water pumping because there must be constant flow through the chiller's evaporation coil or it will become unstable. A constant-volume chilled water system is a simple solution, but it consumes a lot of energy. Instead, a primary/secondary system addresses both issues. **See Figure 8-32.** A primary pump ensures constant flow through the chiller, while a larger secondary pump only pumps what is needed by the cooling coils. Secondary pump motors are controlled with a VFD. (The condenser side of the chiller, between the chiller and the cooling tower, is constant-volume pumping.)

Facts

Transmitters are used to send a signal from a sensor to a controller. Transmitters are selected so that the variable they sense falls near the middle of their range. HVAC mixed air applications have temperatures near 55°F and often use a transmitter with a range of 0°F to 100°F.

When the system is fully loaded, the flow in the primary pump equals the flow in the secondary pump. In effect, the primary and secondary pumps are in series and the chilled water coming from the chiller goes out to the coils.

Control Sequence. A typical sequence for controlling chilled water pumping includes the following:

1. The controller monitors inputs from physical sensors.
   - A digital signal from the chiller controller indicates the enabled/disabled status of the chiller. This signal also indicates the ON/OFF status of the primary evaporator and condenser pumps.
   - A differential static pressure sensor provides an analog input indicating the pressure differential between the chilled water supply pipe and the chilled water return pipe.
2. The controller compares the measured differential static pressure to the setpoint.
   - If the differential static pressure is below the setpoint, the secondary pump speed must be increased.
   - If the differential static pressure is above the setpoint, the secondary pump speed must be decreased.
3. The controller generates an analog signal to the secondary pump VFD to modulate the pump speed.

Primary and secondary flows are not equal when the chiller plant is operating at less than full capacity. Flow through the primary loop (chiller) is constant. Flow through the secondary loop (building loop) varies because of VFD control output changes in response to the differential pressure sensor at the end of the building chilled water source (CHWS) and chilled water return (CHWR) pipes. Flow is balanced through a decoupler. With a building management system, the best practice to control the primary and secondary loop is with temperature monitoring of five points. The five points are the primary chilled water supply, primary chilled water return, secondary chilled water supply, secondary chilled water return, and decoupled loop.

Free Cooling

The industry standard terminology for producing chilled water during the winter is "free cooling." This refers to the fact that the chiller, a large energy consumer, does not need to run when the outside-air temperature is low. **See Figure 8-33.**

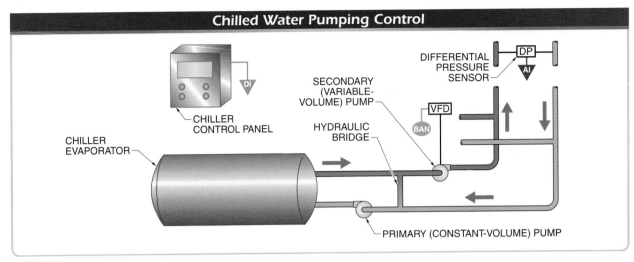

Figure 8-32. Chilled water pumping involves a pair of pumps that provide adequate flow without wasting energy.

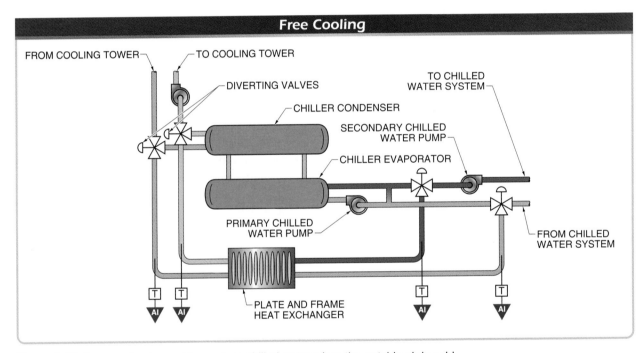

Figure 8-33. Free cooling is used to produce chilled water when the outside air is cold.

When the outside-air temperature falls below the chilled water setpoint by approximately 5°F, diverting valves reroute the chilled water and cooling tower water through a "plate and frame" heat exchanger. Temperature sensors on the heat exchanger only monitor the heat exchanger for diagnostics. The cooling tower outlet water temperature is changed from a condenser water setpoint to a chilled water setpoint.

While chilled water may not be required for space cooling during the winter because the AHUs can use the economizer mode, other equipment in the building may need cooling, such as computer room air conditioning units, medical equipment, or foodservice refrigeration compressors.

In certain climates and building types, free cooling can actually use less energy than an AHU with an economizer. The additional pumping energy may be negligible in comparison to the energy required to keep a building properly humidified when a lot of outside air is used for economizer cooling.

REVIEW QUESTIONS

1. How do some HVAC system sequences affect each other?

2. Explain the function of hardwired safety circuits in HVAC applications.

3. How are static air pressure setpoints related to supply-air fan control?

4. How does economizing save energy?

5. Explain the differences between the three different smoke control modes.

6. How does stairwell pressurization work?

7. What is a perimeter heating system?

8. What is the relationship of building automation systems to the control of package boiler and chiller units?

9. What is one method of sizing steam valves for heat exchangers?

10. Why does chilled water pumping require more control than hot water pumping, and how is this control achieved?

Chapter 8 Review and Resources

Plumbing System Control Devices and Applications 9

All plumbing systems require some degree of control to facilitate the operation of plumbing fixtures and to carry away waste. With growing interest in energy efficiency and water conservation, plumbing systems are becoming increasingly automated. Automated systems can optimally regulate water temperature and flow for various applications and for different parts within the same system, which greatly reduces wasted water and energy. With careful design and implementation, this control can also add extra benefits in convenience, hygiene, and reliability.

PLUMBING SYSTEMS

All buildings that are occupied by people, such as residences, office buildings, and manufacturing facilities, include a plumbing system. A *plumbing system* is a system of pipes, fittings, and fixtures within a building that conveys a water supply and removes wastewater and waterborne waste. These tasks are handled by different subsystems of the plumbing system. Four common systems are water supply systems, sanitary drainage systems, vent piping systems, and stormwater drainage systems.

Water Supply Systems

A *water supply system* is a plumbing system that supplies and distributes potable water to points of use within a building. *Potable water* is water that is free from impurities that could cause disease or other harmful health conditions. Potable water is drinkable.

Potable water is supplied from a municipal water supply source through a water main. **See Figure 9-1.** A water service pipe extends from the water main to each building. Underground cocks (valves) on the service pipe at the water main and near the curb line allow the flow of potable water to a building to be turned on or off.

A water meter is installed on the water service where the water service pipe enters a building. The water meter measures the volume of water that passes through the water service. This allows the water utility to bill the building's owners for their water use. Valves on either side of the water meter allow the meter to be easily serviced or replaced.

The water supply system of a building consists of water distribution pipes, fittings, control valves, and fixtures. These components may be inside or outside of the building, but they must be within the property lines to be considered part of the building's plumbing system. Landscape irrigation systems are considered part of a building's water supply system since they draw potable water from the same water mains.

Water Distribution Pipes. Water distribution pipes convey water from the water service pipe to the point of use. These pipes can be further classified according to their function within the building. A *building main* is a water distribution pipe that is the principal pipe artery supplying water to the entire building. Pipes are connected to the main to distribute the water supply to various areas of the building. A *riser* is a water distribution pipe that routes a water supply vertically one full story or more. A *branch* is a water distribution pipe that routes a water supply horizontally to fixtures or other pipes at approximately the same level. Branches may supply either individual fixtures or groups of fixtures.

Water Supply Systems

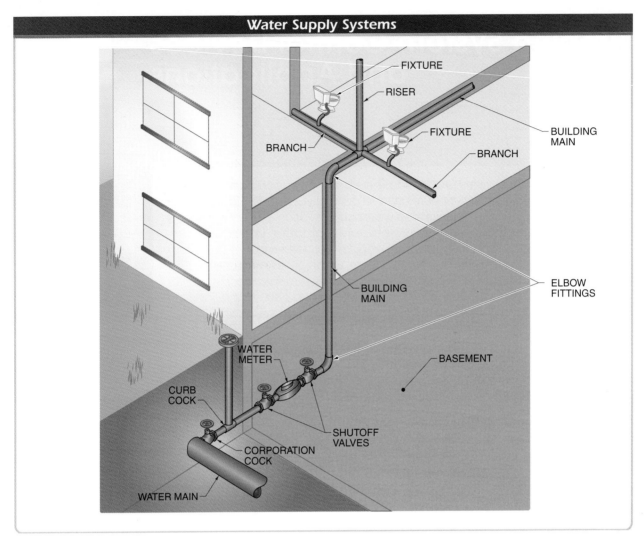

Figure 9-1. A water supply system conveys potable (drinkable) water to plumbing fixtures throughout a building.

A *fixture branch* is a water supply pipe that extends between a water distribution pipe and fixture supply pipe. **See Figure 9-2.** A *fixture supply pipe* is a water supply pipe that connects the fixture to the fixture branch. There is typically a valve between the fixture branch and fixture supply pipe that allows water to individual fixtures to be shut off without interrupting the rest of the system.

Fittings and Valves. A *fitting* is a device used to connect two lengths of pipe. Fittings may be used to extend the length of a pipe only or may serve additional purposes, such as transitioning between different sizes or types of pipes or branching or changing the direction of a pipe.

A *valve* is a fitting that regulates the flow of water within a piping system. There are many types of valves used in plumbing systems. Some are used only when

work is being done on the plumbing system, such as during installation or repair work, while others are used during the daily operation of the system.

Fixtures. A *fixture* is a receptacle or device that is connected to the water distribution system, demands a supply of potable water, and discharges the waste directly or indirectly into the sanitary drainage system. Common fixtures include lavatories (sinks), water closets (toilets), and bathtubs. While most fixtures are permanently connected to the plumbing system, some may be temporary fixtures.

An *appliance* is a plumbing fixture that performs a special function and is controlled and/or energized by motors, heating elements, or pressure-sensing or temperature-sensing elements. Common appliances include water heaters, water softeners, and washing machines.

Controlling the flow of potable water and wastewater within fixtures is accomplished with fixture trim. *Fixture trim* is a set of water supply and drainage fittings installed on a fixture or appliance to control the water flowing into a fixture and the wastewater flowing from the fixture to the sanitary drainage system.

Sanitary Drainage Systems

A *sanitary drainage system* is a plumbing system that conveys wastewater and waterborne waste from plumbing fixtures and appliances to a sanitary sewer. **See Figure 9-3.** *Sewage* is any liquid waste containing animal or vegetable matter in suspension or solution and/ or chemicals in solution. A *sanitary sewer* is a sewer that carries sewage but does not convey rainwater, surface water, groundwater, and similar nonpolluting wastes. A sanitary sewer may be a public sewer, private sewer, individual building sewage-disposal system, or other point of disposal.

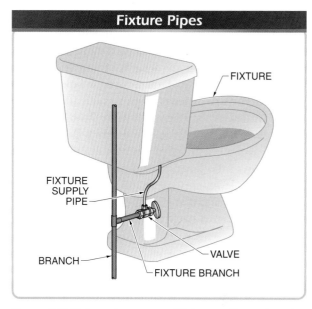

Figure 9-2. Fixture pipes connect fixtures to the main water distribution piping of the water supply system.

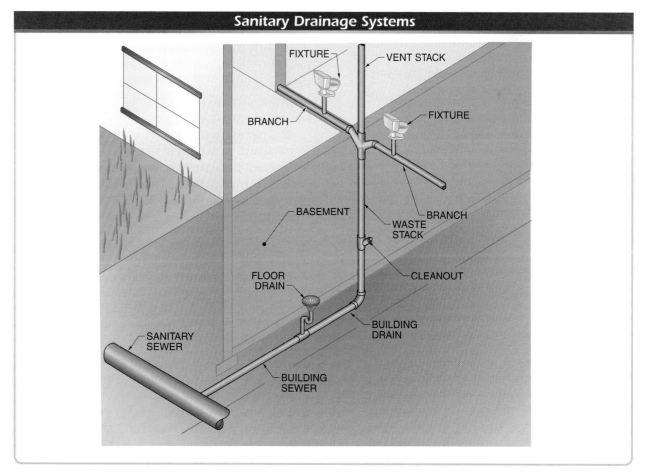

Figure 9-3. A sanitary drainage system collects all the wastewater and waterborne waste from various fixtures and conveys it to the sanitary sewer for disposal.

As wastewater and waterborne waste exit a fixture, such as a lavatory or water closet, gravity causes it to flow toward the sanitary sewer in a system of drainage pipes. Wastewater and waterborne waste first flow through a fixture trap, which provides a liquid seal to prevent the escape of sewer gases without affecting the flow of wastewater or waterborne waste through the trap. **See Figure 9-4.** The waste then enters a fixture drain. A *fixture drain* is a drainage pipe that extends from the fixture trap to the junction of the next drainage pipe. When multiple fixtures are on the same floor, the next pipe is a horizontal branch, which conveys the wastewater to the nearest stack. Individual fixture drains can also connect into the stack.

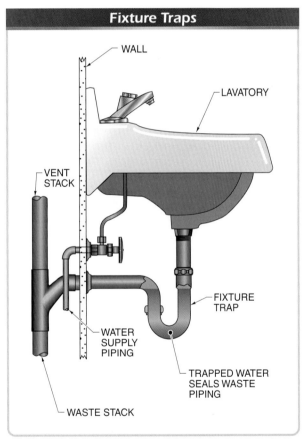

Fixture Traps

WALL

LAVATORY

VENT STACK

FIXTURE TRAP

WATER SUPPLY PIPING

TRAPPED WATER SEALS WASTE PIPING

WASTE STACK

Figure 9-4. A fixture trap retains a small amount of wastewater in the bend of a pipe to seal against sewer gases entering the occupied area.

A *stack* is a vertical drainage pipe that extends through one or more floors. Stacks convey the wastewater or waterborne waste down to the building drain.

A *building drain* is the lowest part of a drainage system, and it receives the discharge from all drainage pipes in the building and conveys it to the building sewer. A *building sewer* is the part of a drainage system that connects the building drain to the sanitary sewer.

The entire system of drainage pipes must be arranged to allow all the wastewater and waterborne waste to flow unimpeded to one large sanitary drainage pipe. Therefore, horizontal branches must be installed with a slight slope to allow for natural flow caused by gravity.

Vent Piping Systems

A *vent piping system* is a plumbing system that provides for the circulation of air in a sanitary drainage system. The sanitary drainage system must be properly vented to ensure adequate removal of sewage and allow sewer gases to properly escape to the atmosphere. *Sewer gas* is the mixture of vapors, odors, and gases found in sewers. Also, ventilating the drainage system prevents trap siphonage and back pressure, which impede the flow of wastewater to the sewer.

As wastewater and waterborne waste flow within the sanitary drainage system, air is displaced within the pipes. In order to keep the displaced air from impeding the flow of waste, the sanitary drainage pipes are vented to the outside air. The air in a drainage system exits the building through a vent pipe in the roof.

Vent pipes are extensions of waste stacks in the sanitary drainage system. One or more vent pipes extend from above the highest connected horizontal drains to the roof. Fixtures on sanitary system branches can also be vented through individual vents that combine into vent branches and connect to the nearest stack vent.

Stormwater Drainage Systems

A *stormwater drainage system* is a plumbing system that conveys precipitation collecting on a surface to a storm sewer or other place of disposal. **See Figure 9-5.** These systems are common for draining rainwater and snowmelt from building roofs and parking lots. A *storm sewer* is a sewer used for conveying groundwater, rainwater, surface water, or similar nonpolluting wastes.

Stormwater drainage systems are similar to sanitary drainage systems in design. The difference is that stormwater drainage systems do not carry sewage. Rainwater enters a stormwater drainage system through roof or surface drains. Drains are covered by a tall strainer basket or flat slotted cover to prevent stones, leaves, and other debris from entering and clogging the system.

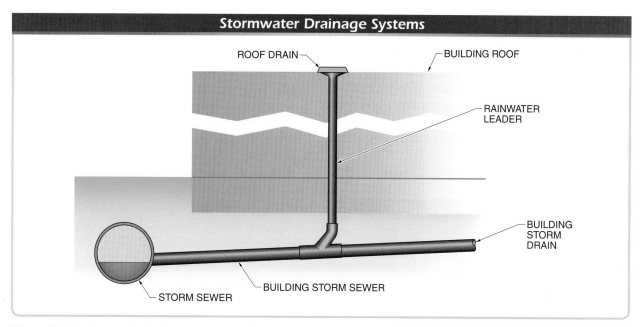

Figure 9-5. A stormwater drainage system collects precipitation from rooftops and other large surfaces and carries it to storm sewers or retention ponds.

The drains convey the water into rainwater leaders. A *rainwater leader* is a vertical drainage pipe that conveys rainwater from a drain to the building storm drain or to another point of disposal. Rainwater leaders inside buildings usually run along columns or in vertical shafts constructed specifically for pipe. Leaders typically extend vertically from the base fitting, run through the floors of the building, and terminate below the roof where they connect to a roof drain.

Rainwater flows from the system of leaders into the building storm drain, which leads into the building storm sewer. If a municipal storm sewer is available, it receives the rainwater from the building storm sewer. If a storm sewer is not available, rainwater is piped to a drainage basin, such as a pond.

A roof drain is used to remove water from a roof.

WATER SUPPLY CHARACTERISTICS

Controls in a plumbing system are typically located in the water supply system. Controls in wastewater systems are less common, though they can be important in certain situations. Vent piping systems are typically left completely open and unobstructed.

Four characteristics of a water supply that are commonly measured and controlled in a plumbing system are the temperature, pressure, flow, and level. This control allows for the most efficient operation of plumbing fixtures.

Water Temperature

Temperature is one of the most important characteristics of the water supplied to building fixtures, but it is also relatively simple to measure and control. Water is supplied to a building at a relatively cold temperature of about 40°F to 60°F, depending on the location and climate. This temperature is primarily influenced by the temperature of the ground, since the water supply is extracted from ground water (via wells) or from underground municipal water main pipes.

Many building fixtures require water at a range of temperatures, so the water supply system is divided into hot and cold water systems. At the fixture, the desired temperature is achieved by mixing the hot and cold water. **See Figure 9-6.** The cold water supply is drawn directly from the building's main water supply, but the hot water supply is directed first through a water heater and then distributed in separate water distribution pipes to the necessary fixtures.

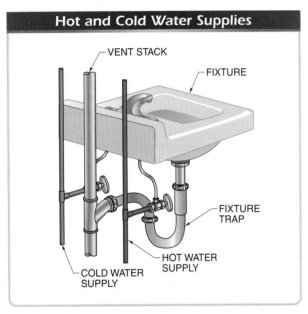

Hot and Cold Water Supplies

VENT STACK

FIXTURE

FIXTURE TRAP

HOT WATER SUPPLY

COLD WATER SUPPLY

Figure 9-6. Some fixtures require both a cold water supply and a hot water supply, which are then mixed together at the fixture to dispense water at a desired temperature.

A *water heater* is a plumbing appliance used to heat water for the plumbing system's hot water supply. Water heaters create a steady supply of hot water that is then distributed to plumbing fixtures throughout the building in a piping system similar, and often parallel, to the cold water supply pipes. Much like boilers, water heaters produce heat either through the combustion of natural or liquefied petroleum gas (LPG) or with electrical resistance elements. A thermostatic control activates the heating function as needed to maintain the set temperature. The water heater accomplishes this without outside control signals.

Not all fixtures require both a cold water supply and a hot water supply. Fixtures that do not require water at a specific temperature, such as water closets (toilets), receive only the cold water supply.

Water Pressure

Water pressure allows water to flow freely into pipes and fixtures when valves are opened. The force of flowing water may make the water more effective for some applications, such as cleaning or rinsing. The pressure may also drive part of the fixture's operation, such as the flush of a water closet.

Water pressure also keeps contaminants from entering the potable water supply by preventing groundwater or wastewater from flowing into the water supply system pipes. A water supply system is a pressurized network of pipes. Water pressure is constantly pushing against all parts of the system. When a part of the system is opened, such as a faucet valve or a pipe crack, water is forced out. This keeps all other gases, liquids, and solids from entering the water supply system, unless forced into it intentionally with pumps.

Pressure is a force distributed over an area. Water pressure is typically measured in pounds per square inch (psi). Water in municipal mains is typically at a pressure of 45 psi to 60 psi. However, some water pressure is inevitably lost in the plumbing system between the main and the final point of use, usually a fixture. Head and friction are the primary sources of pressure loss. Pressure losses must be accounted for in a plumbing design so that adequate water pressure is available at each plumbing fixture for proper operation.

Pressure Loss Due to Head. Water has weight. Therefore, pressure in the water supply system must be high enough to push the water to the topmost plumbing fixture in the building with enough pressure remaining to properly operate the fixture. The pressure required to overcome the weight of water to push it to a certain height is known as static pressure, or head. *Head* is the difference in water pressure between points at different elevations. Head is expressed as a height, commonly in feet.

A column of water 1′ high, with a cross-sectional area of 1 sq in., weighs 0.434 lb. Therefore, water exerts 0.434 psi of pressure for each foot of height (psi/ft). Pipe diameter or shape has no effect on the pressure, since the force is exerted per cross-sectional area. **See Figure 9-7.** Therefore, to raise water, 0.434 psi of pressure is required for every foot of height. Conversely, this can also be thought of as a pressure loss due to height (head).

For example, at a height of 76′, the pressure loss due to head is 33 psi (76′ × 0.434 psi/ft = 33 psi). If the water entering the building from the municipal water main is at a pressure of 45 psi, then there is only 12 psi of pressure remaining at a height of 76′ (45 psi − 33 psi = 12 psi). This may not be adequate for the plumbing fixtures at that level. There would also be a significant difference in the operation of fixtures between the ground floor and the top floor.

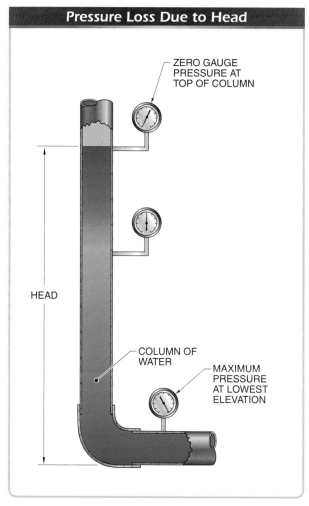

Figure 9-7. The weight of water exerts a pressure at the bottom of a water column in proportion to the height of the column, or head.

Pressure Loss Due to Friction. *Pressure loss due to friction* is the loss of water pressure resulting from the resistance between water and the interior surface of a pipe or fitting. Different pipe materials have different resistances to water flow within the pipe. A coefficient of friction quantifies the pressure loss per length of pipe for a certain diameter, material, and flow rate. **See Figure 9-8.** Flow resistance also occurs when water changes direction as it passes through valves and fittings.

Pressure Loss Due to Friction*

Pipe Size†	Flow Rate‡								
	1	2	3	4	5	10	15	20	25
⅜	0.021	0.075	0.159	not recommended					
½	0.007	0.024	0.051	0.086	0.130				
¾	0.001	0.004	0.009	0.015	0.023	0.084			
1		0.001	0.003	0.004	0.006	0.023	0.049	0.084	
1¼			0.001	0.002	0.002	0.009	0.018	0.031	0.047
1½				0.001	0.001	0.004	0.008	0.014	0.021
2	negligible					0.001	0.002	0.004	0.005
2½							0.001	0.001	0.002
3								0.001	0.001

* psi/ft of pipe
† in in.
‡ in gpm

Figure 9-8. Pressure loss can result from the friction of water flowing through pipes.

Pressure is highest at the bottom of a tank.

The flow resistance effect of various pipe fittings is given as equivalent lengths of pipe. **See Figure 9-9.** For example, a 90° elbow for a 1″ pipe causes the same pressure loss due to friction as a 2.5′ length of 1″ pipe.

The total pressure loss due to friction for a section of piping is the sum of the actual and equivalent pipe lengths (including those for every fitting) multiplied by the coefficient of friction. As more pipe, fittings, valves, and other devices are installed in a water supply system, the pressure loss increases.

Friction Allowance for Fittings and Valves*

| Fitting Size† | Equivalent Tube Length‡ | | | | | | | | | |
| | Standard Elbow | | 90° Tee | | Coupling | Valve | | | | |
	90°	45°	Side Branch	Straight Run		Ball	Gate	Butterfly	Check
⅜	0.5	—	1.5	—	—	—	—	—	1.5
½	1.0	0.5	2.0	—	—	—	—	—	2.0
¾	2.0	0.5	3.0	—	—	—	—	—	3.0
1	2.5	1.0	4.5	—	—	0.5	—	—	4.5
1¼	3.0	1.0	5.5	0.5	0.5	0.5	—	—	5.5
1½	4.0	1.5	7.0	0.5	0.5	0.5	—	—	6.5
2	5.5	2.0	9.0	0.5	0.5	0.5	0.5	7.5	9.0
2½	7.0	2.5	12.0	0.5	0.5	—	1.0	10.0	11.5
3	9.0	3.5	15.0	1.0	1.0	—	1.5	15.5	14.5
3½	9.0	3.5	14.0	1.0	1.0	—	2.0	—	12.5
4	12.5	5.0	21.0	1.0	1.0	—	2.0	16.0	18.5
5	16.0	6.0	27.0	1.5	1.5	—	3.0	11.5	23.5
6	19.0	7.0	34.0	2.0	2.0	—	3.5	13.5	26.5
8	29.0	11.0	50.0	3.0	3.0	—	5.0	12.5	39.0

* allowances are for streamlined soldered fittings and recessed threaded fittings; double the allowances for standard threaded fittings
† in in.
‡ in ft

Figure 9-9. Pipe fittings and valves have a significant effect on pressure loss due to friction. This effect is quantified in equivalent lengths of pipe.

Flow Pressure. *Flow pressure* is the water pressure in a water supply pipe near a fixture while it is wide open and flowing. A minimum of 8 psi of water pressure should be available to each plumbing fixture in order to function properly, though some fixtures require greater pressure. Hose bibbs require 10 psi, and water closets and urinals require 15 psi to 25 psi (depending on the valve type). Insufficient flow pressure prevents water closets from flushing properly and results in inadequate flow rates from faucets.

Flow pressure is affected by pressure losses due to head and friction. The flow pressure at any point in a water supply system is determined by the available pressure at another point (such as the source) minus the pressure lost due to pipes, fittings, and elevation changes between them.

For example, if the available water pressure at the municipal water main is 50 psi, the pressure loss due to head is 33 psi, and the pressure loss due to friction is 2.5 psi, then the remaining flow pressure is 14.5 psi (50 psi – 33 psi – 2.5 psi = 14.5 psi). This pressure is adequate for many fixtures, but if this fixture were a water closet valve, which may require 25 psi of flow pressure, the fixture would not operate properly.

Water Flow

Water flow can be measured as a flow rate or a total flow. *Flow rate* is the volume of water passing a point at a particular moment. *Total flow* is the volume of water that passes a point during a specific time interval. For example, the flow rate of pumping water is typically given in gallons per hour (gph) or gallons per minute (gpm). The total flow is the total number of gallons pumped.

Each plumbing fixture served by a water supply system has a specified flow rate. Multiple fixtures drawing water from the supply system increases the water demand. **See Figure 9-10.** The total of the minimum flow rates for every plumbing fixture within a building yields a total demand. However, plumbing fixtures in a building are rarely all used at the same time, so this estimate is unnecessarily high.

Facts

A potential difference is the driving force that causes material or energy to move through a process. Fluid flows from high pressure to low pressure. Heat flows from high temperature to low temperature. Electricity flows from high voltage to low voltage.

Minimum Flow Rates for Common Plumbing Fixtures	
Type of Fixture	**Flow Rate***
Lavatory faucet, standard	2.0
Lavatory faucet, self-closing	2.5
Kitchen sink faucet	3.0
Bathtub faucet	4.0
Laundry tub faucet	4.0
Shower head	4.0
Water closet flush valve	3.5
Drinking fountain	0.75
Sillcock or wall hydrant	5.0

* in gpm

Figure 9-10. Minimum flow rate requirements for different plumbing fixtures must be considered to ensure that each fixture receives an adequate water supply.

Based on the reasonable assumption that plumbing fixtures are not all used simultaneously, a better estimate of the total demand on a water supply system is determined from water supply fixture units. A *water supply fixture unit (wsfu)* is an estimate of a plumbing fixture's water demand based on its operation. A plumbing fixture, such as a water closet or lavatory, is assigned a wsfu value based on the following:

• fixture flow rate when the fixture is used

• average time water is actually flowing when a fixture is being used

• frequency that the fixture is used

• type of building where the fixture is installed

For example, a domestic dishwasher for private use has a 2 wsfu demand and a commercial dishwasher for public use, such as in a restaurant, has a 4 wsfu demand. There is a difference because a domestic dishwasher uses less water and is not used as frequently as a restaurant dishwasher.

Water Level

There are only a few applications where water level is measured and/or controlled in a building plumbing system, but these applications can be extremely important for preventing undesirable water conditions. Level information is used to determine when to add or remove water from vessels by activating or deactivating pumps or valves. This prevents overflow of a vessel or supplies water to a vessel when it runs low. Level is normally measured in linear units of height, which may be translated into units of volume or weight. Some applications require continuous

measurement, while others need to know only that the level is within the desired limits. **See Figure 9-11.**

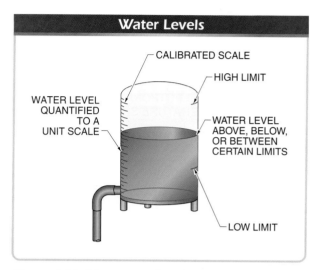

Figure 9-11. Like many other parameters, water level can be either quantified or only confirmed as being within certain limits.

PLUMBING SYSTEM CONTROL DEVICES

Plumbing controls are the devices used to control water in a plumbing system. Most plumbing control is related to water flow, but some plumbing control may incorporate control of the water's temperature or level in certain vessels. Plumbing control equipment includes sensors, pumps, and control valves.

Water Condition Sensors

In order to control the various characteristics of water in a plumbing system, sensors are necessary to measure them. Not all plumbing systems include sensors, but they are increasingly important as plumbing systems are progressively more automated. Automated plumbing systems may include sensors to measure water temperature, pressure, flow rate, and level.

Immersion Temperature Sensors. Water temperature is measured to control water heating and temperature adjustment functions. Temperature sensors used with plumbing systems are very similar to those used with HVAC systems. Water-sensing temperature sensors may either measure an exact temperature or, like HVAC thermostats, provide a digital output based on a comparison of the water temperature to a setpoint. **See Figure 9-12.** Both types must be immersion instruments that are rated for wet environments.

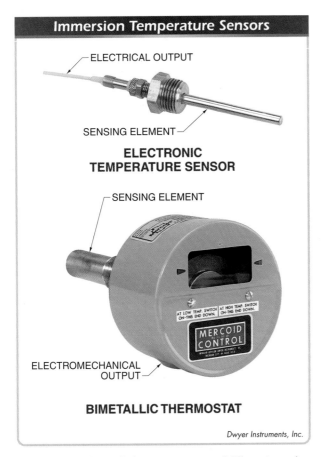

Immersion Temperature Sensors

ELECTRICAL OUTPUT

SENSING ELEMENT

ELECTRONIC TEMPERATURE SENSOR

SENSING ELEMENT

ELECTROMECHANICAL OUTPUT

BIMETALLIC THERMOSTAT

Dwyer Instruments, Inc.

Figure 9-12. Although the appearances of different sensing elements are similar, several types of thermometers are used with plumbing systems.

The primary types of electronic temperature-sensing elements are thermocouples, thermistors, and resistance temperature detectors (RTDs). These sensors change their electrical characteristics in proportion to temperature by either producing a small voltage or changing resistance. These changes may be read directly, or the sensors may include electronics to interpret these changes and convert them into standard analog signals or structured network messages that correspond to the exact temperature in a standard scale, such as °F.

Immersion thermostats may be either electronic or electromechanical and include normally open and/or normally closed contacts. Electronic thermostats are similar to electronic temperature sensors, except that they output only ON/OFF digital signals. Electromechanical temperature sensors incorporate a temperature-sensing element, such as a bimetallic strip, that moves slightly in response to temperature. This movement is used to open or close physical electrical contacts, which provide a similar ON/OFF output signal. The thermostat

is adjusted so that the signal change occurs at a specific temperature, which indicates that the water temperature is either above or below the setpoint temperature.

Regardless of the sensing method, immersion temperature sensors, which are used in wet environments, must have some protection. The sensing element is encased within a thermowell. A *thermowell* is a watertight and thermally conductive casing for immersion temperature sensors that mounts the sensing element inside the pipe, vessel, or fixture containing the water to be measured. **See Figure 9-13.** The thermowell allows a leakproof conduit for signal conductors to pass outside the sensor. Special plumbing fittings may be necessary to mount thermowells.

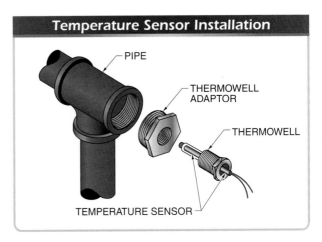

Temperature Sensor Installation

PIPE

THERMOWELL ADAPTOR

THERMOWELL

TEMPERATURE SENSOR

Figure 9-13. Special fittings allow the temperature sensor to be in contact with the water while the rest of the device remains dry and accessible.

Pressure Gauges. Since water pressure is important for proper plumbing fixture operation, especially in tall buildings, the pressure is monitored by pressure gauges and pressure switches. A *pressure gauge* is a pressure-sensing device that indicates the pressure of a fluid on a numeric scale. A *pressure switch* is a switch that activates at pressures either above or below a certain value. Pressure switches are used in some applications where exact pressure is not needed. Pressure gauges for plumbing applications are similar in operation to those used in HVAC systems, except that they are designed for wet environments.

Both plumbing and HVAC pressure gauges include an elastic deformation pressure element, which is a piece of material that flexes, expands, or contracts in proportion to the pressure applied to it. Examples of elastic deformation pressure elements include diaphragms, bellows, and coiled tubes, which move back and forth in response to pressure changes. One side of the element is open to the pipe or vessel containing the water to be measured. The other side is open to the atmosphere, because water pressure is measured in reference to atmospheric pressure. The motion of the pressure element in response to the force exerted on it from the water is proportional to the water pressure. **See Figure 9-14.** A greater displacement indicates a higher pressure.

Most common flow meters for plumbing systems are based on the rotation of mechanical spinners in a housing set in-line with the pipe. **See Figure 9-15.** The spinner is offset slightly so that the force of flowing fluid pushes against one side, causing the spinner to rotate. The spinner is often visible through a clear glass window in the flow meter. The speed of the resulting rotation is proportional to the flow rate of the water.

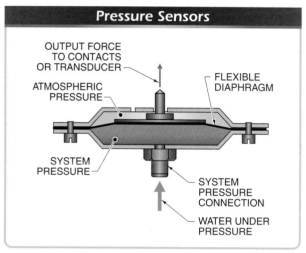

Figure 9-14. Pressure sensors include some type of elastic deformation pressure element and an electronic means to interpret its movement.

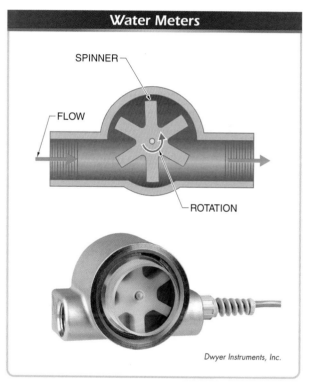

Dwyer Instruments, Inc.

Figure 9-15. Water meters convert rotation induced by flowing water into a flow rate.

For pressure gauges, the motion is transferred to a transducer, where it is converted into an electrical signal. Additional electronics convert the transducer output into standard analog signals or structured network messages. For pressure switches, the motion is used to make or break an electrical contact when the applied pressure reaches a preset level. The switch can be adjusted to activate at a certain pressure. As the pressure rises or falls through the setpoint pressure, the contacts switch positions.

Flow Meters. A *flow meter* is a device used to measure the flow rate and/or total flow of fluid flowing through a pipe. Every plumbing system that draws water from a municipal water main has a flow meter installed on the water service pipe to measure and indicate water flow so that the building owner can be charged for water used. This particular flow meter is typically called a "water meter." However, additional flow meters may be installed at other points in the plumbing system to record water usage within sections of the system.

For electronic monitoring, sensors inside the flow meter detect each revolution. Some flow meters output a pulse for each revolution of the spinner, which a separate device can then translate into a flow rate. For example, if a flow meter is designed to output one pulse for each 0.1 gal. of water that flows through the meter, then a controller receiving 100 pulses within a minute can convert this information to a flow rate of 10.0 gpm. Other flow meters are already calibrated to convert the pulse rate into an equivalent flow rate. These then output a standard analog signal or structured network message based on the resulting flow rate.

Flow Switches. When the presence of flow, rather than the exact flow rate, must be monitored, flow switches are used. Flow switches include a vane mounted inside a pipe that is deflected from the force of flowing water. A certain amount of deflection, and therefore flow, activates the switch. It deactivates when the water flow falls below another setpoint. Switch connections include normally open and/or normally closed contacts. Flow switches in plumbing systems are used to indicate the operation of pumps or valves. **See Figure 9-16.**

Level Switches. Level is not as commonly measured and controlled as other characteristics, even in many automation systems. However, there are some applications, such as drainage sumps, where monitoring water level is important. A *level switch* is a switch that activates at liquid levels either above or below a certain setpoint.

In plumbing systems, the most common level-measuring devices use floats. **See Figure 9-17.** These devices are switches with normally open and/or normally closed contacts. The contacts are used either to directly control another device, such as a pump, via a relay or to input digital signals to controllers.

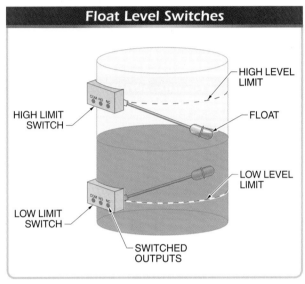

Figure 9-17. Mechanical level switches are actuated by the vertical position of a chamber that floats at the top surface of the water.

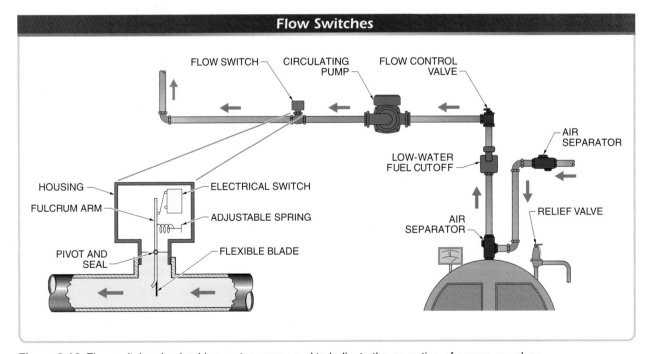

Figure 9-16. Flow switches in plumbing systems are used to indicate the operation of pumps or valves.

A float is a sealed chamber of air that floats at the top of the water, so its position is always a direct indication of the water level. The float is attached to the rest of the level-measuring device. Some use a rigid rod. When the float rises or falls with the water, the rod angle changes. Similar to vane-type flow switches, this movement activates a set of contacts within the switch housing.

Another type of float device uses an oblong chamber with a switch inside that is activated by orientation. The float is attached to the bottom of the vessel with a flexible cord that transmits the switch signal. When the level is low, the cord is slack and the float floats sideways. When the level rises, the float rises until the cord pulls it vertically, activating the switch.

Pumps

A *pump* is a device that moves water through a piping system. A pump moves water from a low pressure to a high pressure. It overcomes this difference by adding energy to the water. Pumps can operate either continuously or intermittently. Continuously operated pumps can be used to circulate water through a closed loop or boost pressure within a water system. Intermittently operated pumps can be used to empty sumps, draw water from wells, or operate plumbing appliances.

Water pumps are normally driven by electric motors. For smaller pumps, the motor is typically integrated into the pump housing. For pumps with a pumping rate above 100 gpm, the motor is normally connected by a coupling between the shaft and motor. The motor can be turned on or off with relays or motor starters. With a variable-frequency drive, the speed of the electric motor can be controlled to achieve the desired pumping rate.

Pumps can be categorized as either centrifugal pumps or positive-displacement pumps. Centrifugal pumps are the most widely used type of pumps.

Centrifugal Pumps. A *centrifugal pump* is a pump with a rotating impeller that uses centrifugal force to move water. An *impeller* is a bladed, spinning hub of a fan or a pump that forces fluid to the perimeter of a housing. **See Figure 9-18.** Water enters the pump through the center of the impeller. The impeller rotates at a relatively high speed and imparts a centrifugal force to the water, which moves it outward. The shape of the casing directs this water to the outlet.

Multiple impellers can be used in series to build higher pressures. In a multiple-stage centrifugal pump, the discharge from the first impeller enters the suction

of the second impeller and increases the pressure even more. As many as four impellers may be on a single impeller shaft.

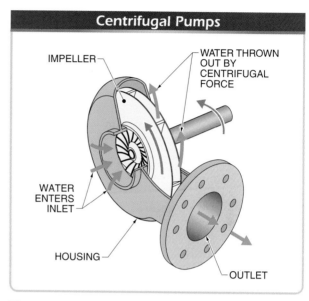

Figure 9-18. The rotating impeller of a centrifugal pump forces water to the perimeter of the pump housing, where it is then directed out through the outlet.

Positive-Displacement Pumps. A *positive-displacement pump* is a pump that creates flow by trapping a certain amount of water and then forcing that water through a discharge outlet. Positive-displacement pumps are further classified as either reciprocating-type or rotary-type, though both types operate using the same principle. **See Figure 9-19.** Each type of pump includes a chamber for water that cycles between a small volume and a large volume. Water is drawn into the chamber from the pump inlet as the chamber volume increases. When the chamber is at maximum volume, the inlet is closed off and the outlet is opened. Then the volume decreases, pushing the water out.

Reciprocating-type pumps manipulate the chamber by moving some portion of it in a reciprocating (back-and-forth) motion. This changes its volume. The movable portion is typically a piston or flexible diaphragm.

Rotary-type pumps create the displacement chamber by enclosing spaces within a circular housing. A rotor near the center includes vanes, lobes, or other projections that seal against the housing's inner wall. As the rotor rotates, the volume enclosed by these projections changes from small to large and back to small, which moves water through the pump.

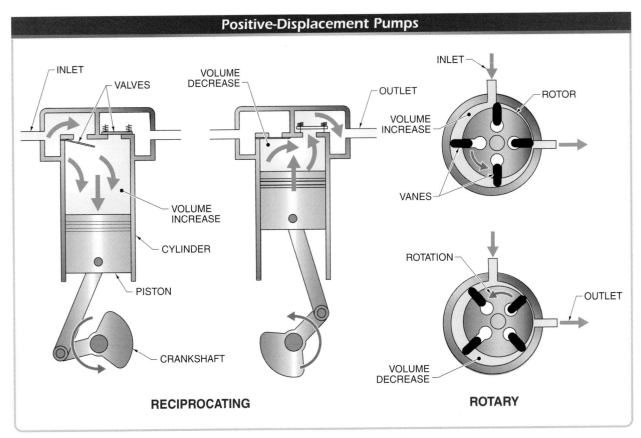

Figure 9-19. Positive-displacement pumps are classified by the method used to move water.

Positive-displacement pumps are typically self-priming and the resulting flow rate is independent of the water inlet pressure. However, some positive-displacement pumps create surges in flow and pressure with each cycle and are limited in capacity.

Troubleshooting Pumps

There are many causes of problems with pumping systems. A technician should be familiar with the most common types of problems in order to assist in troubleshooting. A very common cause of problems with pumps is inadequate supply of liquid to the pump. The supply tank may be empty, a valve may be closed or throttled, or a strainer may be plugged. After these common problems have been eliminated, other typical causes of problems with pumps involve priming, cavitation, mechanical seals, and packing.

Priming. A pump must be able to pull liquid into the suction-side inlet in order to start pumping. Since most pumps move liquids, not air, a pump will not operate if the suction line contains air. Pumps operate under suction head or suction lift, depending on the elevation of the liquid for the pump inlet. For suction head, the water level is above the pump inlet and a liquid naturally flows into the pump.

For suction lift, the liquid normally drains back to the tank when the pump is not operating and the pump can fail from lack of prime. In addition, air leaks in the suction line can cause the water to flow back down the feed line. There must be some means of getting and keeping the liquid in the pump. *Priming* is the process of overcoming suction lift and getting liquid to a pump inlet. The pump prime can be maintained by installing a foot valve to keep the liquid from draining back to the tank. A *foot valve* is a check valve installed at the bottom of the suction line on a suction-lift pump that keeps the suction line primed when the pump shuts down. **See Figure 9-20.**

Vacuum pumps are also used as priming pumps. A *priming pump* is a vacuum pump that ejects air from the suction line of a larger suction-lift pump installation. The vacuum created causes liquid to flow up from below and into the larger pump, priming the larger pump.

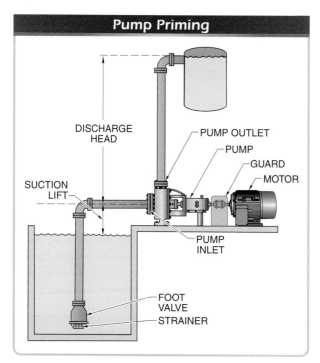

Figure 9-20. The pump prime can be maintained by installing a foot valve to keep the liquid from draining back to the tank.

Cavitation. *Cavitation* is the process in which microscopic gas bubbles expand in a vacuum and suddenly implode when entering a pressurized area. As the pump pulls against a fluid, a vacuum is created. Any microscopic air or gas within the fluid expands. Expanded bubbles on the inlet side collapse rapidly on the outlet side of the pump. The small but powerful implosions can cause great damage to pump parts. **See Figure 9-21.**

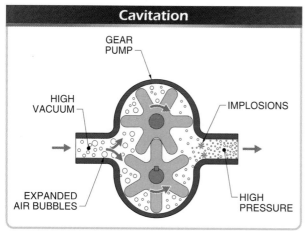

Figure 9-21. Cavitation can cause great damage to pump parts.

Cavitation can cause a high shrieking sound or a sound similar to loose marbles or ball bearings in the pump. Cavitation occurs when the flow into a pump is restricted. Restricted flow occurs when the suction line is damaged, plugged, has collapsed, or when the fluid is too hot and boils when suction is applied by the pump. Cavitation may also be caused by an increase in pump rpm that requires more fluid than the system piping allows or by fluids with an increased viscosity due to lower ambient temperatures.

Mechanical Seals. A *mechanical seal* is an assembly of mechanical parts installed around a pump shaft that prevents leakage of the pumped liquid along the shaft. **See Figure 9-22.** A mechanical seal is used instead of packing around a pump shaft or shaft sleeve. Mechanical seals are particularly useful when leakage of the material could create a fire hazard or noxious vapors, when the liquid is valuable, or in any other case where leakage is unacceptable.

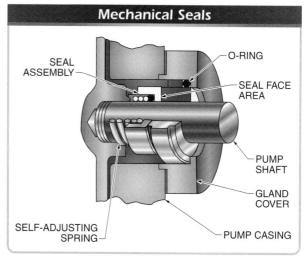

Figure 9-22. Mechanical seals are particularly useful where leakage is unacceptable.

Mechanical seals should not leak more than a few drops per day. Any leakage more than this amount indicates that the seal is about to fail. A mechanical seal should not be adjusted after it is installed. The springs in the seal make the adjustments automatically. Attempts to adjust a mechanical seal will cause premature failure.

Packing. *Packing* is an assembly of compressible sealing material installed around a pump shaft that prevents the pumped liquid from leaking along the shaft. Packing is commonly made of graphite-impregnated square rope, Teflon®, or other similar materials.

Packing is installed in the packing gland area, or stuffing box. The packing is cut to length to create individual rings around the pump shaft. The packing follower compresses the packing so that it throttles the leakage of liquid from around the shaft. If the packing is tightened so much that smoke comes from the packing gland, the packing has overheated, is probably damaged beyond repair, and should be replaced.

Valves

A valve is used to regulate water flow within a system. Valves are used to turn the water flow on and off or to regulate the direction, pressure, and/or temperature of a fluid within the system.

Most valves operate when the valve stem is rotated. When the valve stem is rotated, it either raises or lowers an obstruction to water flow within the valve or rotates a conduit for water flow into or out of position. When used in automation systems, mechanical valves are augmented with electromechanical actuators to allow for automatic operation. Electromechanical actuation can be accomplished with a small electric motor mated to the rotating valve stem or a valve redesigned to include electromechanical parts.

These motors require onboard controls to allow for rotation in both directions (for both opening and closing) and any position in between (for throttling). Complete electronic valves include the necessary electronics to operate the motor and accept input signals that indicate the desired valve position.

There are many designs for valves, which are classified by their intended use as either full-way or throttling valves. Either type can usually be used for the opposite function, but they will likely not perform optimally. There are also special types of valves for controlling water pressure and temperature. All of these valves are available with electronic actuators. Some actuators can be purchased separately and fitted to existing (compatible) manually actuated valves to add automatic operation.

Full-Way Valves. A *full-way valve* is a valve designed to operate in only the fully open or fully closed positions. Full-way valves are also called shutoff valves. When open, the internal design permits a straight and unrestricted fluid flow through the valve, resulting in very little pressure loss due to friction. Full-way valves are installed in the building's water supply so that the water can be shut off to individual rooms or fixtures without interrupting water service to other areas of the building.

Since they are intended for only ON/OFF operation, some mechanical full-way valves are designed to operate with only a 90° turn. A short rotation makes it easier and faster to change between the fully open and fully closed positions.

One type of full-way valve design is the solenoid valve. **See Figure 9-23.** Solenoid valves are valves with completely integrated electromechanical components. A *solenoid valve* is a full-way valve that is actuated by an electromagnet. When supplied with an electric current, the integrated solenoid coil creates a magnetic force that pulls on a plunger. In a normally closed valve, this action opens the valve. This works against the force of a spring, which returns the plunger to the normal position when the coil is deenergized. Since the spring can slam the plunger back to its normal position with great force, solenoid valves may include a slow-closing feature to prevent damage.

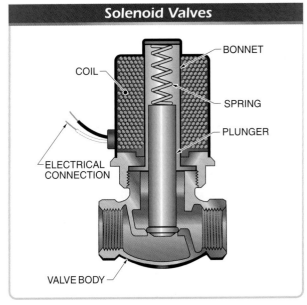

Figure 9-23. When fully open, a full-way valve permits unrestricted flow through the valve.

A full-way valve may require only a digital input, which corresponds to the fully open and fully closed positions. This may be signaled with ON/OFF signals or with structured network messages.

Throttling Valves. A *throttling valve* is a valve designed to control water flow rate by partially opening or closing. Throttling valves are also called control valves. Due to the internal configuration, water flowing through a throttling valve changes direction several times, resulting in flow resistance and a pressure drop. This makes throttling valves ideal for reducing water flow and pressure when required by the application. However, because they have some of this effect even when fully open, they are not recommended for use as full-way valves.

Throttling valves are installed on fixture supply pipes for individual fixtures. Examples of throttling valve designs include globe valves and butterfly valves. **See Figure 9-24.** Many throttling valves are required to be installed with the flow direction arrow pointing in the downstream direction.

Throttling valves require analog valves to determine intermediate positions. Many valve actuators accept standard analog signal types. Some are available for use with structured network message systems.

Three-Way Valves. A *three-way valve* is a valve with three ports that can control water flow between them. A *port* is an opening in a valve that allows a connection to a pipe. The use of the three ports as inlets or outlets may vary. Most often, water enters the valve from one port and the flow of that water can be directed into either of the two outlet ports. **See Figure 9-25.** The valve may also allow a position that stops all flow through the valve. These types of valves typically use actuators that receive digital signals to rotate the valve stem into these fully actuated positions.

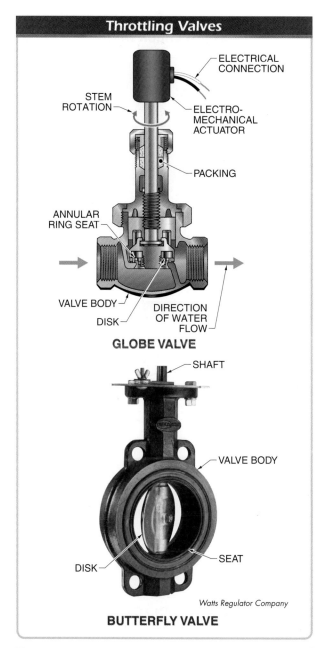

Figure 9-24. Throttling valves are used in positions between fully open and fully closed in order to reduce water flow and pressure at the outlet.

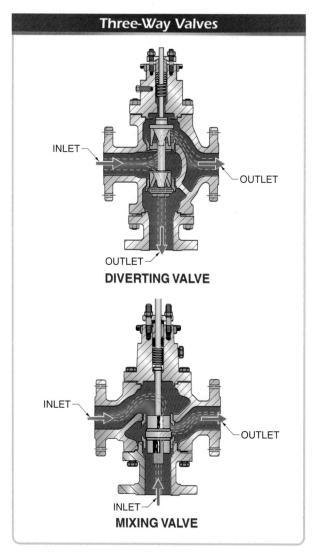

Figure 9-25. Three-way valves direct or mix water flow between three ports.

Three-way valves can also be designed to mix water from two inlet ports and discharge the result into an outlet port. A particular type of valve using this design is a thermostatic mixing valve. A *thermostatic mixing valve* is a valve that mixes hot and cold water in proportion to achieve a desired temperature. **See Figure 9-26.** For many plumbing applications, the hot water supply is too hot and the cold water supply is too cold, but an optimal water temperature is achieved by mixing hot and cold water together.

A thermostatic mixing valve can be set for all cold water, all hot water, or any position in between. For example, a position in the middle mixes 50% hot water with 50% cold water to produce warm water. Because of this range of possible positions, automated thermostatic mixing valves, or other types of three-way valves for mixing water supplies, require either analog signal inputs or structured network messages that include the desired valve position.

Check Valves. A *check valve* is a valve that permits fluid flow in only one direction and closes automatically to prevent backflow (flow in a reverse direction). **See Figure 9-27.** Check valves react automatically to changes in the pressure of the fluid flowing through the valve and close when pressure changes occur.

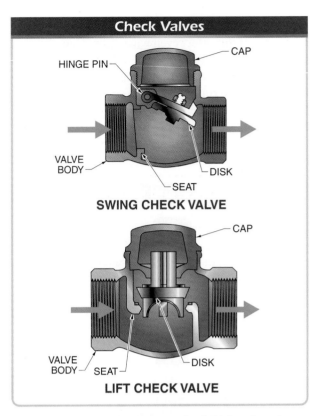

Check Valves

SWING CHECK VALVE

LIFT CHECK VALVE

Figure 9-27. A check valve permits fluid flow in only one direction.

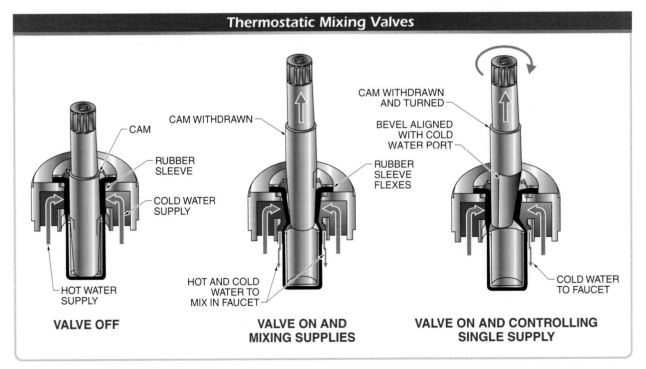

Thermostatic Mixing Valves

VALVE OFF

VALVE ON AND MIXING SUPPLIES

VALVE ON AND CONTROLLING SINGLE SUPPLY

Figure 9-26. Thermostatic mixing valves mix water from the cold water supply and the hot water supply to dispense water at a desired temperature.

Check valves are common in several parts of all plumbing systems. Check valves are used in sanitary drainage lines to prevent sewage from flowing back into a building. Check valves are also needed for several automation applications that use water pumps. Check valves prevent water from backflowing and hold the pressure the pump built up on its outlet side.

Two common types of check valves are swing check valves and lift check valves. A *swing check valve* is a check valve with a hinged disk. In its normal operating condition, fluid flows straight through the valve and holds open the hinged disk. When backflow occurs, the hinged disk swings down into position, blocking flow. A *lift check valve* is a check valve with a disk that moves vertically. In its normal operating condition, fluid pressure forces the disk from its seat, allowing fluid to flow. When backflow occurs, the disk drops onto its seat, preventing backflow from occurring.

PLUMBING SYSTEM CONTROL APPLICATIONS

Control devices are integrated into a building's plumbing system to provide control and automation for certain aspects of its functionality. There are many possible control applications, depending on the specific system requirements. However, a few applications are relatively common.

The water pressure in some plumbing systems can be low or highly variable. This can be the case if the building draws water from a well or is at the end of a long municipal water main (caused by pressure loss due to friction in the mains). Fluctuating pressure may be a problem for overtaxed municipal water supply systems or for buildings located near other facilities with significant water usage. For the plumbing system to operate adequately, some means is needed to boost and/or stabilize the water pressure.

> **Facts**
>
> The only way a check valve can be cleaned is by removing it from the system and cleaning it manually. To save time and cost, a dirty check valve can be replaced with a new one.

Plumbing System Control Application: Boost Pumps

Boost pumps are the simplest method for increasing water pressure. A *boost pump* is a pump in a water supply system used to increase the pressure of the water while it is flowing to fixtures. **See Figure 9-28.** Boost pumps are fairly small and quiet and do not require additional equipment. However, they cannot keep up with large water demands, so they are typically suitable only for residences or small office buildings.

Operating Conditions. Boost pumps are installed directly in the water supply system between the water supply (municipal main or well) and the rest of the water distribution system. Pressure switches monitor the water pressure on either side of the boost pump, control the pump to provide a certain water pressure as fixtures are opened and closed, and protect the water source from excessively low pressure.

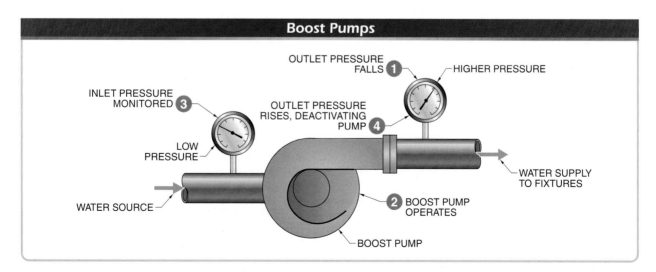

Figure 9-28. Boost pumps are installed in water distribution systems to increase the pressure of low-pressure water supplies.

Control Sequence. The following control sequence is used to activate a boost pump as needed:

1. As fixtures draw water from a building's water supply system, the water pressure falls. A pressure switch on the outlet side of the boost pump activates if the water pressure falls to a low-pressure setpoint.

2. The activation of the pressure switch activates the boost pump via a relay or controller. The boost pump operates, increasing the pressure in the water supply system.

3. If the pressure of the water source falls to a low-pressure setpoint, the pressure switch on this side of the pump deactivates the pump. This protects sources such as wells from collapsing due to excessive water drawing.

4. As fixtures are closed, the water pressure in the water supply system increases. If the pressure rises to the high-pressure setpoint, the downstream pressure switch deactivates the pump.

Plumbing System Control Application: Elevated Tanks

Water pressure changes according to height (head). In a column of water, the pressure at the bottom is greater than the pressure at the top. This principle is important for understanding why water pressure is lost when water rises in tall buildings, but it can also be used to solve this problem.

Water is typically supplied from underground mains. From this low reference point, all fixtures in the building are at a higher elevation, so the water pressure at those points is lower than the water pressure in the main. However, if water is supplied from the top of the building, then all the plumbing fixtures are below the supply and receive water at a pressure higher than that of the supply. The pressure still varies between fixtures near the top and bottom of the building, but if the fixtures near the top receive a minimum adequate pressure, the higher pressures at lower elevations can be controlled with pressure-reducing valves. This scenario is accomplished with an elevated tank.

Operating Conditions. A large water tank is on or near the roof of a building and includes a pair of level switches to monitor the level of water within a specified range. **See Figure 9-29.** These tanks are not pressurized or sealed, other than measures needed to protect the potable water from outside contaminants. A pump moves water up to the tank from the water main, and a check valve prevents the water from flowing back down through the pump. A control sequence is required to maintain an adequate supply of water in the tank.

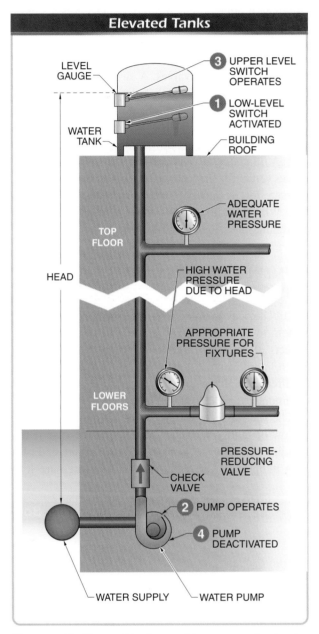

Figure 9-29. Elevated tanks provide a pressurized water supply to fixtures at lower elevations due to head (height).

Facts

Cavitation is a condition in a pump in which some of the inlet water flashes to vapor. Cavitation of a pump can be caused by low water pressure on the inlet side of the pump.

Control Sequence. The following control sequence is used to control water supply system level in an elevated tank:

1. Water usage in the building draws water from the tank, and the water level falls until the low-level switch is activated.
2. The activation of the low-level switch activates the pump via a relay or controller. The pump operates, adding more water to the elevated tank.
3. The water level in the tank rises until the upper level switch is activated.
4. The activation of the upper level switch deactivates the pump via a relay or controller. This prevents the tank from overfilling.

Plumbing System Control Application: Hydropneumatic Tanks

When a large roof tank is undesirable, pressure-boosting systems can employ a hydropneumatic tank. These tanks accomplish the same result as elevated tanks, but they can be installed at any elevation, even in a basement. A *hydropneumatic tank* is a water tank with an air-filled bladder that raises the pressure of water as it is pumped into the tank. **See Figure 9-30.**

As water is pumped into the tank, it displaces some of the space that was occupied by the flexible air bladder. Unlike water, air is compressible, so the air is forced to occupy less space. This increases the pressure exerted by the air, which increases the water pressure. When a fixture draws water from the system, the air pressure pushes water out of the tank to supply the fixture. As more water is drawn, the pressure gradually falls until it must be boosted again by the pump.

Systems with hydropneumatic tanks have similarities to both boost pump systems and elevated tank systems. Like boost pump systems, the water pressure is monitored to control the pump. Like elevated tank systems, a water supply is stored that can be used for some time after the pump shuts off, reducing the operating time of the pump.

Operating Conditions. Hydropneumatic tank systems require a pump to charge (fill) the tank. A pair of pressure gauges or switches is installed on either side of the pump. The downstream sensor controls the pump to maintain the water supply pressure within a specified range. The upstream sensor protects the water source, such as a well, from excessively low pressure.

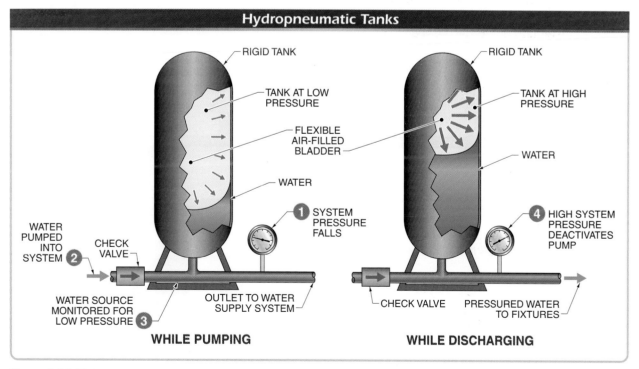

Figure 9-30. Hydropneumatic tanks can be installed at low elevations because they have an air-filled bladder that is compressed by water, which exerts pressure on the water supply system.

Control Sequence. The following control sequence uses an elevated tank to control water supply system pressure:

1. As fixtures draw water from a building's water supply system, the water pressure falls. A pressure switch on the outlet side of the hydropneumatic tank activates if the water pressure falls to a low-pressure setpoint.

2. The activation of the pressure switch activates the pump via a relay or controller. The pump operates, adding water to the hydropneumatic tank and supplying water to fixtures.

3. If the pressure of the water source falls to a low-pressure setpoint, the pressure switch on the low pressure side of the pump deactivates the pump. This protects sources such as wells from collapsing due to excessive water drawing.

4. As fixtures are closed, the pressure within the hydropneumatic tank increases. If the pressure rises to the high-pressure setpoint, the pressure switch deactivates the pump.

Facts

A pump is said to be deadheaded when a discharge valve is closed while the pump is operating.

Plumbing System Control Application: Hot Water Loops

Hot water in a plumbing system's distribution pipes will cool to room temperature when it is not drawn regularly. Since the cooled water must pass through the fixture first before recently heated water from the water heater is drawn, it usually takes several moments for water drawn from a faucet to get hot. **See Figure 9-31.** This is not only inconvenient, but it also wastes water.

One solution for this problem is to implement a hot water loop. A *hot water loop* is a closed circuit of hot water supply distribution pipes, including the water heater, through which hot water is circulated. A small pump circulates hot water through the loop from the water heater to the farthest fixture and back. **See Figure 9-32.** Therefore, the water in the pipes is always hot, and little water is wasted waiting for the water to become hot at the fixture.

Some hot water loops operate the circulation pump continuously. However, when the heat loss through the pipes is low enough, the pump may only operate intermittently, conserving electricity. This requires that the system be automated.

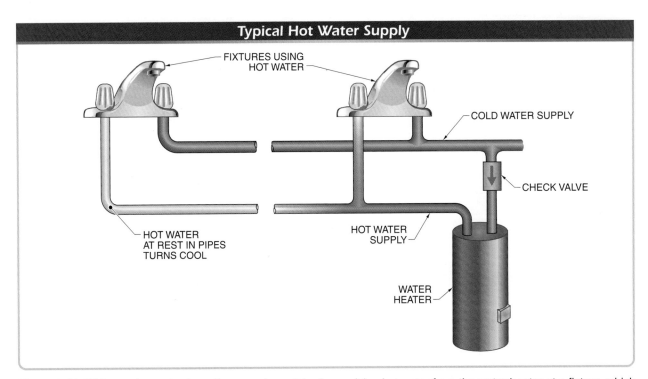

Typical Hot Water Supply

FIXTURES USING HOT WATER

COLD WATER SUPPLY

CHECK VALVE

HOT WATER AT REST IN PIPES TURNS COOL

HOT WATER SUPPLY

WATER HEATER

Figure 9-31. Without a hot water loop, there can be a delay in receiving hot water from the water heater at a fixture, which wastes water.

Hot Water Loops

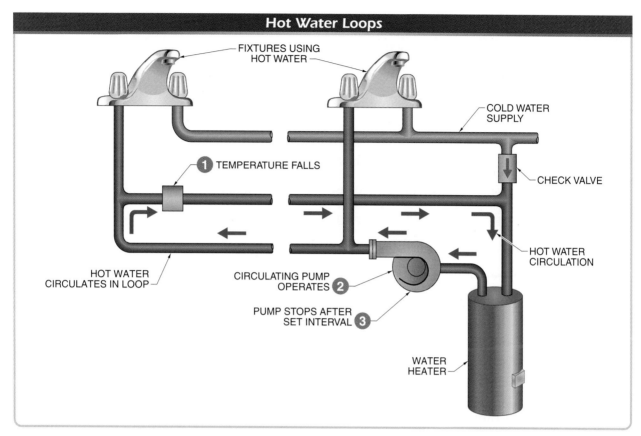

Figure 9-32. Hot water loops maintain a supply of hot water in a system of piping that is close to the fixtures that require a hot water supply.

Operating Conditions. An immersion thermostat is installed at the farthest part of the loop from the outlet of the water heater, which is where the water is likely to be the coolest. A timer may be used to run the pump for a certain amount of time for each cycle. Schedules can also be used to limit the operation of the water heater and circulating pump to parts of the day when occupants are likely to use hot water, such as during business hours for office buildings.

Control Sequence. The following control sequence is used with a hot water loop:

1. If the building is in occupied mode and hot water has not been drawn by a fixture for some time, the hot water in the loop begins to cool. The immersion thermostat activates when the temperature falls to the low-temperature setpoint.

2. The activation of the thermostat turns the pump on via a relay or controller. The pump operates, circulating hot water through the loop. The activation of the pump may be used to also activate a cycle timer.

3. The pump is deactivated when either the thermostat registers a temperature above the high-temperature setpoint or the pump has operated for a set amount of time, depending on the system design.

Safety valves are used to relieve pressure on water heaters if the pressure gets too high.

REVIEW QUESTIONS

1. How is a water supply system different from all other portions of a building's plumbing system?

2. What are the four characteristics of a water supply that are controlled in a plumbing system?

3. Describe the two primary sources of pressure loss in a water supply system.

4. Why are some sensors used in plumbing systems divided into two parts?

5. How can the pumping rate be controlled with motor-driven pumps?

6. Why are throttling valves not recommended for use as full-way valves?

7. List the common types of problems with pumping systems.

8. What are the differences between building mains, risers, and branches?

9. What is an advantage of including hot water loops in a plumbing system?

10. Explain the three primary techniques for increasing water pressure in a water supply system.

 Chapter 9 Review and Resources

Automated Building Operation 10

The number of potential ways in which various building systems can be integrated is practically limitless. Most automated buildings include only a few applications. Some include many applications, but there are always additional possibilities. There are also different protocol systems that can be used to achieve automation goals. In many situations, a building automation system must be designed to integrate more than one protocol. Various strategies can be used to fully integrate these systems together, including strategies that include gateways and intermediary frameworks that bridge the gap between the different information structures. A walk-through of an example automated building highlights only some of the common ways in which building systems are integrated and automated.

BUILDING SYSTEM OPERATION

Building automation systems use the abundant sharing of system information to provide a highly functional and flexible "living" environment for building occupants. This exchange occurs on an infrastructure that connects the control devices for HVAC, lighting, fire, security, and other systems together on a common network. **See Figure 10-1.** Additionally, with energy management capabilities, these systems can reduce overall energy use and costs.

Figure 10-1. Building automation systems are becoming increasingly common in modern commercial buildings.

Although the initial cost for a building automation system is typically higher than for conventional construction, total life-cycle costs are generally lower due to energy savings, reduced maintenance costs, and increased value to potential tenants. The cost of implementing a building automation system is minimized when it is incorporated into new construction. Retrofitting a building for automation is usually possible, but it is at a greater expense, inconvenience, and potential loss of system flexibility.

System Interdependencies

Building systems have been self-contained and independent in past generations. As multiple building systems are connected together into a "whole building" automation system, control devices are somewhat dependent on each other. The actions of one system may affect another. For instance, a controller may use sensor information from other networked devices in order to complete its sequences and share its output data for other devices to use. Critical interactions must be carefully designed for reliability, but the advantage is a significant reduction in redundancy. For example, only one outside air temperature sensor may be needed if its data is shared with all of the controllers that require this information.

Control Device Interoperability

A major concern with traditional automation systems is that they are typically limited to a single building system or equipment manufacturer. This severely limits the potential ability of automation to seamlessly and reliably involve multiple systems.

Most modern automation devices are built on a framework that allows complete compatibility and reliable and accurate communication between nearly any building automation devices, regardless of the manufacturer or the system in which they are found. For example, information from an occupancy sensor can be used to control both the lighting and the HVAC functions within a room. This interoperability of the system components allows for nearly limitless flexibility in system design and component sourcing. Interoperability is the ability of diverse systems and devices to work together effectively and reliably.

The communication requirements of such an interconnected system include a robust network infrastructure. **See Figure 10-2.** It is becoming increasingly common for building automation systems to merge with traditional information technology (IT) infrastructures. The sharing of these networks requires the IT professionals to take an active role in building automation system operations. Integration with Internet standards allows the scope of a building automation system to extend beyond the walls of the building.

Location Applications

The flexibility of building automation systems is evident in the wide variety of possible applications. Building automation system location applications can be classified as building-wide applications, campus-wide applications, and worldwide applications. **See Figure 10-3.**

Because of security concerns, many buildings allow internal monitoring only. For example, owners and managers of federal facilities, homeland security buildings, and law enforcement buildings want to prevent unauthorized access and generally do not allow any external monitoring. Facility maintenance staff members are connected through a secure network to address issues within the building.

Building-Wide Applications. Building-wide applications include those operations within a discrete facility. The most common implementations of building automation systems are building-wide. The scope of the system extends no further than that of the building systems themselves. Most systems are monitored, controlled, and modified only from within the building.

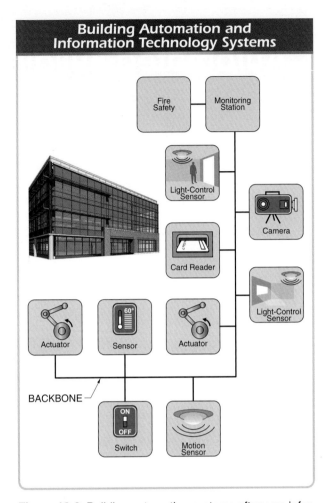

Figure 10-2. Building automation systems often use information technology (IT) networks similar to those used for computer systems.

Campus-Wide Applications. With a suitable communications network, automation systems can extend beyond a single building. Campus-wide applications include operations between and within a group of buildings in relatively close physical proximity.

For example, a campus consisting of a library, a student union, classroom buildings, and an athletic building may be managed by an automation system consisting of devices installed throughout the campus. The same systems managed as a building-wide application can be managed as a campus-wide application. This is especially important for campuses with centralized plant facilities, such as boilers and chillers. The campus-wide automation system can manage these resources in the most effective way across the entire area served by these plants.

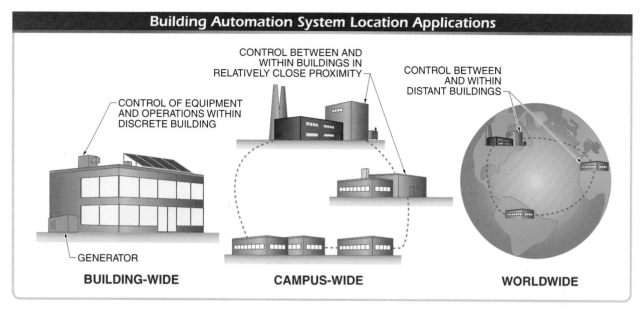

Figure 10-3. With a suitable network infrastructure, building automation systems can extend beyond the walls of a single building.

Another example of a campus application is an airport. Facility systems in areas such as terminals, attached hotels, and the administration building can be managed from an automation system. Direct control and system monitoring is critical for addressing homeland security issues.

Worldwide Applications. The worldwide Internet is a readily accessible network for applications beyond local areas. Many automation system device manufacturers include network tools needed to access and control automation function from anywhere in the world.

Most worldwide automation applications involve the monitoring of system operations remotely. System technicians, maintenance personnel, or building owners can access building operation via an Internet connection and either special software or a standard web browser. Some systems even allow authorized users to change system information, such as operating modes or setpoints, remotely. Remote monitoring is particularly helpful when outside consultants are needed to troubleshoot system problems. Some issues may be resolved by monitoring system information via the Internet, thus reducing the need for expensive on-site visits.

Systems can also be designed to automatically share control information between distant locations, though this requires careful design. For example, upon activation of a security alarm at a satellite office, personnel in the headquarters building can remotely control the surveillance CCTV system to investigate the trouble.

Building-wide applications use control devices throughout the entire building.

Automation systems are also used to monitor steam plants remotely, allowing some routine monitoring and control functions to be performed by off-site personnel. **See Figure 10-4.** A boiler system's microcomputer burner control system (MBCS) is the local controller that analyzes data from boiler-specific sensors and accessories and makes the necessary data available for other controllers or remote monitoring via the building automation system. The remote monitoring of boilers or other potentially dangerous equipment does not replace the need for qualified personnel in close proximity to a facility. However, it provides additional options for monitoring and control.

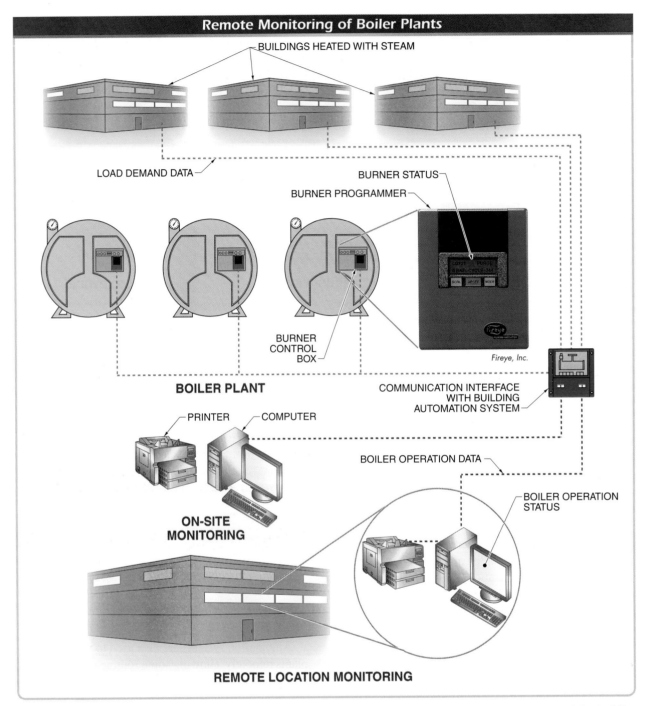

Figure 10-4. Remote access to a building automation system allows off-site personnel to monitor and control the building systems.

CROSS-PROTOCOL INTEGRATION

A building automation system based on a single protocol is usually the best option, but this is not always possible or practical. It is often necessary to design control systems using a mixture of LonWorks, BACnet, and even proprietary protocol devices. There is no one cross-protocol solution that fits every building or control situation. Often, multiple system designs can be integrated, though some may have advantages in scale, efficiency, ease of use, cost, or other aspects.

When multiple protocols are integrated together, the primary concern is the translation of control information from one protocol to another. **See Figure 10-5.** Designing systems that translate information accurately and reliably between two or more protocols can present significant challenges. However, since this is not an uncommon scenario, solutions are available.

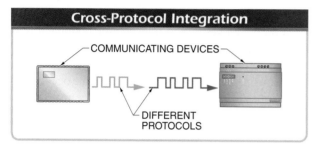

Figure 10-5. Cross-protocol integration strategies are solutions for facilitating communication between devices using different protocols.

Information Translation

The translation of information encoded in a protocol is similar to language translation. The complexities of meaning and context must be considered when designing a cross-protocol solution for building automation systems.

For example, the English word "temperature" can easily be translated into the Italian word "temperatura" with the meaning intact. **See Figure 10-6.** However, the translation does not necessarily address every aspect of the original meaning in English, such as scale and resolution.

Translation can be further complicated when information is retranslated in the other direction. For example, if the English word "hello" is processed by a translator into Italian, it may be interpreted as "ciao" on the other side. However, if "ciao" is inserted on the Italian side, it may emerge on the English side as "hello," "goodbye," or "at your service" because it has multiple meanings. Perfect translation is not always possible because of the idiosyncrasies and origins of different languages' words. Likewise, translating from degrees Fahrenheit to degrees Celsius, and then back into degrees Fahrenheit, could result in a loss of accuracy (77.61000°F = 25.33888°C = 77.60998°F).

This seemingly minor change can be significant in certain situations. For example, if a chiller unit is enabled at 77.60°F, a value of 77.61°F is sent from a temperature sensor through a gateway. The gateway maintains the value at two decimal places (disregarding the rest, instead of rounding) as it is converted to 25.33°C. Converting back to degrees Fahrenheit, the chiller receives a value of 77.59°F. Therefore, it is not enabled, though it should be.

Gateways

A gateway can be used to integrate two or more protocols into one information communication system. A *gateway* is a network device that translates transmitted information between different protocols. **See Figure 10-7.**

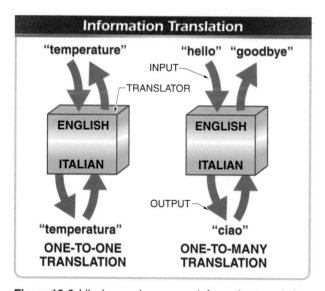

Figure 10-6. Like human languages, information translation can create problems if there is not a one-to-one relationship between words or concepts.

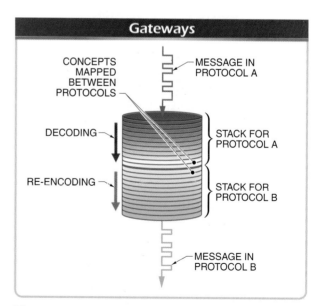

Figure 10-7. Gateways decode messages through one protocol stack and re-encode the same concept through another protocol stack to produce an equivalent message.

Gateways and routers are not the same. Routers manage the forwarding of message traffic based on the destination address and may connect segments of different media types. They simply move information from one network to another without changing the content. However, gateways manipulate message traffic at an application level by translating certain information.

A gateway must have a one-to-one mapping of the information to be shared from one side to the other. *Mapping* is the process of making an association between comparable concepts in a gateway. Each concept in one protocol must have one, and only one, equivalent concept in the other protocol. If the two protocols have different features or levels of complexity, mapping can be extremely difficult. Gateways can be used for translating content or translating protocols, though often both are involved. **See Figure 10-8.**

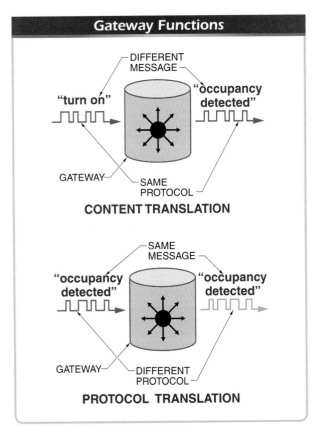

Figure 10-8. Gateways provide content translation and protocol translation services, though applications often require both simultaneously.

Translating content may be required when devices communicate with the same protocol but share incompatible information. This is often the case when the

devices were not originally intended to be integrated together. For example, an occupancy sensor that is designed specifically for a lighting system may not easily share its information with another system. If the occupancy sensor commands a set of lights to "turn on," this output cannot be interpreted by a security system, which expects an "occupied" or "unoccupied" status. The information is different, the context is different, and the desired response of the other system may be different. A gateway can be used to translate the command into something that can be understood by the security system.

Translating protocols involves altering the structure and encoding the message. For example, the occupancy sensor may report a status, such as "occupancy detected," which can be used by both the lighting system and the security system. However, if the two systems use different languages (protocols), the meaning of the message is not conveyed. This may also involve changes to the lighting system, which must now decide if action is required for "occupancy detected" instead of being commanded directly to "turn on." In this situation, a gateway can be used for protocol translation. This kind of translation almost always involves some content translation as well, since different protocols have different models for organizing information.

Perhaps the most common use of a gateway is to connect a building automation network to an intranet or the Internet. The control data can then be monitored and recorded by another system for any purpose.

Facts

The purpose of the OSI model is to create a framework to be used as the basis for defining standard communication protocols. This framework is used to divide the problem of computer-to-computer communication into several smaller pieces.

Intermediary Frameworks

Gateways may not be adequate for very complex systems, especially if there are multiple protocols. In response, some companies have developed intermediary frameworks. An *intermediary framework* is a complete software and/or hardware solution for integrating multiple network-based protocol systems through a centralized interface. **See Figure 10-9.**

For example, Tridium, a subsidiary of Honeywell International Inc., has developed a software platform and supporting hardware solution that integrates LonWorks,

BACnet, Modbus®, and other protocols simultaneously. Their Niagara^AX platform transforms data from protocol formats into common software components, which bridges the gap between the control networks and enables protocol-to-protocol communications.

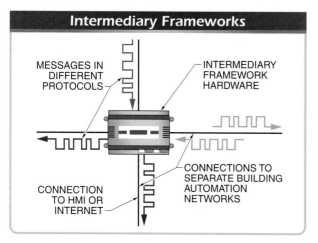

Figure 10-9. Intermediary frameworks translate all incoming messages into a common format, which can be used for system monitoring or management or translated into another protocol for outgoing messages.

Intermediary frameworks provide a variety of integration options, eliminating the need for protocol-specific gateways and custom software. The disadvantage is that the entire system is tied to that framework, relying on a single solution. However, relying on a gateway may create a similar situation. An integrator must weigh the advantages and disadvantages with the end user to determine whether an intermediary platform makes sense for an installation.

Extensible Markup Language

A development of the intermediary framework idea is the standardization of the information exchange, which allows vendors to provide competing and complementary solutions for the same framework platform. Extensible Markup Language (XML) is a leading solution.

Extensible Markup Language (XML) is a general-purpose specification for annotating text with information on structure and organization. XML is used in a variety of data communication applications as a standardized way to define and structure shared data. XML is known as "extensible" because it allows the elements to be defined by the user and extended as needed.

An XML element is nothing more than a pair of tags—composed of plaintext words, values, and symbols—that surround the data being structured. **See Figure 10-10.** Each tag is enclosed by the "<" and ">" symbols. A start-tag, which is in front of the data, must be matched by a corresponding end-tag. End-tags always include the "/" character.

For example, the temperature 25.33° could be represented in XML coding as the number "25.33" surrounded by the tags "<temperature>" and "</temperature>". The number "25.33" is the raw value that is being conveyed, and the tags provide descriptive information about the meaning of the value. The placement of the tags clearly indicates the beginning and end of the value and the entire element.

If further descriptive information about the value is needed, attributes can be added to the start-tag. Each attribute consists of a name and value. For example, if the temperature 25.33° is in the Celsius scale, that information can be added within the start-tag as an attribute. Other attributes in this example could include the resolution, source, sensor type, or timestamp of the measurement.

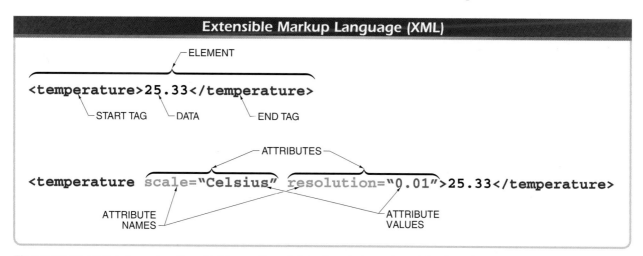

Figure 10-10. XML is based on the strict formatting of data descriptions, though it allows flexibility in description choices.

Nearly any type of information can be represented in this way. With only a few basic rules, XML allows any descriptive word (or hyphenated phrase) to be used as an element. Elements can also be empty (with no value) or include other elements. Nested elements can be used to build hierarchical structures. **See Figure 10-11.**

The flexibility and simplicity of XML have made it the most widely used, interoperable information-sharing model for building automation systems. XML element structures can be used to fully describe data that must be passed between systems using different protocols.

However, XML allows multiple ways to represent the same information. **See Figure 10-12.** XML elements can have exactly the same meaning with very different wording or attributes. This can be a benefit, but it can also be a potential problem. Programmers can use the most appropriate elements and attributes for the application, but they can easily introduce inconsistencies. Within a system, all senders and receivers of XML-formatted information must be programmed with the same element types or the data could be interpreted incorrectly or not at all.

Custom programming adds complexity and cost and can make a solution proprietary. Instead, standard XML formats that use the advantages of XML for gateway translation but do not use custom programming have been developed. A standard XML data-sharing model includes standardized definitions for structuring information for specific applications, such as building controls. This allows manufacturers to market off-the-shelf gateways and software interfaces that work together seamlessly without custom programming.

XML Structures

```
<lighting-system>
    <switch type="manual" location="Room 101">ON</switch>
    <switch type="occupancy" location="Room 107">OFF</switch>
    <ballast location="Room 101">
        <status type="dim-level">2</status>
        <priority type="demand-limiting">low</priority>
    </ballast>
    ...
</lighting-system>
```

Figure 10-11. XML elements that contain other XML elements form hierarchical structures that can be used to form logical relationships of building automation system devices and information.

XML Equivalent Elements

```
<temperature scale="Celsius" resolution="0.01"
        location="outside air">25.33</temperature>


<value type="OA temperature" units="Celsius"
        resolution="0.01">25.33</value>


<outside-air variable="temperature" units="Celsius"
        resolution="0.01">25.33</outside-air>
```

Figure 10-12. Because of its flexibility, XML can be used to create elements with very different wording that have the same meaning.

oBIX. The *open building information exchange (oBIX)* is an XML-based model for conveying control information between any building automation protocol, including enterprise and proprietary protocols. This model was developed within the Continental Automated Buildings Association (CABA) and is now a responsibility of the Organization for the Advancement of Structured Information Standards (OASIS). Several companies and corporate groups, including LonMark International, have adopted oBIX as the preferred XML model for control information.

While oBIX can share information from one system to another, it does not allow one control system to control another. This is because the different protocols have different management methods and requirements.

BACnet XML. *BACnet XML* is a proposed XML-based model for conveying control information between BACnet and other protocols. Unlike oBIX, BACnet XML is specific to one underlying protocol. This model has been fully developed and proposed as an addendum to the BACnet standard, though it is not yet approved. However, some member companies of BACnet International already provide products based on this format. The biggest benefits of BACnet XML are in the enterprise-level collection of information from one or more BACnet systems and in the sharing of XML-encoded BACnet messages between BACnet systems for control.

Both the oBIX and BACnet XML implementations require gateways or intermediary frameworks to translate information and receivers (computers, systems, and/or additional gateways) that understand the respective formats. However, a specific XML format can be self-describing by using additional files that explain, in a standard XML way, how to interpret the formats. Therefore, custom programming is usually not required if off-the-shelf tools are available.

It must be noted that some programming is still required at the gateway level in order to instruct the gateway what data to send and in what level of detail. However, this should be far less programming than what would be required in gateway-customization scenarios.

Enterprise System Integration

An *enterprise system* is a software and networking solution for managing business operations. These systems are often custom applications that include specialized services developed specifically for unique aspects of a particular company's operations. **See Figure 10-13.** For example, a large company may have an enterprise system for handling its accounting and project scheduling needs. Software applications are commonly referred to as "enterprise level" if they are part of this solution. This software is typically hosted on servers where it simultaneously supports a large number of users over a network.

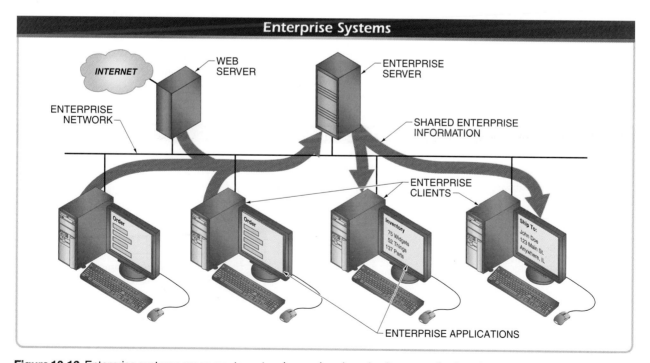

Figure 10-13. Enterprise systems are computer networks running shared software applications focused on business operations.

Businesses may use multiple enterprise systems that may need to be integrated. It may also be desirable to integrate control information from a building automation system into enterprise systems. For example, energy usage data from the building automation system can be shared directly into the accounting system for billing building tenants.

The means of sharing information with enterprise systems is similar to that for different control systems. Either a common, standardized platform is used by all systems, or an interface is required to translate and transport information between the different systems. Similar to building automation solutions for integration, gateways and intermediary frameworks are available for bridging an enterprise system and a building automation system.

Interfacing a building automation system with multiple protocols to an enterprise system can be accomplished in two different ways. Either each building automation system protocol is integrated separately, or the enterprise system is integrated with only one protocol system that handles the through-traffic to and from the others.

Likewise, there are efforts to promote XML as an integration standard, thus eliminating the limitation of single-vendor intermediary frameworks. One such standard from OASIS is the electronic business using Extensible Markup Language (ebXML). It not only enables intracompany exchange of business information but also intercompany exchange.

Gateway Connection. A gateway can be used to translate control information. The implementation of XML for translating control information to the enterprise level may be directly specified, which can significantly affect the feasibility of integration, or it may be influenced by the capabilities of the enterprise applications. Ideally, enterprise applications must be able to accept at least one of the control system's available XML implementations. Otherwise, additional programming must convert one XML format to another that is compatible with the enterprise system. **See Figure 10-14.**

Direct Connection. If a control system includes a supervisory control device, it can be used similarly to a gateway, translating information into a format that can be read by the enterprise system. **See Figure 10-15.** The control information is shared with the central device in the native protocol. This device then interfaces directly with the enterprise system using its data exchange method, such as ebXML.

Alternatively, the enterprise system servers could be equipped with the hardware necessary to interface directly to the control system, such as an internal interface card. Software drivers translate the hardware interface card information into a data exchange format that is native to the enterprise system.

Tunneled Connection. If the enterprise system is designed to understand control information formats, either natively or through additional protocol driver software, then the control information can be received directly. **See Figure 10-16.** This is not considered to be a gateway solution because there is no protocol translation. The control information is tunneled through the IP infrastructure into the enterprise system network.

> **Facts**
>
> The LEED® Green Building Rating System is a method of quantifying the measures taken during the construction and operation of a building to save energy and use sustainable materials.

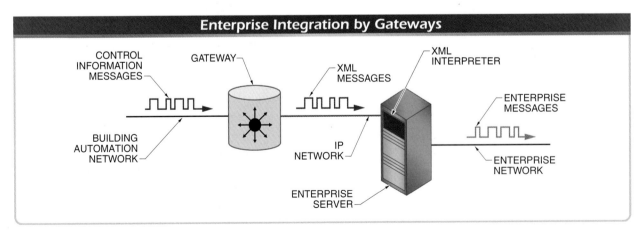

Enterprise Integration by Gateways

Figure 10-14. Gateways can be used to translate control information from a building automation network into XML data, which can then be processed for use in an enterprise system.

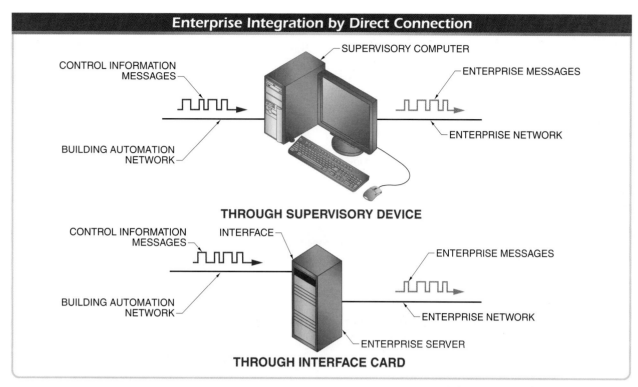

Figure 10-15. Building automation networks can be directly integrated to enterprise systems through supervisory devices or simple interfaces.

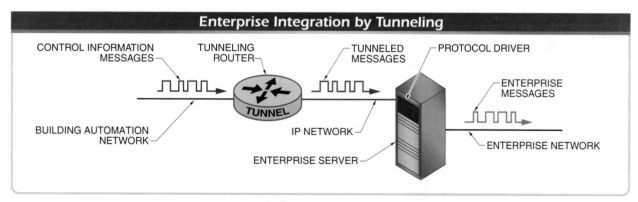

Figure 10-16. Control information messages can be tunneled through an IP network directly into an enterprise system protocol driver or interface.

Departmental Cooperation. Designing the best integration solution involves understanding the responsibilities and concerns of the building's information technology, accounting, and facility management departments. Many IT departments resist sharing their infrastructure with the control system. Their primary concern is the security of the facility's networks from both the outside and the inside. Adding control information traffic to the IT infrastructure may complicate the department's responsibilities, especially if they know very little about the automation system.

The accounting department can use control information for departmental/tenant billing purposes and for energy efficiency (cost savings) monitoring. However, this group may object to connecting the control system directly to the accounting enterprise system out of concern for unauthorized access to their financial information. The facility management department uses control information to maintain building operations and relies on the IT and accounting departments to help deliver and process the information.

All three departments, and any others involved in a particular automation solution, must work together to satisfy all the concerns while meeting the demands of integration. Control information must be communicated in a reliable and secure way.

Facts

BACnet® and LonWorks® are industry standards that allow manufacturers to design their system protocols and connecting media for communication compatibility with other manufacturers' systems.

EXAMPLE BUILDING AUTOMATION SYSTEM

Building automation systems are increasingly popular in both new construction and existing buildings. The most common automation applications are related to HVAC and life safety operations. In some buildings, other applications such as security, access control, and other building functions are also included. Most buildings do not integrate all possible building systems. However, an overview of a building automation system example can be used to emphasize several applications that are possible with existing technology. **See Figure 10-17.**

Example Building

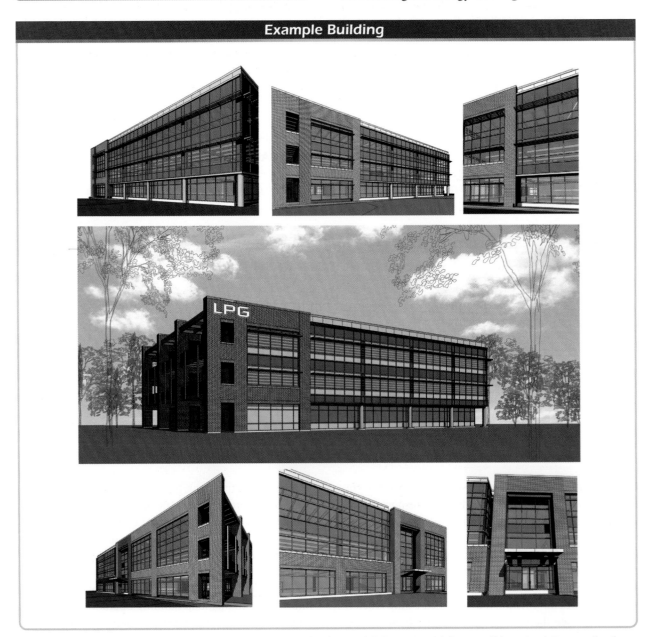

Figure 10-17. A walk-through of an example automated building highlights some of the possible automation applications.

Example Building Owner Profile

The Lincoln Publishing Group (LPG) is a 100-employee organization that publishes several trade magazines, journals, training products, and government documents. The company works with government officials, authors, advertisers, and other vendors located throughout the country. A variety of media is used in these products, including print, audio, and video. Some publications involve homeland security and classified government documents.

Because many of the publications cover technology and its applications in industry and government facilities, the building was designed with many automation features that showcase technology and energy savings. Additionally, the building was designed as a sustainable building and includes provisions for energy efficiency, reduced maintenance, and enhanced occupant well-being and performance. The building was also designed and built to be Leadership in Energy and Environmental Design® (LEED®) certified by the U.S. Green Building Council.

Example Building Description

At the early planning stage, a conscious decision was made by the management of the company to construct a building with sustainable building features. Input was solicited and received related to features and amenities that provide efficiency and comfort in the working environment. In addition, future expansion for increased production capacity was considered in the facility requirements.

The building site is 4.25 acres of previously farmed land in an area that has seen limited development. The area is zoned as office/research/light commercial. Because the property is not part of a subdivision, improvements for roads, utilities, sewer, and water detention were required inside and adjacent to the property boundaries. Grading was completed to facilitate the flow of water into a dry bottom detention area and away from the property by earth swales. **See Figure 10-18.** A crushed rock walking path around the detention area provides an opportunity for moderate outdoor exercise. Landscape lighting is provided on the path in intervals.

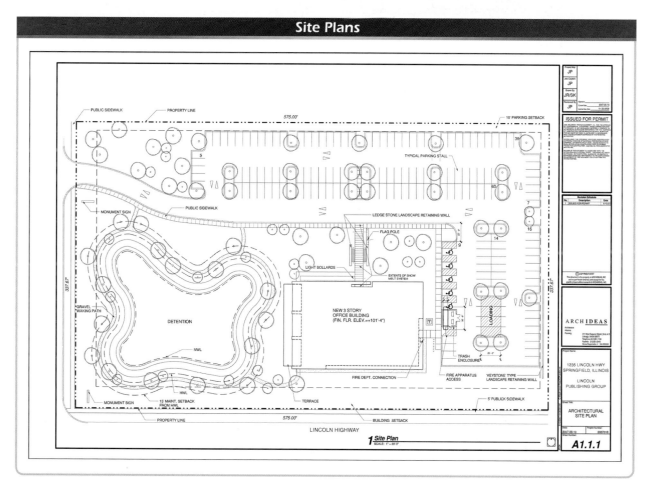

Figure 10-18. Grading facilitates the flow of water into a dry bottom detention area.

The building has 45,000 sq ft of total usable space on three floors. It has a structural steel core and glass curtain walls with aesthetic brick facades. Building orientation and architectural features provide shading for maximum lighting and heating benefits from the sun. A patio is located off the west wall of the first floor. Balconies are provided above the patio on the second and third floors. A sidewalk heating system circulates hot water from a boiler through heat exchanger loops in the sidewalk to prevent slippery conditions from ice and snow in the winter. A soil moisture sensor and temperature sensor activate the system.

A 400 kW diesel-powered generator serves as a back-up power source for the fire pump, elevator, computer servers, and other critical load circuits. **See Figure 10-19.** Critical loads are grouped into a common panelboard.

> **Facts**
>
> *When power is lost, controllers can shut down. Therefore, many controllers have back-up power supplies that last long enough for the power to come back on or for an orderly shutdown.*

Each floor has 10′ ceilings to maximize natural light penetration and an open environment. Package HVAC units are located centrally on the roof. A trunk and branch HVAC system provides conditioned air into building spaces. The supply air is distributed by branches extending from the trunk (main supply duct) to each register. Airflow from the register is controlled by the vanes on register panels. Return air is drawn through grills back into the return-air ductwork. **See Figure 10-20.**

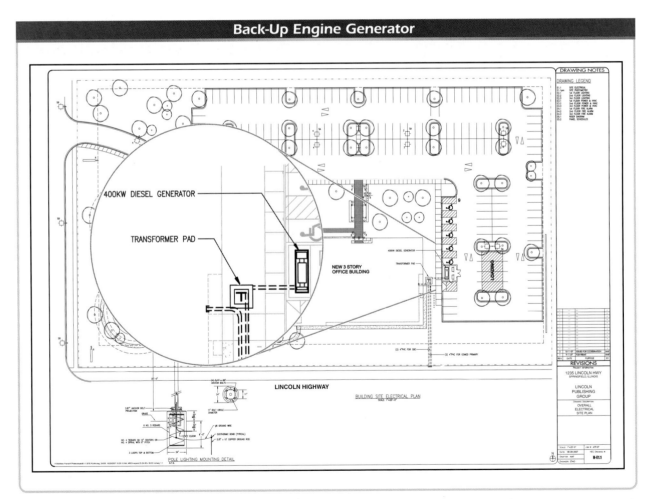

Figure 10-19. Automation of the electrical system may include managing the transition to secondary power sources, such as diesel engine generators.

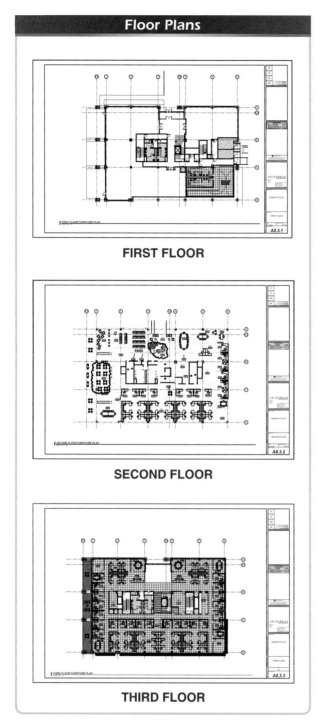

Floor Plans

FIRST FLOOR

SECOND FLOOR

THIRD FLOOR

Figure 10-20. Building interiors made pleasant and comfortable with automated lighting and HVAC systems improve the productivity of the occupants.

The first floor is designed for approximately 9000 sq ft of tenant space, with a 3000 sq ft exercise area in the southeast corner for LPG employees. The first floor tenants determine the floor plan required for their specific needs. The building electrical, HVAC, and VDV equipment is designed for flexibility in meeting tenant requirements. The second floor is occupied by LPG's administrative and marketing departments. The third floor is occupied by the editorial department. The lobby area on the first floor secures access to all building spaces. Employees can access the second and third floor via an elevator or stairs.

The reception area for LPG is located on the second floor across from the elevator. The space above the reception area opens to the third floor. The perimeter of the lobby space occupying the second and third floor is enclosed by glass in compliance with fire department code for smoke isolation and containment during a fire.

The second floor contains private offices and workstations associated with the administrative and marketing functions of the company. In addition, the second floor also houses the lunchroom, training rooms, library, business center, and corporate boardroom. The third floor contains private offices and workstations associated with the product development functions of the company. In addition, the third floor also houses a photography studio and multimedia suite.

A Voice over Internet Protocol (VoIP) system is used to transmit telephone calls over the data network and Internet. Combining voice and data allows long-distance calling without telephone company charges. The system requires VoIP telephones.

The common areas are heated to a temperature setpoint based on the outside air temperature, sun position, and number of occupants. During office hours on weekdays, electrical demand and consumption are reduced during peak utility demand periods by adjusting temperature setpoints by a few degrees. This small change is usually imperceptible to building occupants but results in significant savings from peak demand utility rates.

The fire alarm system is monitored by an off-premises supervising station, and alarms are also automatically transmitted to the local fire department. This alerts monitoring and emergency personnel immediately and communicates building information prior to their arrival. The fire alarm system takes over the management of other building systems to facilitate safe evacuation and limit the spread of fire and smoke. For example, commands are activated for HVAC shutdown, elevator recall, door unlocking, and smoke door closure.

Facts

Control transformers often operate at 24 V and are supplied from 120 V or 240 V sources. These control transformers supply low voltage to the controllers that manage heating, ventilation, air conditioning units, and other electrical devices for building automation systems.

Example Building LEED® Certification

The LEED® Green Building Rating System is the nationally accepted standard for the design, construction, and operation of green buildings. LEED® serves as a guide for building owners and operators for realizing the construction and performance of a building. **See Figure 10-21.** The LEED® certification process addresses sustainable site development, water savings, energy efficiency, materials selection, and indoor environmental quality throughout the construction and operation of the building. Two major criteria in the LEED® certification process are the energy performance measures and the renewable energy standard. Both of these criteria are affected by the building automation system.

The example building was designed to meet the criteria for gold certification by the U.S. Green Building Council. This level of certification requires a minimum of 39 points. Each of the categories was analyzed during the designing and construction phase for compliance. **See Figure 10-22.** For example, the building structural steel is recycled, and the bricks were supplied by a manufacturer located within a 500-mile radius. In addition to design and construction requirements, the building must also comply with the standards for maintenance and operation, including cleaning procedures, pest control methods, and office waste recycling. **See Appendix.**

Example Building Automation System Use

Understanding the potential interactions of various building systems is critical to the design and implementation of a building automation system. Selected, potential capabilities of a sophisticated automation system can be highlighted by describing a day in the life of employee Pat Smith of LPG. Pat is the leader of the new product development staff and has worked for 11 years as Senior Development Editor.

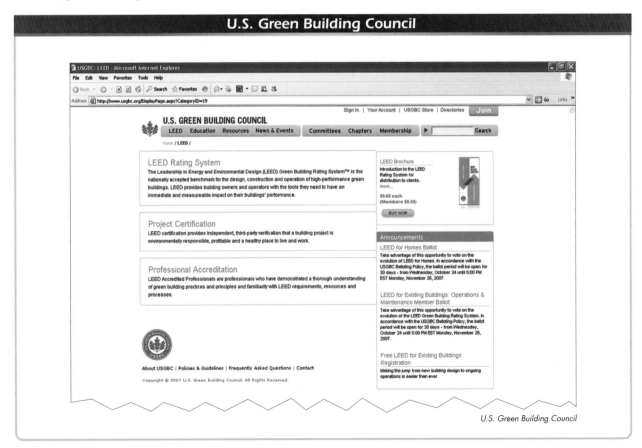

U.S. Green Building Council

Figure 10-21. The LEED® program establishes criteria for the certification of green buildings.

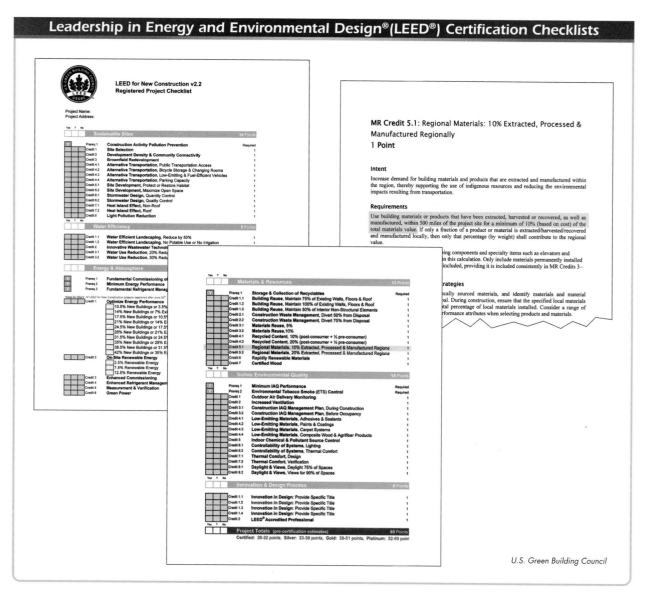

Figure 10-22. The LEED® project checklist defines certification levels that can be achieved through the accumulation of points.

Pat leaves for the office early on Saturday morning in mid-October. It is still quite dark at this time in the morning. LPG headquarters is located just off the interstate highway in Springfield, Illinois. The building is the first of several buildings being constructed in the new Capitol City Industrial Campus. Construction equipment for other new buildings is located on adjoining properties.

Turning into the drive, Pat's vehicle is sensed by the in-ground loop. The gate arm rises and allows access to the parking facility. **See Figure 10-23.** A CCTV security camera installed in the gate records the entrance of the car and the license plate. Monitors located in the security office display the video footage captured by these and other CCTV cameras in the facility. The footage is recorded by digital video recorders (DVRs) if needed, but it is otherwise overwritten at scheduled intervals.

A new software release has necessitated a training seminar for key people in the development team. There will be a guest presenter and 10 more staff members arriving later at 8:30 AM. The expected staff members have been logged for preauthorized access to common areas during the seminar. Since the building normally operates in efficiency mode during the weekend, an exception schedule is set to override the Editorial Department HVAC zone operation for Saturday. This was entered into the system via the building management website.

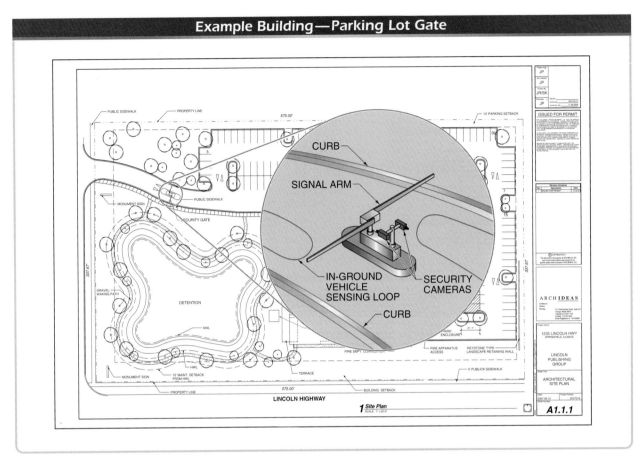

Example Building—Parking Lot Gate

Figure 10-23. Automated security and access control systems provide seamless integration with other building systems.

Approaching the front entrance door, Pat notices that the irrigation system is activated for the morning landscape watering. Stormwater runoff from the building roof is collected in a cistern and supplied to the sprinkler control and distribution unit. A water level sensor in the cistern provides water level readings indicating the amount of available irrigation water. The sprinkler control unit uses a soil moisture sensor so that the landscaping is only watered when necessary.

Pat swipes her access card. An overhead CCTV camera captures her image and the card data is sent through the network to the access control system. Access is granted to the exercise area on the first floor and the second and third floor spaces for management-level employees. In her daily tasks, Pat requires access to most building spaces. Having identified Pat, the building automation system is programmed to turn on the lights in her office and the adjoining work areas. **See Figure 10-24.**

The lobby doors to the elevator and second and third level stairway are locked. Typically, the receptionist activates the release of the entrance doors. Pat swipes her card and the doors unlock, allowing access. Anticipating her need for an elevator, the access control system signals the elevator system to send the car to the first floor.

On the third floor, Pat exits the elevator. Fire doors in the glass lobby perimeter are held open by electromagnetic door holders. If there were a fire, the alarm signal from the fire alarm control panel would deenergize the magnetic holders, allowing the doors to close. This is a code requirement that prevents the spread of smoke throughout the building. The elevator is also programmed to be immediately recalled to a designated level in the event of a fire alarm.

The third floor lights are on, and the space is comfortable at 72°F and 40% rh, according to the exception schedule. The sun has begun to illuminate the area. Pat walks to her office near the telecommunications room and swipes her access card. The door is unlocked and she enters and hangs up her coat. The office is slightly cooler than the common areas—just how she likes it.

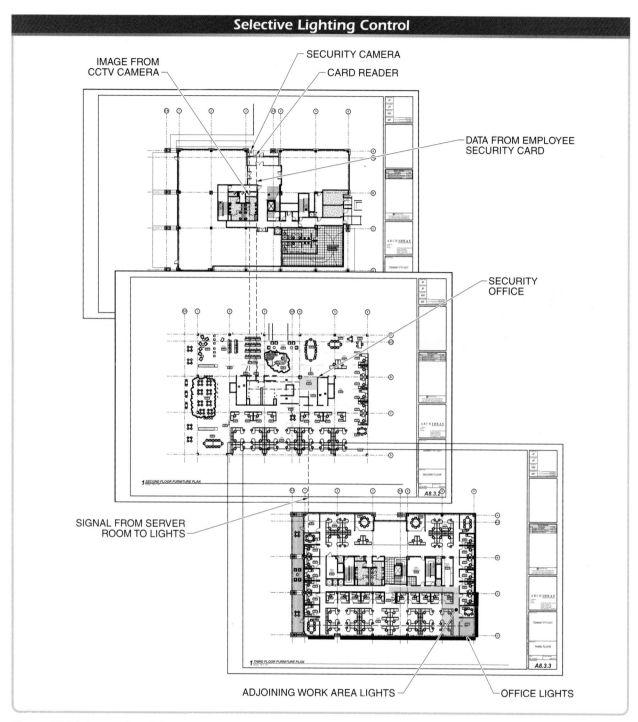

Selective Lighting Control

IMAGE FROM
CCTV CAMERA

SECURITY CAMERA

CARD READER

DATA FROM EMPLOYEE
SECURITY CARD

SECURITY
OFFICE

SIGNAL FROM SERVER
ROOM TO LIGHTS

ADJOINING WORK AREA LIGHTS

OFFICE LIGHTS

Figure 10-24. Integration of the access control and lighting systems allows certain lighting circuits to be turned on based on the identification of a person entering the area.

After responding to e-mail messages, Pat is ready for a cup of coffee and she walks down the stairway between the second and third floor. Light from the brightening sky has caused the photo control sensors to turn off the parking lot lights. Pat opens the lunchroom door and an occupancy sensor turns on the lighting. After making a large espresso, Pat returns to her office to continue reviewing the new software menus and capabilities.

Back on the third floor, Pat observes that the perimeter lights (nearest the exterior walls) in the common areas are off. Sufficient sunlight entering the building has caused the daylighting sensors to turn those lights off. **See Figure 10-25.** Back in her office, a security system pop-up window shows an image of the front entrance. It is 8:10 AM and Chris Tate, Production Editor, has swiped his access card. Pat catches a glimpse of lights turning on in Chris's office, indicating that the building systems detected his presence. The rest of the seminar attendees will be arriving soon.

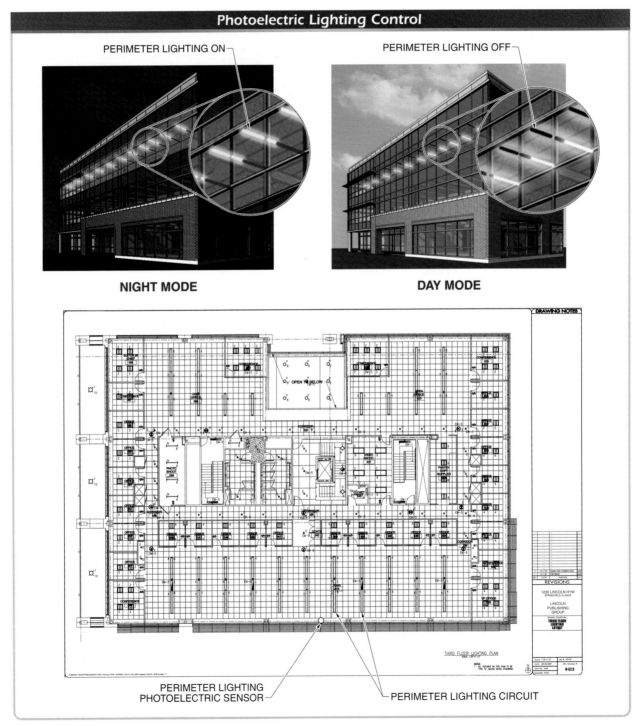

Figure 10-25. Daylighting controls minimize the use of artificial lighting when natural sunlight is available.

A few minutes later, another security window pops up. Jack Morton, Copy Editor, left his access card at home and uses the keypad at the front door to enter the building. Sue Barton, IT Coordinator, arrives with Gary Mandro, the guest presenter and Technical Support Specialist for his software company. The CCTV security camera in the lobby records Gary and Sue entering the elevator hallway. Sue gives Gary a guest security card to use throughout the day. His card provides limited access to building areas and services.

In preparation for the software upgrade training class, Chris runs copies of the handouts and prepares the training room. A PowerPoint® presentation will provide an overview, and the handouts will include a sample application to be completed by the group. The training room has a scene lighting control scheme. The scene settings for the training room correspond to the room functions. In presentation mode, the blinds close and the lighting is dimmed.

> **Facts**
>
> *Bilevel switching is a technique that controls general light levels by switching individual lamps or groups of lamps in a fixture separately.*

The overall lighting is provided by a three-lamp fluorescent fixture with bi-level switching. **See Figure 10-26.** The outer two lamps are switched separately from the middle lamp, allowing the user to switch on one, two, or all three lamps. One lamp will be used for the multimedia presentation, two lamps will be used for the question-and-answer session, and all three lamps will be used for the group activity.

Chris begins the presentation at 8:30 AM with all 10 attendees present. Chris introduces the presenter, Gary Mandro, and uses the scene controller to reduce the lights to the lowest intensity. After an hour, the room lights and the multimedia projector suddenly quit. Pat hears the emergency back-up generator come on. **See Figure 10-27.**

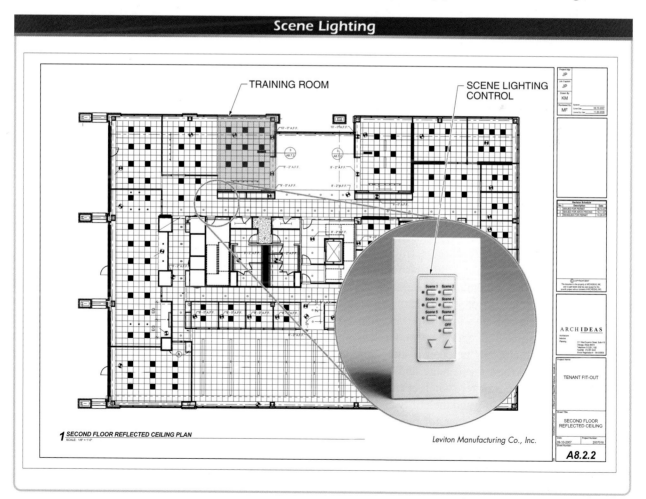

Leviton Manufacturing Co., Inc.

Figure 10-26. Scene lighting controls allow for special lighting schemes optimized for certain tasks.

Back-Up Power

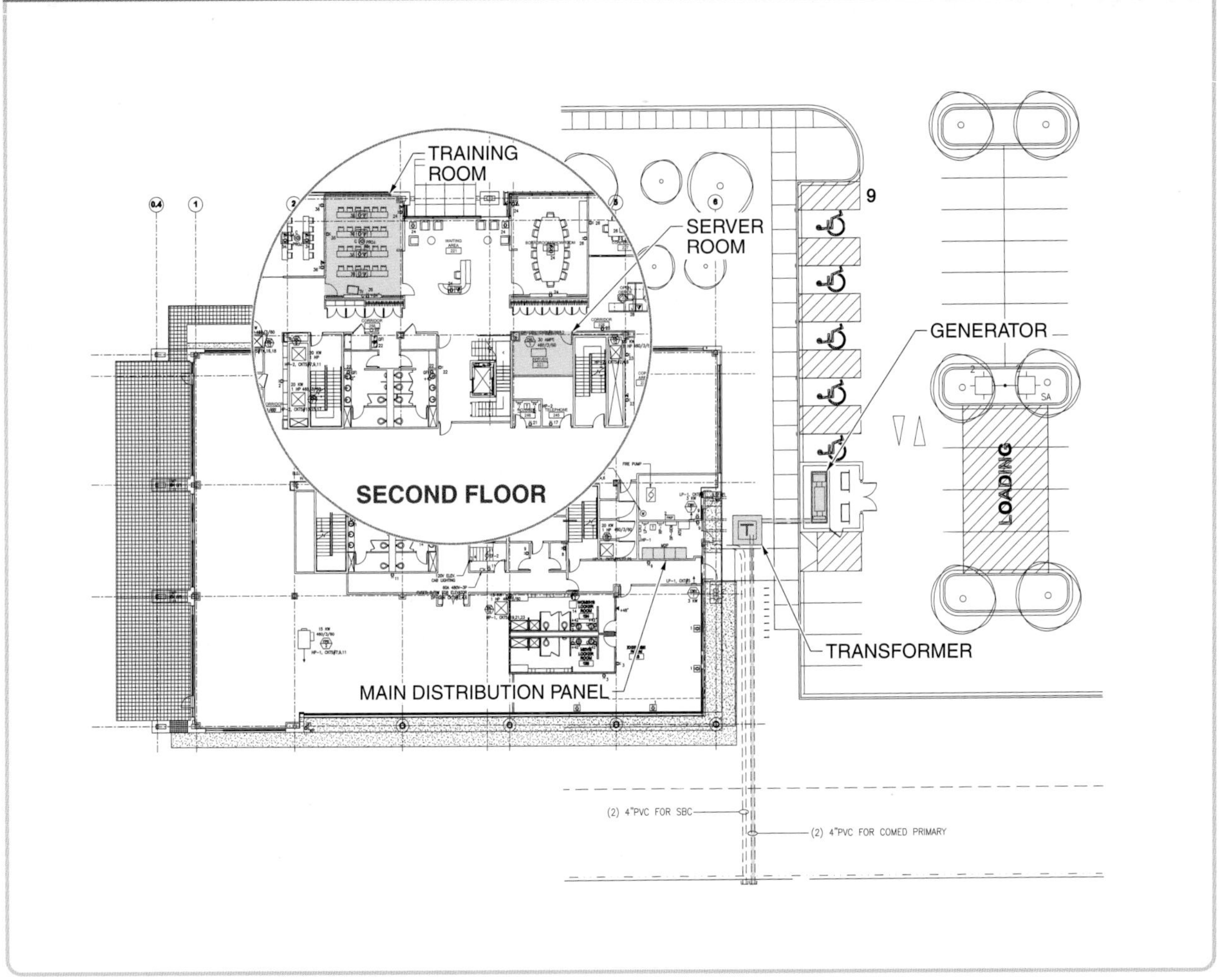

Figure 10-27. Automatic transfer switches and uninterruptible power supplies ensure continuous and reliable power to the building in the event of a utility power outage.

Contractors working at a nearby construction site inadvertently breached the building's lateral electrical service with their construction equipment. The building's transfer switch sensed a loss of power, delayed 10 sec, and then initiated an engine generator start-up sequence. The generator engine takes a short time to come up to speed. When the voltage and frequency stabilize, the automatic transfer switch opens the connection between the building and the utility and closes the connection between the generator output and the building's electrical system. The uninterruptible power supply (UPS) powers the computer network system, building automation controls, security system, and other critical systems during this interruption, ensuring no data is lost or corrupted.

The presenter, Gary Mandro, sees Chris pointing to his watch, indicating that the power loss may be a good time for a break. A joke is made that "it is good that the coffee maker is on a backed-up circuit." The attendees proceed to the lunchroom on the second floor. Some of the lights along the way are being powered by the back-up generator. On the way, Gary realizes that his cell phone is low and will not make it through his five new voicemail messages. A conference room ahead has a phone on the table. He swipes his guest access card and turns the door latch. It does not open. From down the hall, Chris calls out, "Use the business center—your card will work there."

The power is restored in 20 min, and the group reconvenes for the rest of the presentation. The next part of the presentation involves how sensitive documents are accessed with the new software using a password. Access to secure files is controlled from the server room. Sue leaves the training room and heads to the server room.

A fingerprint scanning biometric device on the wall by the door prevents security breaches such as "card swapping" or other unauthorized access. **See Figure 10-28.** The physical (fingerprint) characteristics unique to Sue Barton are required for authentication and are not easily falsified. Sue accesses the password file, creates a temporary password, and allows limited access to certain secure files for the training room presentation.

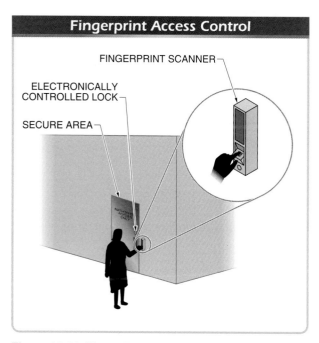

Figure 10-28. Biometric access control devices ensure the identity of the user for access to secure areas.

The training seminar continues into the afternoon with the participants successfully completing the hands-on activities specified on the handouts. The training seminar adjourns with closing comments by Chris and Pat at 4:30 PM. Pat still has several tasks to complete before leaving, so she returns to her office to work.

After everyone else has left the building, Pat resets the motion sensors located in the common areas away from her office. **See Figure 10-29.** These sensors will now be active input devices for the security system (instead of just for the lighting system) in those security system partitions.

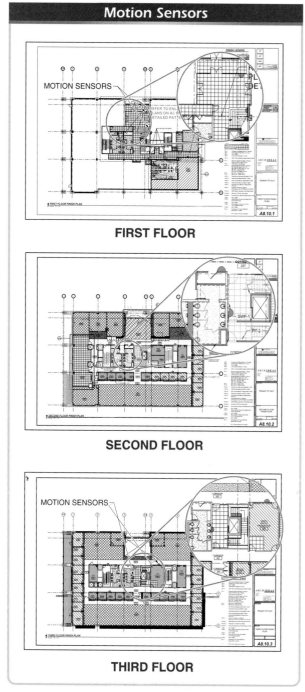

Figure 10-29. Motion detectors throughout a building provide detection of intruders for the security system.

Facts

Security systems can be set up to provide temporary access cards for guests. The cards are typically limited to common areas and any special areas assigned to the guest.

The lobby area, entrance door to the workout area, and hallways on the first and second floor have dual-technology motion sensors to detect intruders. If an individual were present, the passive infrared (PIR) sensor would sense the heat energy of the individual within its field of view in a detection area, and the microwave sensor would sense movement. The detector would only initiate an alarm if both sensors were tripped. Motion sensors would activate a building alarm, and a pop-up notice on Pat's computer screen would appear.

With the final tasks completed, Pat disables the motion sensors and shuts down her computer. She turns off her lights, locks the door, and proceeds to the elevator. In the lobby, Pat inputs the night mode security code on the keypad, which activates the exit delay countdown warning. A light on the security system control panel blinks, increasing in frequency to indicate the amount of exit delay time remaining before the system is armed. The programmed exit delay time is 45 sec. This is the amount of time between when a security system is armed and when the control panel begins registering alarm signals from sensing devices.

The exit delay allows Pat time to leave the detection area before her presence causes an alarm. With the alarm system now set for an unoccupied building, the building automation system uses this information to signal the HVAC and lighting systems to return to the weekend energy-saving mode. Pat leaves the building. While leaving the parking lot, the security CCTV camera and DVR record her rear license plate at the gatepost.

REVIEW QUESTIONS

1. How can automated building systems become interdependent when integrated?

2. What is the primary advantage of interoperability?

3. What is the primary factor in extending an automation system beyond a single building?

4. How can remote building automation system access via the Internet help with monitoring and operating building systems?

5. What general areas does LEED® certification address?

6. What is the open building information exchange (oBIX)?

7. What is the primary concern of cross-protocol integration?

8. What is a gateway?

9. What role can Extensible Markup Language (XML) play in intermediary frameworks?

10. What is an enterprise system and how can it be integrated with a building automation system?

Chapter 10 Review and Resources

System Integration  11

The ultimate goal of a comprehensive building automation system is the integration of multiple building systems into a common information-sharing network. Events or changes in one system can be used to trigger changes in the operation of otherwise unrelated building systems. The possible combinations of system interactions are practically infinite, although some control scenarios are relatively common. Regardless of the required interactions, it is likely that the automation system can be designed using a variety of hardware combinations or protocol systems. It may be necessary to consider multiple possible implementations when designing a building automation system.

BUILDING AUTOMATION SYSTEM EXAMPLE

Building automation systems are increasingly popular in both new construction and existing buildings. The most common automation applications are related to HVAC, lighting, and life safety operations. Some buildings also integrate additional systems, such as security, access control, elevator control, or other special functions.

A study of an example building automation system can be used to emphasize possible applications of existing building automation technology. Common building systems are integrated to show how various systems can be combined under a single protocol. The design focuses on implementing a robust system that limits the possibilities of a single source of failure and allows communication between various systems.

Facts

Commercial buildings with a large number of electric baseboard heaters or small exhaust fans often use duty cycling as a control strategy. Because a duty cycle alternates the areas it heats, loss of comfort may occur when the load is off.

The example building belongs to the Lincoln Publishing Group (LPG). The building has 45,000 sq ft of usable space on three floors. **See Figure 11-1.** The architectural design, including the building's orientation, optimizes light penetration and shading for maximum lighting and heating benefits.

Example Building

Figure 11-1. The integration requirements for an example automated building are useful for comparing the possible implementations of different protocol systems.

LPG occupies the second and third floors. The first floor includes tenant space, plus common areas for entrances, an elevator lobby, and an exercise facility. **See Figure 11-2.** The lobby area secures access to all building spaces. The building's electrical, HVAC, fire protection, and voice-data-video (VDV) equipment is also located on the first floor.

Example Building Floor Plans

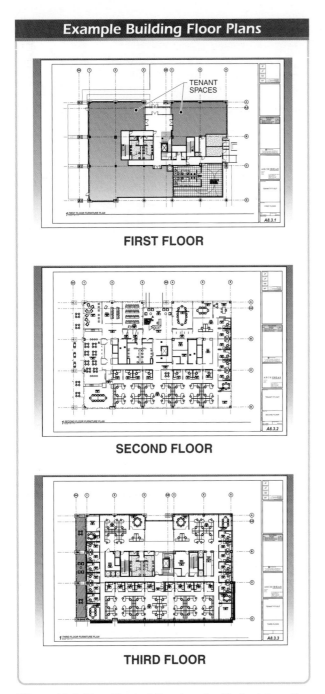

FIRST FLOOR

SECOND FLOOR

THIRD FLOOR

Figure 11-2. The office building described in the integration example consists of three floors that include open offices, individual offices, multiuse space, a training room, a lunchroom, an exercise room, and tenant spaces.

Building System Requirements

The company recognizes the importance of building automation technology in accomplishing the goals of energy efficiency, reduced maintenance, and enhanced occupant comfort and performance. A number of possible control scenarios were developed, though only some were ultimately implemented, based on a balance of benefits and costs. All building automation system designs use an open protocol, such as LonWorks® or BACnet® protocols, and all systems are interoperable with this protocol. The resulting building automation system integrates many of the building's systems.

Electrical System. The building includes a 400 kW diesel-powered generator to serve as a back-up power source for the fire pump, elevators, computer servers, and other critical load circuits, as well as for demand shedding during peak load situations.

Lighting System. Outdoor site lighting is primarily controlled according to a schedule, which turns the lights on from dusk until late at night and again during a period early in the morning. Light level sensors are also used to automatically turn the lights on if needed during the scheduled off times, such as during a thunderstorm.

Indoor lighting is controlled by a lighting control system that uses schedules, occupancy sensors, manual override switches, and control sequences specific to the lighting system. Light fixtures incorporate dimmable ballasts or bi-level switching to adjust lighting levels. Work areas near the perimeter use daylight harvesting. All life-safety-designated lighting, such as lighting in egress stairwells, is on emergency power.

HVAC System. The building is divided into HVAC zones. The HVAC system manages zone temperature with variable-volume air terminal units (ATUs), which use information from temperature sensors to control their operation. The ATUs are served by a rooftop air-handling unit (AHU) with a direct expansion (DX) cooling coil. The air conditioner (with DX evaporator coil) has an air-cooled condenser. The AHU and air conditioner are stand-alone components, not a package unit.

Setback temperature setpoints are used during unoccupied periods. The system uses an optimization strategy to provide adequate heating/cooling in anticipation of scheduled occupancy. All components of the HVAC system that are critical to life safety are on emergency power.

Plumbing System. Automated plumbing subsystems include pressure boosting and hot water circulation equipment. The critical plumbing subsystems, such as the lift station and sump pump, are on emergency power. The building automation system does not enable/disable the lift station, but it monitors the lift station for alarms.

Fire Protection System. The fire protection system must be reliable during life safety situations. Therefore, the fire protection system can share information with other building automation systems, but direct control of the fire protection system is not allowed. During an alarm condition, the fire alarm system controls critical functions such as door unlocking, HVAC shutdown, and elevator recall. The fire protection system is UL®-listed and fire-marshal-approved for compatibility with BACnet and LonWorks protocols. The fire protection system is on an emergency power circuit.

Security System. The security system is separated into four partitions: one partition for each of the three floors and a fourth partition for the common areas (elevator lobbies, front entrance, employee entrances, and exercise facility). The security system monitors the closure of main entrances and windows on the first floor and patio windows and doors on the second and third floors. Common areas on each floor are monitored with motion detectors, which are also integrated with the lighting control system to indicate occupancy.

Like the fire protection system, the security system must ensure reliability during alarm events. Therefore, critical system programming functions and critical operations, such as zone bypassing, are only allowed to be performed by authorized personnel. The security system is on emergency power.

Access Control System. All building entrances have electric locks and card readers for employees to gain access to the building. The front doors automatically unlock and lock on a schedule programmed for normal business hours. Doors leading into tenant spaces also have electric locks and card readers that are programmed to unlock and lock according to the tenants' schedules. Stairwell doors are locked using magnetic locks and require a valid card read to enter or exit, regardless of the time of day. For safety purposes, stairwell doors use a touch-sensitive panic bar that releases the magnetic lock, allowing free egress from the occupied area. This event triggers an alarm condition if a valid card read was not received.

Like the security system, the access control system must be protected from unauthorized operation. The access control system can be monitored by building maintenance staff, but critical system programming and critical operations, such as remote door release, are only allowed to be controlled by authorized personnel. The access control system is on emergency power.

VDV System. The VDV system operates on its own network, as opposed to the common building automation network. The VDV system is on emergency power.

Elevator System. During normal occupancy periods, the elevator controller allows unrestricted use of the elevator. During scheduled unoccupied periods, the elevator system requires access card authentication before the elevator operates. The identity of the cardholder also determines which floor the elevator provides access to, based on privileges programmed into the system. The elevator system is on emergency power.

> **Facts**
>
> *Building automation systems are typically used to verify equipment operation as part of the commissioning process.*

Building Automation System Implementations

There are multiple ways to implement an effective building automation system that meets the automation and integration requirements of the various building systems. In most cases, a control scenario can be designed with any of a number of different protocol-based systems, including the two most common open protocols, LonWorks and BACnet.

The choice of protocol is typically based on cost, contractor experience, ease of programming, available compatible control devices, existing infrastructure and equipment, and other considerations. For comparison, it is particularly useful to produce preliminary implementation designs of the same control scenarios in multiple protocol systems. For the example building, the control scenarios are studied from both a LonWorks and a BACnet perspective.

LonWorks Implementation. The building automation system for the Lincoln Publishing Group building may be based on the LonWorks technology. A flat network architecture integrates the lighting, HVAC, plumbing, fire protection, security, and access control systems. Node communication is peer-to-peer without the need for programmable gateways or network supervisor nodes. A gateway is needed only to interface with the proprietary elevator control system.

The network uses twisted-pair cabling. A TP/XF-1250 channel provides a high-speed backbone that is arranged in a bus topology and connects each floor. **See Figure 11-3.** TP/FT-10 free-topology node channels are used for each building system on each floor. Multiport routers on each floor isolate network traffic for system channels. Channel terminators are installed at router ports.

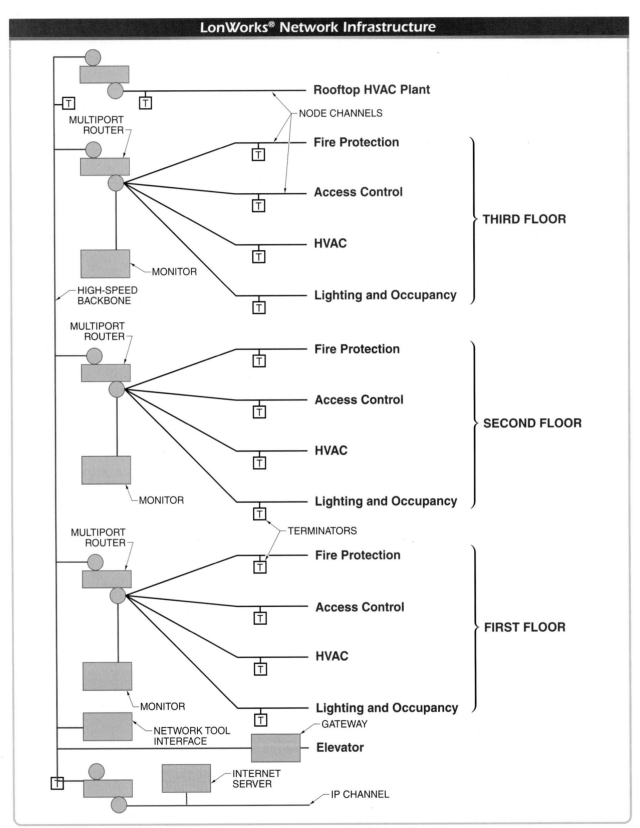

LonWorks® Network Infrastructure

Rooftop HVAC Plant

NODE CHANNELS

MULTIPORT ROUTER

Fire Protection

Access Control

HVAC

THIRD FLOOR

MONITOR

HIGH-SPEED BACKBONE

Lighting and Occupancy

MULTIPORT ROUTER

Fire Protection

Access Control

HVAC

SECOND FLOOR

MONITOR

Lighting and Occupancy

TERMINATORS

MULTIPORT ROUTER

Fire Protection

Access Control

HVAC

FIRST FLOOR

MONITOR

Lighting and Occupancy

NETWORK TOOL INTERFACE

GATEWAY

Elevator

INTERNET SERVER

IP CHANNEL

Figure 11-3. The network architecture of the proposed LonWorks® implementation uses a high-speed backbone to connect to many node channels, which are divided according to building floor and system type.

Nodes adhere to LonMark interoperability standards as well as local safety jurisdictions. Wherever possible, application-specific nodes use LonMark standard functional profiles, which provide a high degree of vendor interchangeability. The network management platform is LonWorks Network Services (LNS). Node configuration, where possible, uses LNS-compatible software plug-ins.

LonMark International maintains a master list of standardized variables and descriptions. Each variable and description is called a standard network variable type (SNVT). Typically, each SNVT is defined by a maximum and minimum value. In addition, the master list defines the data resolution and measurement units. For example, the variable "temperature" is represented by a number from 0 to 65,535 that represents a temperature range of –274°C to 6279.5°C. The data has a resolution of 0.1°C.

All devices that use this master list use the same standard values for the variable.

The network human-machine interface (HMI) is browser-based and provides graphical representations of equipment operating conditions and access to data logging, scheduling, and alarm-reporting functions. These monitoring and control functions are provided by an Internet server node. The access control management and database software runs on a secure computer that is accessible by authorized personnel only.

BACnet Implementation. Alternatively, the control system network for the Lincoln Publishing Group building may be based on the BACnet standard. The network architecture provides for peer-to-peer communication within each subnet. **See Figure 11-4.**

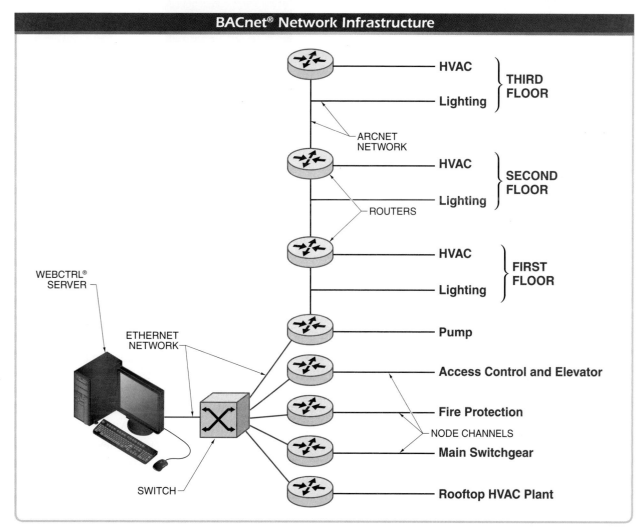

Figure 11-4. The proposed BACnet® network architecture uses a fast Ethernet network to connect to system-based sub-networks, which use bus topology ARCNET networks to connect between nodes.

The network uses twisted-pair cabling. Integration between multiple building systems is accomplished primarily with Ethernet and ARCNET. An Ethernet network provides a high-speed connection to each floor. There, routers connect to subnets based on the building system. The lighting and HVAC systems are integrated on ARCNET networks. Integration with proprietary systems, including the fire protection, security, access control, and elevator systems, is accomplished via a Modbus RTU protocol. The electrical demand limiting and plumbing monitoring and control are accomplished with ARCNET controllers.

Nodes adhere to BACnet interoperability standards as well as local safety jurisdictions. Wherever possible, application-specific nodes use BACnet standard object types, which provide a high degree of vendor interchangeability. The network management platform is an Internet server running Automated Logic Corporation's WebCTRL® software.

The network HMI is browser-based and provides graphical representations of equipment operating conditions and access to data logging, scheduling, and alarm-reporting functions. These monitoring and control functions are provided by an Internet server node. The access control management and database software runs on a secure computer that is accessible by authorized personnel only.

CONTROL SCENARIO: OPENING THE BUILDING ON A REGULARLY SCHEDULED WORKDAY

A very basic control scenario concerns the actions needed by the building system when it changes to an occupied mode, such as at the beginning of a normal workday. The control scenario of opening the building on a regularly scheduled workday is initiated by an employee presenting a proximity card at either building entrance. **See Figure 11-5.** Opening scenarios on non-workdays, as determined by calendar schedule in the building automation system, may behave differently.

When the proximity card is read, the access control system detects the access request of an employee that works on the second floor. The card access system compares the credential to the access control database. Upon authorization of the request, the building automation system executes actions to make the building ready for occupancy. This scenario affects the lighting, HVAC, plumbing, access control, security, and elevator systems.

Opening the Building on a Regularly-Scheduled Workday

Figure 11-5. The action of a person requesting access to the building at the beginning of a regularly scheduled workday initiates changes in the building systems in order to prepare for occupancy.

Lighting System Response

The lighting system takes several actions to ensure adequate lighting for the occupant entering the building. This includes lighting in general work areas plus additional areas specific to the employee's location, such as the following:

- Interior lighting in the common areas of the respective employee's floor (second floor) is turned on and will remain on until shut off by the building closing scenario.
- Interior lighting in the work areas, in both enclosed office and open office areas, is controlled with occupancy sensors and turns off after a time delay. Alternatively, the employee can override the occupancy sensors with a local override switch.
- If the employee enters an unassigned area, such as a conference room, lighting is controlled by occupancy sensors and turns off again after a time delay.
- Interior work areas near the building perimeter use daylight harvesting and dimmable ballasts.
- Interior life safety lighting is turned on and stays on until shut off by the building closing scenario.

All lighting actions are determined by programmed schedules of times of day and days of week. For example, if an employee enters the building at 3 AM, the lighting system actions are based on programming for unoccupied periods, and lighting in each area is controlled with occupancy sensors. Outdoor lighting is controlled by anticipated occupancy schedules and light level sensors that may turn the site lighting on for dark mornings before the expected arrival of the first employee.

LonWorks Lighting System Implementation. The interior lighting system includes dimmable fluorescent fixtures operated by LonWorks lighting controllers. Common area fixtures are energized from a central lighting control panel that also acts as a LonWorks node. **See Figure 11-6.** Individual offices use a multifunction LonWorks node that provides occupancy detection and temperature and ambient light level measurement. A manual override switch is used to bypass the timed off function.

Nodes involved with life safety systems and demand-limiting functions also share data with lighting nodes. This is needed for performing energy-reduction strategies while maintaining safety and security.

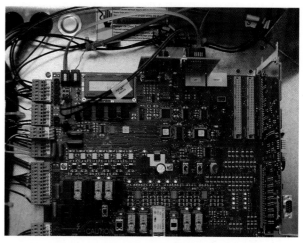

Elevator control panels may provide connection terminals for integration with a building automation system.

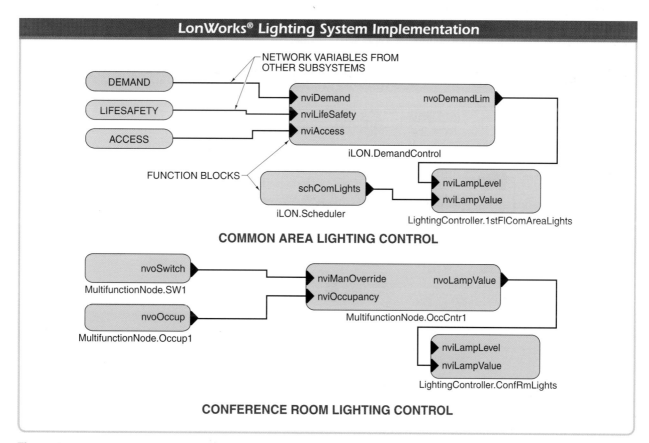

Figure 11-6. A LonWorks®-controlled system controls lighting according to schedules, electrical demand, access, occupancy, and life safety network data.

BACnet Lighting System Implementation. The interior lighting system includes dimmable fluorescent fixtures operated by BACnet-based lighting controllers. Common area fixtures are energized from a central lighting control panel that also acts as a BACnet node. **See Figure 11-7.** Individual offices use a multifunction BACnet node that provides occupancy detection and temperature and ambient light level measurement. A manual override switch is used to bypass the timed off function.

The first floor and second floor lighting control panels turn on the common area lighting and emergency lighting. When the employee gains access to the second floor, the access control system sends a signal to the second floor lighting control panel. The lighting control panel turns on the lighting in the open office area on the second floor.

Nodes involved with life safety systems and demand-limiting functions also share data with lighting nodes. This is needed for performing energy-reduction strategies while maintaining safety and security.

HVAC System Response

The control of the HVAC system at the beginning of an occupancy period assumes that temperature setpoints are at setback levels during unoccupied times as an energy-conservation measure and that a preheating/precooling sequence in anticipation of occupancy has not already been initiated via schedule.

Both central and employee-assigned HVAC systems are initiated and stay on until shut off by the closing scenario. During scheduled unoccupied periods, occupancy sensors are used to indicate occupancy, which is used to turn on the HVAC system equipment serving the specific occupied area.

Some systems use a standby temperature mode that is programmed to respond to the occupancy sensor when a zone is unoccupied during a normally occupied period. The system can establish a new temperature setpoint between the normal occupied and unoccupied setpoints. This allows for energy savings by adjusting the setpoint when temperature control is not needed. It also allows for quick response when the zone becomes occupied.

LonWorks HVAC System Implementation. The HVAC subsystems include a rooftop mechanical plant and variable-air-volume (VAV) terminals for each floor. LonWorks nodes from nine different manufacturers share information on the network in order to perform their local applications and the specified system sequences of operations.

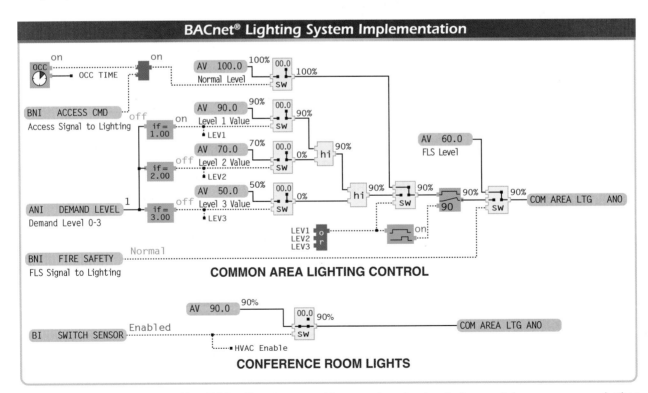

Figure 11-7. Lighting control with a BACnet® system uses binary and analog inputs from switches, sensors, and other building systems to determine the output of common area and individual room lighting.

LonWorks VAV controllers for each HVAC zone receive space-temperature values and occupancy status from a ceiling-mounted multisensor node. **See Figure 11-8.** Carbon dioxide levels from an indoor air quality node are also sent to VAV controllers in order to implement demand-controlled ventilation sequences. The rooftop mechanical plant includes a LonWorks programmable node that adjusts fan speed based on static pressure, mechanical cooling stages, outside-air temperature, and hot water boiler operations.

The programmable plant controller is bound to each VAV zone controller's terminal load output, which indicates the level of cooling or heating required. The plant controller then resets the supply air temperature according to the zone of greatest demand and adjusts outside-air dampers in response to indoor carbon dioxide levels.

Alarm contacts in proprietary fire and smoke detection devices in life safety systems provide system shutdown and smoke evacuation during alarm events.

During peak energy conditions, a network variable from the demand-limiting subsystem initiates a load-reduction sequence that includes increasing setpoints for individual zones and reducing supply-fan speed.

BACnet HVAC System Implementation. The HVAC subsystems include a rooftop mechanical plant and VAV terminals for each floor. BACnet nodes share information on the network in order to perform their local applications and the specified system sequences of operations.

BACnet VAV controllers for each HVAC zone receive space-temperature values and occupancy status from a ceiling-mounted multisensor node. **See Figure 11-9.** Carbon dioxide levels from an indoor air quality node are also sent to VAV controllers in order to implement demand-controlled ventilation sequences. The rooftop mechanical plant includes a BACnet programmable node that adjusts fan speed based on static pressure, mechanical cooling stages, outside-air temperature, and hot water boiler operations.

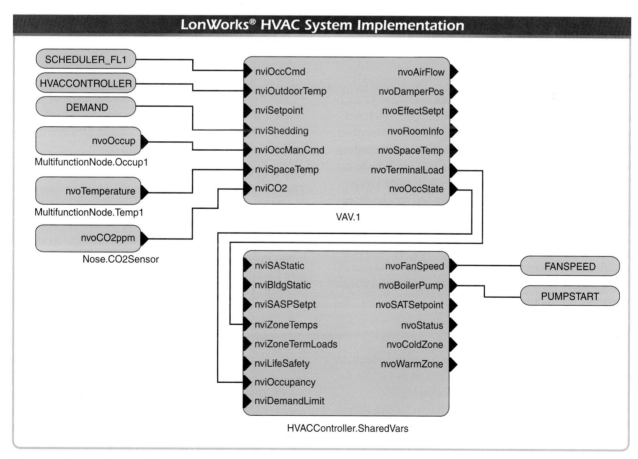

Figure 11-8. LonWorks® HVAC control integration includes occupancy, temperature, indoor air quality, and demand limit variables as inputs into function blocks governing central plant controllers.

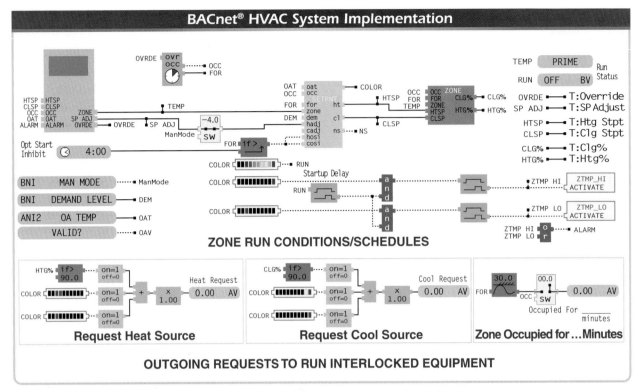

Figure 11-9. VAV logic can be programmed with BACnet® objects and their properties.

The programmable plant controller is bound to each VAV zone controller's terminal load output, which indicates the level of cooling or heating required. The plant controller then resets supply air temperature according to the zone of greatest demand and adjusts outside-air dampers in response to indoor carbon dioxide levels.

Alarm contacts in proprietary fire and smoke detection devices in life safety systems provide system shutdown and smoke evacuation during alarm events. During peak energy conditions, a signal from the demand-limiting subsystem initiates a load-reduction sequence that includes increasing setpoints for individual zones and reducing supply-fan speed.

Plumbing System Response

The plumbing system takes actions to ensure adequate water supply for the plumbing fixtures in use while the building is occupied, such as the following:

- The electrical circuit serving the hot water circulating pump is switched on and stays on until shut off by the closing scenario.
- The lift station pump package is fully operational 24/7/365 with self-contained controls.

- The fire protection system provides a signal that is used to initiate an emergency mode, which shuts down gas-fired boilers in the event of a fire alarm. Other devices may also be shut down.

LonWorks Plumbing System Implementation. The plumbing subsystems use general-purpose LonWorks input/output nodes as well as analog control and schedule functions provided by the network Internet server. **See Figure 11-10.** Transducers wired to analog inputs in the node measure loop water pressure and energize the booster pump when the pressure drops below a setpoint. The scheduler controls equipment operations, including pumps and domestic hot water boilers, based on expected building occupancy.

BACnet Plumbing System Implementation. The plumbing subsystems use general-purpose BACnet input/output nodes as well as analog control and schedule functions. **See Figure 11-11.** Transducers wired to analog inputs in the node measure loop water pressure and energize the booster pump when the pressure drops below a setpoint. The scheduler provides time-of-day control of equipment operations, including pumps and domestic hot water boilers.

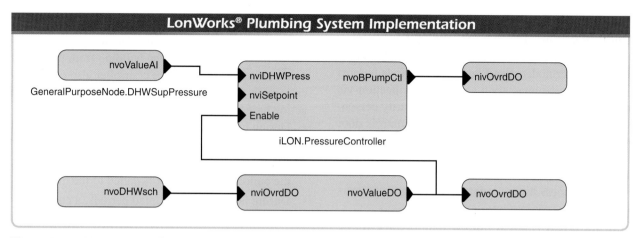

Figure 11-10. The LonWorks® implementation of the plumbing system automation uses a general-purpose controller node, interacting with an Internet service, to control when pumps and the boiler are turned on.

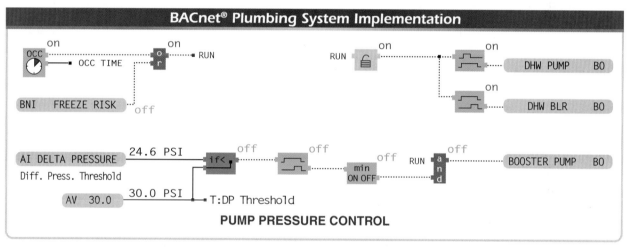

Figure 11-11. Plumbing system control with a BACnet®-based system determines when to activate pumps and the hot water boiler.

Access Control and Security System Responses

The security system receives information from the access control system and scheduler. The access control and security systems can be a combination system covering both functions, but they may have the ability to act as two individual stand-alone systems. Access control and security system responses are as follows:

- The employee's credential is authenticated by the access control system and allows the door to open. The access control panel compares the access event to the programmed schedules and transmits information to the security system.
- The security system turns off the alarm partition of the common zone and the occupant's assigned area.

- If there is a network communication problem, the security system remains armed. The door is allowed to open, but the security system panel just inside the entrance annunciates that the security system is still armed, which requires a code to be entered during the time-delay period.

Facts

Building automation control systems often include exercise functions. When the exercise function of a hydronic heating system is enabled, the control ensures that each pump of the system is operated at least once during a specified time period, such as every few hours, days, or weeks.

LonWorks Access Control and Security System Implementation. The security system uses information from LonWorks occupancy sensors, card readers, and scheduler nodes. **See Figure 11-12.** Node outputs include alarm annunciation and event notification via cell phone text messaging or email to security personnel. A LonWorks access control system controls door entries, detects unauthorized entries, and provides network data used to initiate building opening events in HVAC, plumbing, and lighting systems. Access-control software records access events and manages the users of the door access system.

BACnet Access Control and Security System Implementation. The security system uses information from BACnet-based occupancy sensors and scheduler nodes. Node outputs include alarm annunciation and event notification via cell phone text messaging or email to security personnel. A card access control system with a BACnet gateway controls door entries, detects unauthorized entries, and provides network data used to initiate building opening events in HVAC, plumbing, and lighting systems. Access-control software records access events and manages the users of the door access system.

Elevator System Response

The elevator system receives information about the identity of the person entering the building from the access control system. The elevator controller also communicates with the scheduler to determine the appropriate action based on the programmed schedule as follows:

- During normal occupied periods, the access control system enables unrestricted use of the elevator.
- When an employee assigned to an upper floor enters the building, the elevator is called to the first floor.
- During normally unoccupied periods, a card reader located in the elevator restricts elevator operation to authorized employees. When authenticated, the access control system then enables the elevator call button for the floor assigned to the employee.

LonWorks Elevator System Implementation. A LonWorks node integrates with the access control system and scheduler and interfaces with the proprietary elevator control system. Dry relay contacts are used to enable/disable call buttons according to access events and building schedules. Access-control software records access events and manages the users of the elevator access system.

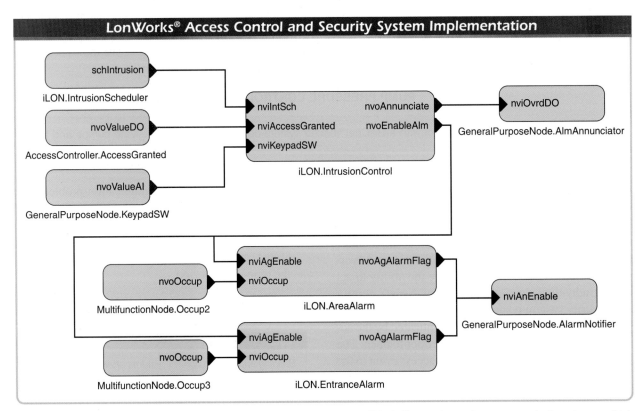

Figure 11-12. Occupancy sensors also provide information to the LonWorks® security and access control systems, which initiate alarms when intrusion is detected.

BACnet Elevator System Implementation. A BACnet node integrates with the access control system and scheduler and interfaces with the proprietary elevator control system. Dry relay contacts are used to enable/disable call buttons according to access events and building schedules. Access-control software records access events and manages the users of the elevator access system.

CONTROL SCENARIO: DEMAND LIMITING

Independent of the utility electric meter, a facility energy meter measures electrical demand and consumption for monitoring purposes. These electrical demand values are calculated over a 15-minute sliding window and evaluated according to the utility's time-of-day demand rate schedules. During peak periods, demand-limiting strategies are initiated if electrical demand is at or above a setpoint.

When the building automation system receives the signal to initiate the demand-limiting sequence, certain noncritical loads are commanded to turn off or reduce their duty cycle. This consists of changes to lighting and HVAC system operations, such as dimming light fixtures, adjusting temperature setpoints, and reducing supply-fan speed, in order to reduce the overall electrical demand of the building.

LonWorks Demand Limiting Implementation

A LonWorks electrical meter node measures electrical power demand and consumption in parallel with the utility electric meter. The node also measures and reports power-quality parameters such as voltage, current, frequency, and power factor for each phase. The meter node calculates average demand over a sliding 15-minute time window and shares this information with a LonWorks-programmable node. **See Figure 11-13.** If the demand is at or above the setpoint, the node initiates a demand-limiting sequence by instructing HVAC, lighting, and plumbing control devices to turn off or reduce their duty cycle. Energy consumption is logged and compared with utility billing for accuracy.

> **Facts**
>
> Lighting in certain areas may be dimmed or turned off for demand-limiting purposes. However, in the event of a life safety alarm, such as a fire alarm, all lighting is restored to its full level to aid building evacuation.

BACnet Demand Limiting Implementation

An electrical meter node measures electrical demand and consumption in parallel with the utility electric meter. The meter also measures and reports power-quality parameters such as voltage, current, frequency, and power factor for each phase. The meter transmits this information to a BACnet node via the Modbus RTU protocol. The BACnet node calculates the average demand over a sliding 15-minute time window and compares it to the setpoint. If the demand is at or above the setpoint, the node initiates a demand-limiting sequence by instructing HVAC, lighting, and plumbing control devices to turn off or reduce their duty cycle. Energy consumption is logged and compared with utility billing for accuracy.

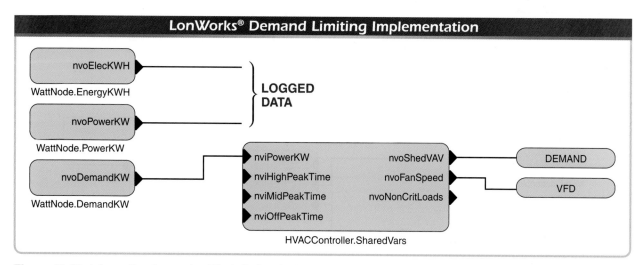

Figure 11-13. Information from a LonWorks® electrical meter node is either logged or shared with a function block that determines when to initiate demand limiting.

Lighting System Response

When it receives a signal to initiate demand limiting, the lighting system takes several actions to reduce the lighting levels in several areas of the building. The scheduler's programmed information may be overridden to accomplish some of the demand limiting. Lighting system responses are as follows:

• Life-safety-designated lighting remains on.

• Lighting levels in the interior common areas are reduced but stay on until shut off by the closing scenario.

• Lighting in the work areas, in both enclosed and open office areas, is controlled with occupancy sensors and turn off after time delay.

• The lighting control system overrides the daylighting sequence to lower the target lighting levels.

• If lights have been enabled and demand limiting becomes active, lighting levels are reduced to one of three different presets. Abrupt changes to lighting levels are avoided via a ramping function.

LonWorks Lighting Demand Limiting Implementation. LonWorks-enabled dimmer controllers are used to reduce lighting levels. Upon receiving a signal to reduce electrical demand, LonWorks lighting nodes lower lighting energy usage while maintaining high levels of safety and security. **See Figure 11-14.**

BACnet Lighting Demand Limiting Implementation. BACnet-enabled dimmer controllers are used to reduce lighting levels. Upon receiving a signal to reduce electrical demand, BACnet lighting nodes lower lighting energy usage while maintaining high levels of safety and security.

HVAC System Response

When it receives a signal to initiate demand limiting, the HVAC system takes several actions to reduce the heating/cooling loads in several areas of the building by adjusting the zone temperature setpoints. HVAC system responses are as follows:

• Every thermostat setpoint is raised (when in cooling mode) or lowered (when in heating mode) by 2°F.

• Occupancy sensors are used to raise or lower setpoints as much as 5°F in areas where occupancy has not been detected for a minimum period of time.

LonWorks HVAC Demand Limiting Implementation. Upon receiving a signal to reduce demand, VAV terminal devices adjust heating/cooling setpoints accordingly to reduce energy consumption. Through a browser-based user interface, individual occupants can voluntarily participate in additional demand-limiting measures by specifying an acceptable setpoint offset during periods of high energy demand.

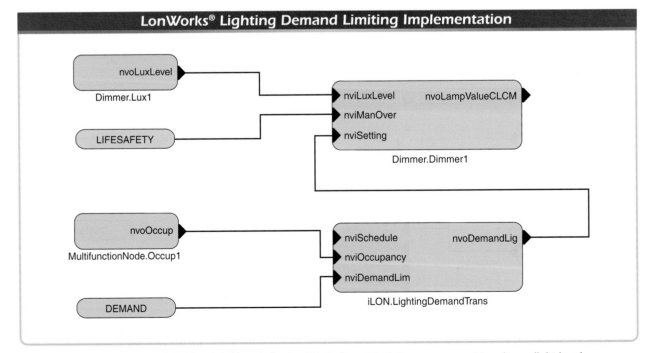

Figure 11-14. When demand limiting is initiated, the LonWorks®-enabled dimmers are set to a lower light level.

BACnet HVAC Demand Limiting Implementation.
Upon receiving a signal to reduce demand, VAV terminal devices adjust heating/cooling setpoints accordingly to reduce energy consumption. **See Figure 11-15.** Through a browser-based user interface, individual occupants can voluntarily participate in additional demand-limiting measures by specifying an acceptable setpoint offset during periods of high energy demand.

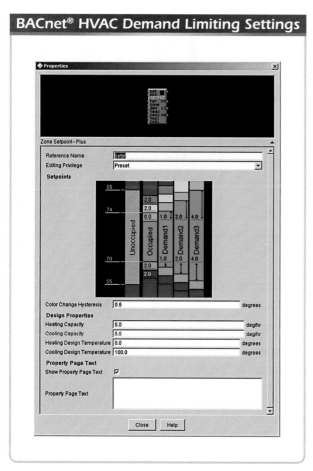

Figure 11-15. HVAC system setpoints for different demand-limiting scenarios can be set with BACnet® system software.

CROSS-PROTOCOL IMPLEMENTATIONS

A building automation system based on a single protocol is usually the best solution, but it is not always possible or practical. For example, an older building may have had systems installed or upgraded at different times and different protocols may have been used. It is often necessary to design control systems using a mixture of LonWorks, BACnet, and even proprietary protocol devices.

As an example of one cross-protocol integration solution, a control scenario for the Lincoln Publishing Group building with multi-protocol requirements is explained below. The resulting design is a commonly implemented solution utilizing industry-available products. This example outlines some of the strategies and implications of cross-protocol integration.

> **Facts**
>
> *The opening of a building at the beginning of a workday is often programmed for a certain time. However, many occupants arrive early, and their entry may be used to trigger the building opening sequence.*

System Description

The system and building requirements are identical to those in the previous examples, except for the following:
- The lighting system uses BACnet nodes.
- The HVAC system uses LonWorks nodes.
- The plumbing system uses Modbus TCP nodes.
- The fire protection system operates on a separate proprietary network for life safety reliability reasons, but it can share information with the building automation network via a BACnet/IP integration board.
- The security system uses a VYKON® Security controller.
- The access control system uses a VYKON Security controller.
- The elevator system operates on a separate network, but it can share information with the building automation network via an integration board.

VYKON is a suite of Niagara-based products designed to integrate networked control devices into a unified, Internet-enabled, web-based system. VYKON controllers integrate LonWorks, BACnet, Modbus, oBIX, the Internet, and web services protocols. They include network management tools to support the design, configuration, installation, and maintenance of interoperable networks. **See Figure 11-16.**

The VYKON JACE® (Java Application Control Engine) is a line of controller/servers that combines control, supervision, data logging, alarming, scheduling, and network management functions. These devices can control and manage external control devices over the Internet and present real-time information to users within web-based graphical interfaces. Optional input/output modules can be added for local control applications.

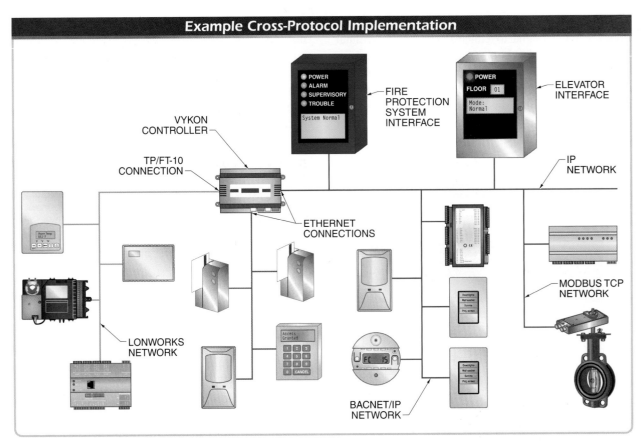

Figure 11-16. An example cross-protocol implementation may use intermediary framework controllers to connect separate networks using different building automation protocols.

VYKON Security is a security management solution that integrates with a building automation system to enable any building system to react to access events and alarm conditions. A Security JACE is used to integrate many of the building systems in the example building. The device includes several connections for integrating networks of different protocols. A TP/FT-10 network port is connected to the LonWorks network for the HVAC equipment. A primary Ethernet port is connected to operator workstations or additional JACE devices, which somewhat isolates the critical security and access control systems. A secondary Ethernet port is connected to the building's IP infrastructure that hosts BACnet/IP and Modbus communication. This integrates many of the other building systems into the overall system.

Control Scenario: Opening the Building on a Regularly Scheduled Workday

An employee swiping a proximity card at either building entrance initiates the control scenario of opening the building on a regularly scheduled workday.

Opening scenarios on non-workdays, as determined by calendar schedule in the building automation system, may occur differently.

The access control system detects the access request of the employee and, upon authorization of the request, shares the identity of the employee with other building systems. The building systems react in various ways to this event.

Electrical System. The electrical system receives no signals from this event. The electrical system is in standard operating mode, receiving electricity from the utility's service entrance.

Lighting System. The lighting system takes several actions to ensure adequate lighting for the occupant entering the building. This includes lighting in general work areas and the employee's assigned location. The Security JACE controller sends commands to the lighting system via BACnet/IP messages. The lighting controllers interact with the scheduler to determine the appropriate actions as follows:

• Interior life safety lighting is turned on and stays on until shut off by the building's closing scenario.

- If not already turned on by system schedules (within the JACE controller), the interior lighting in the common areas of the employee's assigned floor is turned on and remains on until shut off by the building closing scenario.
- Interior lighting in other work areas, both enclosed offices and open office areas, is controlled with occupancy sensors and turns off after a time-delay period. Alternatively, the employee can override the occupancy sensors with a local override switch.
- Work areas near the building perimeter use daylight harvesting and dimmable ballasts.

HVAC System. The Security JACE controller uses schedules to automatically adjust heating and cooling setpoints during unoccupied periods. The Security JACE controller uses control algorithms to determine the optimal start time for the HVAC equipment based on current space temperatures, setpoints, scheduled event times, and dynamically adjusted recovery rates.

If the employee accesses the building prior to the optimized start event, the Security JACE controller commands the HVAC equipment via the LonWorks network to initiate both common-area and employee-assigned HVAC systems. These systems stay on until shut off by the closing scenario.

Plumbing System. The plumbing system takes several actions to ensure adequate water supply for the plumbing fixtures in use while the building is occupied. The Security JACE controller uses internal control programming and remote input/output modules for controlling the plumbing subsystem as follows:
- The electrical circuit serving the electric water heater, heating elements, and their self-contained controls is switched on and stays on until shut off by the closing scenario.
- The electrical circuit serving the hot water circulating pump is switched on and stays on until shut off by the closing scenario.
- The pressure-boosting pump package controller is enabled and stays on until shut off by the closing scenario.

Fire Protection System. The fire protection system receives no signals from this event. All fire detection, alarm notification, fire protection system monitoring, and fire safety function subsystems are fully operational 24/7/365.

Security System. The Security JACE controller closely integrates security and access control functions. When an intrusion is detected, the Security JACE controller is programmed to call the police, fire department, building security, or other authorities. Additionally, remote input/output connections are used to activate sirens and strobes. The security system receives information through an interface with the access control system, since both of these critical systems are networked separately from the rest of the building automation system. Security system responses are as follows:
- The alarm partitions for the common areas and the occupant's assigned floor are disabled.
- If the network connection is not available, the alarm system remains armed. The door is allowed to open, but the security panel just inside the entrance annunciates the armed status, requiring an additional disarming code to be entered during a time-delay period.

Access Control System. During normal occupied periods, the Security JACE controller signals the credential reader module to unlock the exterior doors. If accessed during the normally unoccupied period, the Security JACE controller signals the door to be unlocked for the configured unlock period. The Security JACE controller logs the access control event based on the outcome as follows:
- If the door is opened, an "access granted" event is logged.
- If the door is not opened, an "access granted, but not used" event is logged.
- If the door does not properly close, the Security JACE controller initiates a door-held-open alarm.

VDV System. The voice (telephony) and data portions of the system take no actions from this event. These subsystems are active 24/7/365. The video surveillance portion of the system takes no additional actions beyond those defined in the scenario controlling activation of the site when someone enters onto the property.

Elevator System. During normal occupied periods, the Security JACE controller enables unrestricted use of the elevator. When an employee assigned to an upper floor enters the building, the elevator is called to the first floor. During normally unoccupied periods, a proximity reader located in the elevator restricts elevator operation to authorized employees. When authenticated, the Security JACE controller, through distributed input/output modules, enables the applicable elevator call buttons.

REVIEW QUESTIONS

1. What types of automation applications can be integrated with a building automation system?

2. Can an effective building automation system be implemented to meet the requirements of various building systems? Briefly explain.

3. How is a protocol system typically chosen?

4. In the example building scenario, how is the opening of the building initiated on a typical workday?

5. In the example building scenario, what is the HVAC system's response to the opening of the building?

6. In the example building scenario, what is the elevator system's response to the opening of the building?

7. In the example building scenario, how is the demand-limiting sequence initiated?

8. In the example building scenario, what is the HVAC system's response when it receives a signal to initiate demand-limiting?

9. Why might cross-protocol integration be required?

10. In the example building scenario, how is the opening scenario with cross-protocol implementation initiated and how do some of the building systems respond?

Chapter 11 Review and Resources

Trends in Building Automation **12**

The building automation industry is a relatively new participant among the building system trades, especially the portion dealing with electronic controls and networking technologies. Therefore, the industry is still maturing in a climate where building controls are becoming increasingly important, both for new construction and existing buildings. Industry watchers notice a number of trends from the past several years that are expected to influence the continuing evolution of building automation in the near future.

INDUSTRY TRENDS

The building automation industry supports a number of companies and organizations involved in choosing, specifying, designing, integrating, maintaining, and otherwise working with building automation systems. Changes in the building automation industry have affected the roles of these industry members and how the systems are implemented.

Integrated Building Automation Teams

The traditional building automation team includes individuals and groups that take the requirements, concerns, and ideas for a new or existing building and create an automated building. Traditional teams include building owners, architects, consulting-specifying engineers, and controls and electrical contractors. Building officials often are required to approve the plans before construction begins. As building automation systems become more integrated, the roles and responsibilities of the traditional team members change and new members are added. Newer teams include systems integrators and information technology staff.

Building Owners. Building owners decide which building automation strategies to implement based on the desired results. The concerns of building owners include initial costs, return on investment, maintenance, impact on occupants, system flexibility, system upgradeability, and environmental impact. These factors are complicated when retrofitting an existing building that may be partially occupied because of the impact of renovations on the operations of existing tenants.

Architects. An *architect* is a professional who designs and draws plans for a structure. The architect determines the parameters of the necessary building space based on property size and shape, desired interior and exterior spaces, and available project budget. Additional items discussed with the owner include intended usage, necessary life span of the structure, and design and style preferences. Materials, equipment, finishes, and fixtures are also discussed. The architect also works with the owner and planners to ensure zoning requirements and building codes are met.

Consulting-Specifying Engineers. A consulting-specifying engineer defines the scope of a building automation system from the owner's list of desired features and coordinates these desires with the architectural, mechanical, and electrical systems requirements. The engineer produces the contract documents used by the controls contractor to bid on a project. The engineer then reviews and approves the shop drawing prepared by the controls contractor to ensure that the intent of the contract documents has been met. The engineer also reviews and approves the project closeout information, such as commissioning reports.

Controls and Electrical Contractors. Previously, in traditional installations, mechanical engineering contractors were responsible for controls involving HVAC, plumbing, and piping systems. Electrical contractors were responsible for controls involving power distribution, emergency power generation, fire protection, and lighting. There was little coordination between the two disciplines, or even among elements of the same discipline, because everything stood alone in the bid documents. Now, with integrated building controls, both engineering disciplines must work more closely together and produce a coordination document, in addition to the stand-alone components. The coordination document details various automated operational scenarios and how each independent system participates in and responds to these scenarios.

This type of coordination document is not recognized by the Construction Specifications Institute's (CSI's) MasterFormat® standard specification format or ARCOM MasterSpec® software, which put building elements in distinct categories for the convenience of general contractors and construction managers. Therefore, contractors continue to define systems in distinct categories and also reference the requirements of the coordination document. This requires the general contractor to bundle all of the affected sections into one bid package to be assigned to the systems integrator or master systems integrator.

Access control systems are used to allow personnel to restrict access to specific areas of a building.

Building Officials. After the drawings, prints, and specifications have been reviewed and approved by the owner, architect, and engineers, they are submitted to the appropriate governmental agencies that have jurisdiction for a particular geographic location. Building officials review the construction documents to confirm that they are in compliance with all local building codes and zoning regulations.

Systems Integrators. Controls contracting has traditionally been considered a subset of mechanical work, especially since HVAC systems are the most commonly automated building systems. Electrical contractors used to handle the remainder of the low-voltage building systems such as fire detection and lighting control. However, with systems integrated together in automated buildings, a master systems integrator is needed. Just as the mechanical and electrical design engineer work together to produce a coordination document, the master systems integrator works with both the controls contractor and the electrical contractor to create an integrated system.

Integrators are individuals and companies who engineer, implement, and maintain building automation systems, especially those using control devices from multiple vendors. They select the appropriate network architectures, choose control devices, perform installation tasks, program the network, and provide network maintenance services.

It must be noted that each project team will likely approach the coordination of their system's integration in a unique way due to the many factors involved in each scenario. For example, in retrofit installations, the owner may decide to work directly with a systems integrator and bypass the coordination provided by a consulting-specifying engineer because the other system components are already in place. The simplest solution is having an electrical contractor install both low-voltage and medium-voltage equipment and the low-voltage control systems.

Information Technology Staff. In traditional installations, facility managers were responsible for the mechanical and electrical systems without much technical input from other areas of the organization. But with the ability to connect building automation system controllers to the building's IP network, the information technology (IT) staff is becoming involved in the decision-making process. Input from IT staff is needed to manage the integration of data communication networks and ensure reliable operation afterward.

Building Information Modeling (BIM)

Contractors should use standardized methods of incorporating all drawings into their plans. Building information modeling (BIM) is an integrated, electronically managed system that aligns all working drawings, structural drawings, and shop drawings into a consistent system. BIM, also known as virtual design and construction (VDC), allows for integration of CAD-generated documents or building models from various sources to create a complete set of construction documents.

Contractors, subcontractors, and suppliers can all import the electronically produced construction documents or models into a common system that recognizes all of the individual construction components in a fully integrated platform. Prints are generated from the integrated model that takes all construction components into account. BIM technology also allows for the integration of estimating and scheduling software. **See Figure 12-1.**

Controls Vendors

The product lines of controls vendors vary greatly, as do the needs of building owners looking to implement building automation. The choice of controls vendors involves matching their most appropriate products with the building specifications, client expectations, budget, and other considerations. The building automation industry has responded to the needs of clients by presenting a choice of delivery options.

One-Stop Shops. A one-stop shop is a vendor that markets most of the control devices and equipment that a building owner needs. For example, some controls vendors have working agreements with fire detection manufacturers who manufacturer their own security and access control systems. Other vendors have developed a "super vendor" approach for markets such as health care. In this instance, the vendor sells everything to a hospital client, including MRIs, switchgear, and building controls.

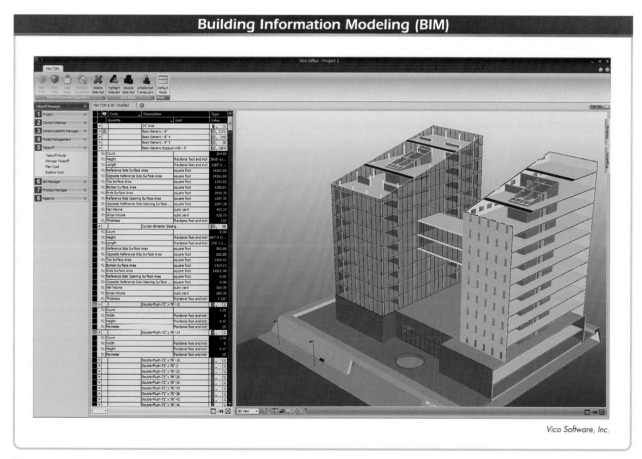

Building Information Modeling (BIM)

Vico Software, Inc.

Figure 12-1. BIM documents contain the geometric information needed to build a structure and may include particular construction details, job costs, and scheduling data.

A variation on the one-stop shop strategy is the inclusion of factory-mounted building controls on traditional building equipment. For example, an HVAC equipment manufacturer can put controls on a package air-handling unit so that the unit is ready for service immediately after connecting power and the communication network.

Best of Breed. The best of breed strategy takes advantage of the individual strengths of various controls vendors to design an all-around robust system. One vendor may produce particularly good HVAC controls but may be weak in access control equipment. A different vendor may have the ideal access control devices for the building owner's needs. The owner or contractor chooses each major component individually and then a systems integrator connects the individual components together to implement an operational scenario. With the interoperability of open protocols, this is a feasible strategy that can make use of the best or most appropriate devices in each building system.

Global Differences in Controls Incentives

The capabilities of building automation systems in the United States are not very different from those in other countries, especially since most controls manufacturers are multinational companies that market their products worldwide. However, building automation systems in other countries differ in the extent of implementation and integration. They are more common and typically achieve higher levels of integration by bringing together more building systems.

The biggest influence on this disparity is the desire for greater energy efficiency. This is generally not driven by the cost of energy, which is determined at the global level, but by government energy policy. For example, German utilities are required by law to buy back excess energy from renewable energy sources at prices that can be more than twice the retail rate. In this situation, it makes sense to add additional control system features that save energy because the payback profile is substantially different. The United States does not currently have any similar federal laws or incentives that encourage this trend.

NETWORKING TRENDS

Trends in networking technologies are aimed at making it easier to transmit control information quickly and reliably. Factors that influence these characteristics are the media, addressing and routing schemes, and network traffic volume. Open protocols are developed with certain network types in mind, though they are subject to addenda that expand their supported types. Improvements are also being made for migrating control data onto existing networking technologies in ways that do not degrade the performance of either system.

Using Existing IP Infrastructures

A master system integrator works with a building's IT department to integrate a building automation network with a data network and IP infrastructure. This network integration is not necessary, but it is becoming more common.

An IP infrastructure is typically already available throughout a building and may have excess message traffic capacity. This intranet allows control data to be distributed quickly and reliably throughout a building. For sharing data between separate campus buildings, or even between remote locations, an Internet connection provides access to a global information network. This requires security measures for both obscuring the data (so that no unauthorized person can listen in) and ensuring that the data is transmitted unaltered (ensuring that the information sent is the information received). The availability of data via IP networks helps integrators fine-tune automation systems for efficient and effective control.

Many industries have a geographic structure that is ideal for adopting the Internet as a core part of their control systems. Examples include business chains, manufacturers whose parts are constructed in different areas of the country and world, and property-management or ownership firms that monitor buildings in different cities from a central location.

Wireless Networking Technologies

The growth of wireless networking technologies is one of the most significant trends in electronics. Wireless networks have become options for building automation, both as LAN choices for traditionally wired protocols and as the sole transport technology for wireless protocols. **See Figure 12-2.** However, wireless networks have still not made strong inroads in building automation. Wireless networking has always been considered a specialty technology, or at least limited in application. Several factors have kept wireless networks from capturing a large share of the building automation market, but as improvements in technologies mitigate these limitations, the future of wireless networks in building automation may improve.

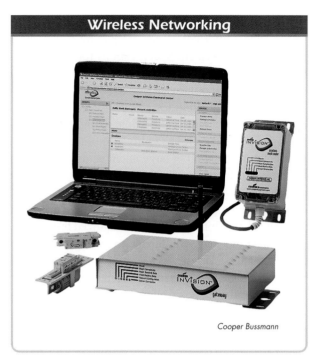

Cooper Bussmann

Figure 12-2. Wireless systems may be particularly suited for certain applications, such as monitoring of maintenance components that cannot be connected by physical media.

Power Sources. A large part of the problem with wireless networks is sourcing power for the control device nodes. In the past, the power limitations and large size of NiCad batteries were too significant for the power-demanding signaling technology.

However, newer battery compositions have allowed for smaller and longer-lasting power sources. The signaling methods have also improved, allowing the nodes to "sleep" (power down to a minimal operating state) when not sending and "wake" upon receiving a certain signal. Additionally, mesh networking (the ability of nodes to relay signals for other nodes) reduces the radio power output needed to deliver a message from one end of a building to the other. These advancements have greatly improved the lifetime of nodes between battery charges or replacements.

Further, there are energy-harvesting technologies that allow for nodes that do not require batteries. *Energy harvesting* is a strategy where a device obtains power from its surrounding environment. Energy-harvesting technologies can use mechanical actions, vibrations, light, thermal gradients, or other energy sources in their areas to operate their electronics or charge short-term batteries.

Interference. There are many concerns about the reliability of wireless sensors and actuators. Wireless networks are vulnerable to disruptions from a variety of electromagnetic sources. Accidental disruptions are caused by stray radio waves in the same frequency, perhaps due to new equipment or transients from people and objects. Network disruption may also be caused by changes in the environment, such as the installation of foil-based wallpaper, the moving of large metallic equipment or furniture, or poorly isolated equipment power supplies. Awareness of the potential causes of wireless network interference contributes to improving the success rate of these implementations.

Security. A major concern for wireless networks is the possibility of intentional network disruptions. A person can intercept wireless signals and either attempt to jam network communications or gain control of nodes or systems. It may be possible for someone outside the facility to infiltrate the network. With a wireless access point, a physical connection to the network is not needed. Depending on the building systems involved, a hacker could cause serious equipment damage. Encryption, authentication, and other security measures are used to minimize the potential for security breaches. Common types of encryption include WEP, WPA, and WPA2. **See Figure 12-3.**

Encryption Comparisons		
Encryption Method	Description	Security
WEP	Weak key; can be hacked with low security knowledge	Poor
WPA	Same encryption as WEP with added authentification; can be hacked with high security knowledge	Good
WPA2	Highest level of security; unhackable with today's tools	Best

Figure 12-3. Different encryption methods have different levels of security.

Infrared. Most wireless nodes use radio frequency (RF) as the media, though this has disadvantages. Infrared (IR) is another, perhaps more secure, wireless platform. IR requires line-of-sight arrangement of communicating nodes, which limits its application to within certain walls. While this may be a disadvantage in most applications, it can be an advantage from an information security point of view. IR communication channels can be used to complement wired and RF-wireless channels for use in particular applications within small areas.

OPEN PROTOCOL TRENDS

Today, building automation installations with open protocols are still a small minority. The majority of installations are still using proprietary technologies, either completely closed or with available application programming interfaces (APIs). Increasingly, however, more controls specifications are demanding open protocols and specifying the limits of exceptions to that rule. **See Figure 12-4.** The various open protocols today will continue to expand while proprietary protocols become less popular. As each open protocol grapples for a larger customer base, the battle is more against proprietary systems than against other open protocols.

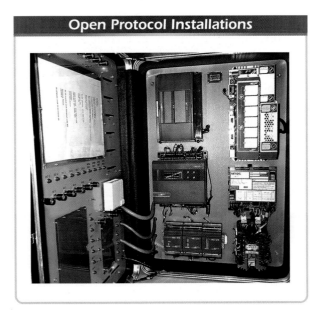

Figure 12-4. Modern building automation systems can use standard open protocols.

However, proprietary protocols will continue to exist and be installed in projects where specifications do not dictate otherwise. The often lower up-front costs may be more attractive to the building owner or construction contractor despite the long-term maintenance contracts, especially if the company constructing a facility is doing so for a different company.

Perceptions of Open Protocols

What makes a protocol "open" is subjective. For manufacturers, an open protocol allows them to focus on their core competency, which is developing products, rather than having to define the platform. The types of data structures used to carry information and the mechanisms by which that information gets from point A to point B are already established. Specifications for open protocols also allow the manufacturer to bid for jobs from which they might otherwise be excluded.

It is the specification for an open protocol that discourages a manufacturer from dominating the installation and maintenance of a facility's automation system. This counters the goal of large manufacturers, which is to involve their product lines in as much of the facility as possible. However, given an increased penetration of only semi-open products in a facility, which allow for specialized additions under the standardized framework, a manufacturer may still be able to lock a facility into a maintenance contract.

For integrators, an open protocol allows the integrator to bring together nodes from one or more manufacturers to create an interoperable system. The integrator is concerned with having the nodes share the same network and not only coexist, but interact intelligibly without additional programming. This keeps system complexity and integration costs low. Also, the increasing implementation of open protocols means that integrators must learn only a few open protocols, rather than many proprietary protocols. This requires fewer employees and less training and allows integrators to focus on actual job installations instead of the details of different product lines from multiple manufacturers.

A building owner's definition of an open protocol focuses on interchangeability of nodes on the network without the requirement for custom solutions by the manufacturer or the original systems integrator. This allows a variety of products from different manufacturers to be easily installed and integrated by a wide selection of integrators and supported by many maintenance organizations.

A building owner typically views an automation system as a long-term, capital asset. The potentially higher short-term costs and complexity of an open system are balanced with the long-term benefits: interoperability, choice of maintenance contractors, and future expandability without additional gateways, programming, or original-manufacturer interaction. However, building owners are rarely concerned with which protocol is used for the facility, as long as it is an open protocol.

IP Addressing Changes

When the present-day IP addressing scheme was invented, 4,228,250,625 (2^{32}) unique addresses seemed adequate for all future needs. However, large groups of IP addresses were assigned to organizations or companies in the early 1990s, making many addresses unavailable, even if unused. This also leaves large gaps in the numbering that prevents unique IP addresses from being doled out sequentially. For example, Apple, Inc. was granted the entire address space under 17.xxx.xxx.xxx, a total of 16,777,216 addresses. Many other companies, institutions, and organizations were granted similar blocks, known as "Class A" or "/8" addresses. Even if the numbers were sequentially assigned, there are not enough addresses for everyone in the world today.

This current IP addressing scheme, known as IP version 4 (IPv4), will eventually be replaced by IPv6, which has a far more equitable address distribution scheme and allows for 2^{128} unique addresses. This allows for more than 48,000,000,000,000,000,000,000,000,000 unique addresses per person (assuming a world population of about 7 billion). Just as this upgrade will significantly change computer-based networks, it will improve addressing opportunities for building automation nodes that also use IP.

However, the migration is complicated and will require years to complete. Support for IPv6 is being added to updates for computer operating systems, updates for new and existing home network routers, Internet service provider equipment, and major IP traffic routing systems. Upgraded equipment and systems are designed to support both IPv4 and IPv6 to facilitate a smooth transition sometime in the future.

LonWorks System Trends

Nodes in a LonWorks network are designed to interoperate and communicate in a peer-to-peer fashion without the need for a supervisory controller. Peer-to-peer communication is the strength of a LonWorks network and makes it a good choice for both small networks and commercial building integrated networks.

Standardization of data structures and LonMark functional profiles ensures a certain level of interoperability between products from multiple manufacturers. A node created today can communicate and interoperate with a node installed in the network a decade prior without the need for additional hardware or glue logic.

The LonTalk protocol continues to be enhanced using increased addressing options, improved IP tunneling standards, and the Interoperable Self-Installation protocol, which standardizes a method for installing nodes in a network without the use of a network management tool. Further, LonMark International continues to develop specifications to profile the functions of existing and emerging building systems, allowing LonWorks systems to be a platform for integrating systems together in a facility or campus.

BACnet System Trends

While BACnet nodes also have the ability to communicate peer-to-peer, their strength is in their programming flexibility, allowing manufacturers to assemble different data points in a node to achieve a particular functionality. The assembly is represented in a machine-readable format that allows a supervisory controller to integrate different nodes into the same network.

Through addenda, BACnet International continues to enhance the protocol by integrating new data structures and control features, including security and authentication. BACnet networking is also being expanded to enhance BACnet/IP accommodation of remote operator access through network address translation (NAT) firewalls and includes the ZigBee® protocol as a BACnet wireless data link layer.

Facts

For years there was no compatibility among computerized control system manufacturers. Then two standards emerged: BACnet® and LonWorks®. These industry standards allowed manufacturers to design their system protocols and connecting media for communication compatibility with other manufacturers' systems.

Wireless Networking Protocol Trends

Wireless networks are becoming more reliable. This trend will continue since batteries last longer and transceiver designs are getting smaller and more efficient. Implementation of wireless networks for building automation is expected to increase considerably. However, specific wireless networking protocols will not entirely replace the need for other transport media, including other wireless media. For example, a specific wireless networking protocol may be appropriate for integration and data sharing within a building, but a wired protocol with a longer range is typically needed for transmission between buildings in a campus setting.

Wireless networking protocols are also expected to continue growing in proximity service applications. Applications for radio-frequency identification (RFID) include asset tracking (including salable product, movable equipment, and even personnel tracking), proximity activation (the presence of something or someone causes another action to automatically occur), and proximity security (the allowance or restriction of persons or things based on their proximity). In this regard, wireless protocols are a connection between the physical world and the control network, which may use a very different building automation protocol.

Therefore, the future of wireless networking protocols is as a complement to other building automation protocols that use various media. The success of wireless protocols will benefit automation protocols as a media channel option but not as a complete building automation solution due to their media-dependent design.

Features Added to Existing Open Protocols

In many cases, technologies being developed to enhance the existing set of protocols are beyond the known desires of the user community. These features provide flexibility and control beyond what is contained in any buildings today, but they have not yet been identified as desirable features by the user community. Following the development of these protocol features is training and education supporting the need for such features. The burden is on integration and design firms to educate building owners about new features because it is ultimately the owners' knowledge about an operation and its control that drives a feature to become one of their requirements. It is the application of the available toolsets rather than the tools themselves that drives their implementation.

Media Availability. With the exception of strictly wireless platforms, open building automation protocols are beginning to offer a variety of media choices. The choice of media allows the physical extension of the network into building areas that might not otherwise be reachable. In addition to the typical copper conductor and wireless media, some of the popular protocols use fiber optic media, coaxial cable, power line mains, and infrared. Built-in support for multiple media types allows data transmission from one medium to another through routers without gateways or translators.

Also, the popular building automation protocols now add support for IP-based networks. This allows control data to leave dedicated control networks and be transmitted on IP-based networks, which are shared by personal computer and enterprise systems. Nearly all of the associations that maintain these protocols have embraced both the carrying of their messages over an IP-based technology (though perhaps limited to Ethernet in some cases) and the translation of their data structures to XML.

Integration of Multiple Building Systems. Building automation protocols with origins in HVAC applications are branching into lighting, security, elevator control, and energy management, the last of which is the latest important field in building automation. In many existing control devices, these systems are still proprietary and require low-level input/output connections, high-level gateway interfaces, or IP-based application programming interfaces to be integrated with other building systems. The integration of these applications will encourage development of new control products "native" to these protocols and bring together other distinct building systems. These protocols will be the most desired pathway for cross-system integration projects.

Advanced Cross-Protocol Integration. Both governmental and nongovernmental organizations are trying to meet mandated restrictions on carbon-output limits and energy consumption, which is creating the forum for supporters of different protocols to come together to reach these common goals. Much of the effort in cross-protocol integration is taking place outside the protocol-supporting associations. The competitive spirit between the protocol advocates and their continued desire for market penetration and dominance speeds the respective companies and organizations toward solving the cross-protocol issues as quickly as possible. Innovations that were once being driven by a push from inside the organizations are now being driven by mandates and parties outside those organizations.

The combined effects of organizational efforts and the energy goals will fuel cross-protocol integration at the application level, which will then cause the organizations to work more closely together at defining the interfaces for these new initiatives.

CONTROL STRATEGY TRENDS

Control strategies are used to maintain the desired indoor environment while operating the building systems in an appropriate and efficient manner. There are a number of capable control strategies that are chosen based on the particular requirements and conditions of a building and its occupants. There are also opportunities to improve control strategies in both application program algorithms and system equipment.

Individualized (Local) Controls

Individualized controls allow the occupants of a building to adjust setpoints themselves, usually within an allowable range of values, to create their own optimal indoor environment. Individualized control is typically used for HVAC systems, but it is also sometimes used for lighting systems. As occupants may have different perceptions of comfort, this strategy requires much finer control than most systems. For example, an occupant of one office may prefer a warmer temperature, while occupants at adjacent workstations may prefer cooler temperatures. Controls must also be user-friendly so that any occupant can use them effectively.

The turning point that really created an awareness of individualized control was the creation of a credit within the LEED® rating system that recognized the value of individual control for temperature and ventilation. The credit has two requirements: providing controls for at least half of the occupants to adjust their personal

environment for their own comfort and providing controls for all shared multi-occupant spaces to enable adjustments to suit group needs and preferences.

At the time, this was considered an aggressive mandate, and engineers and manufacturers tried to meet the requirements of this credit in a variety of ways. An increase in the number of air terminal units and thermostats proved to be cost prohibitive. A cost-effective variation of this technique is to place a temperature sensor in every space and then use an averaging sequence to provide the input to the controller.

Underfloor Air Distribution Systems. An underfloor air distribution system uses a raised floor as a plenum space to distribute conditioned air throughout a building space. **See Figure 12-5.** This system has the distinct advantage of placing cooling where it needs to be. Airflow into the occupied space is controlled by the placement and adjustment of floor diffusers. The use of floor diffusers makes it possible for occupants to manually control the amount of airflow into their personal space. However, the conditioning of the air is still controlled centrally with automated controls. The downside to such a system is that, because part of the system is manually controlled, it has a limited ability to set back when unoccupied.

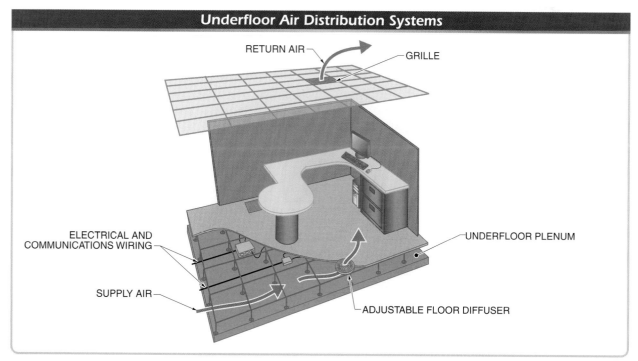

Figure 12-5. Underfloor air distribution systems allow individuals to adjust the flow of conditioned air into their personal space using accessible floor diffusers.

Integrated Furniture Comfort Control Systems. An integrated furniture comfort control system is a variation of the underfloor air distribution system concept. It uses ducts integrated into the occupant's furniture to channel, mix, and filter conditioned air from an underfloor (or wall or column) plenum into the occupant's personal space. **See Figure 12-6.** These systems often also control lighting and background noise. This system allows for a great deal of individualized control but requires significant up-front costs for the furniture-integrated equipment.

Software as a Service

Supervisory software has cycled through different strategies. Early digital control systems were dependent on centralized intelligence and robust networks. After that, intelligence was distributed to the local controllers. It now appears that the next major step in the evolution of automation intelligence will be supplementing existing distributed systems with Internet-scale computing and advanced data collection.

Cabling can be installed during building construction or renovation to allow for both computer networking and building automation needs.

To create a data format that enables systems to work together, a gateway collects the data from the various automation and control systems in a building and pushes it out over the Internet for storage on a server in a data warehouse. **See Figure 12-7.** This data can then be used by analysis systems.

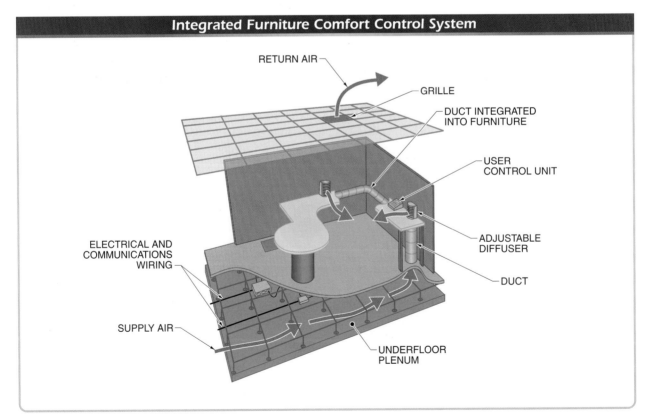

Integrated Furniture Comfort Control System

RETURN AIR

GRILLE

DUCT INTEGRATED INTO FURNITURE

USER CONTROL UNIT

ADJUSTABLE DIFFUSER

DUCT

ELECTRICAL AND COMMUNICATIONS WIRING

SUPPLY AIR

UNDERFLOOR PLENUM

Figure 12-6. By integrating ducts, diffusers, and controls into the furniture, conditioned air from underfloor, wall, or column plenums can be directed into the occupant's workspace.

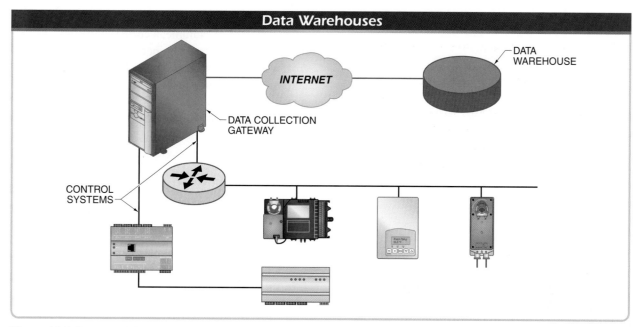

Figure 12-7. Data can be collected from building control systems, even separate or proprietary systems, and sent for storage in a data warehouse.

Data Warehouses. A *data warehouse* is a dedicated storage area for electronic data. The structure of data within the storage area is organized to facilitate accessing and filtering the data at a later time for analysis, trending, and reports. Storing large quantities of data is important for trending analyses. Control points sampled at short intervals over a long period can reveal important historical trends. Data warehouses offer several advantages over trend logging from within a building automation system including the following:

• Data storage is not limited by the capacity of the building automation system. The initial storage capacity of a data warehouse system is typically much higher than a local automation system, and it can be much more easily upgraded as needed. A data warehouse can store the value of every control point for every minute potentially for the life of the building at a cost-effective price point.

• Off-site storage offers security for the stored data, which can become an extremely valuable resource when used for analyzing ways to reduce building energy consumption and other operational costs.

• The standardized cataloging and organization of the data from multiple sources by a single system facilitates data retrieval and analysis.

Fault Detection and Diagnosis. Since symptoms often precede many types of failures, automated fault detection and diagnostics software is able to predict many failures by analyzing data trends. Analysis software sorts through data warehouse archives and applies fault detection and diagnosis algorithms to uncover building operation problems. This automated system looks for trends in the data to determine if equipment is beginning to fail, because failure is often preceded by measurable symptoms. **See Figure 12-8.** The availability of extensive stores of data allows an operator to determine the financial impact of the system failure.

System Optimization. A data warehouse can also be used to calibrate an energy model. When a building is designed, the mechanical engineer typically generates a building model that simulates energy consumption of various building configurations. The system optimization software model can be extended to post-construction analysis. The model's input and output files are modified by current data from the building control system and the data warehouse, rather than being based on preconstruction assumptions. Then, operational scenarios such as "what is the least disruptive method of reducing demand?" are simulated for various desirable outcomes. Since the new simulations use real control data from actual building operation, the optimization accuracy is significantly higher and increases as the available input data grows over time.

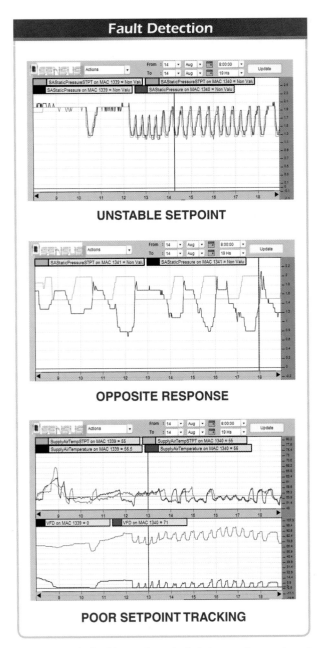

Figure 12-8. Archives of control data can be analyzed by software to find problems in the automation system algorithms.

Enterprise Integration

A building automation system can be integrated with enterprise systems to improve productivity or sales. In a retail environment, the system can use building and environment information to control ambient music, lighting, electronic advertisements, and product displays. This kind of integrated system could be sensitive to the weather, time of day, customers, sales goals, or other factors. For example, a coffee shop could display hot beverages more prominently on cooler days and frozen drinks on warmer days. Music and lighting can automatically change to create a different mood, which may influence buying habits. The result not only benefits the retailer but also gives the consumer an enhanced buying experience.

AUTOMATING EXISTING BUILDINGS

The automation of existing buildings is far more complicated than designing building automation systems for new construction. However, the increasing focus on energy use and environmental impact is encouraging building owners to upgrade their existing system infrastructure, including equipment, building materials, and electronic control.

Upgrading the Existing Infrastructure

Retrofitting existing buildings requires prioritization of the possible measures for improving energy efficiency. Materials and equipment that are relatively accessible are typically replaced, while it is rarely feasible to modify the building design. Adding building automation to new systems can also be challenging. For example, it is relatively simple to replace old lighting ballasts and HVAC equipment with modern, efficient equivalents, but their controls must also be replaced with versions that are operable with other building systems.

Integrating new controls with daylight-harvesting sun blinds and added solar and wind energy solutions can help reach the lower energy-consumption patterns desired. With the emerging trend of adjustable electrical rate tiers from utility companies, energy contracts can be made that benefit building owners if peak electricity demand can be reduced through integrated solar and wind power systems. If the building's electrical system can communicate expected energy requirements with the utility, better rates can be negotiated based on predictability measures.

Merging New and Existing Systems

The first decision in upgrading an existing facility is whether existing systems should be merged with new systems or completely removed. This is a decision that must be made while considering the life-cycle stage of the facility, the depreciation value of the existing equipment, and the cost to migrate versus the cost to remove systems. Costs are measured in terms of labor, downtime, disruption, and capital expenditures.

If the decision is made to migrate from old to new systems over the course of time, finding the right players to support the migration path may be difficult. The question of "whose work is it?" arises not only in terms of installation but also in terms of merging the physical networks, merging the databases and user interfaces, and merging (or rather separating) the maintenance agreements and areas of responsibility—especially where responsibility overlaps within vertical disciplines.

REVIEW QUESTIONS

1. How does a coordination document help with integrating building controls?

2. Why is it now important for information technology (IT) staff to be involved in planning building automation systems?

3. How have controls vendors developed ways to differentiate themselves?

4. What are the advantages of integrating a building automation system with an IP infrastructure?

5. Why have wireless networks not yet been significantly implemented in building automation applications?

6. How do perceptions of open protocols differ between industry participants?

7. How will the migration to IPv6 help the utilization of building automation applications?

8. What are the common features being added to existing open protocols?

9. How does an individualized controls strategy improve occupant comfort?

10. How is data warehouse information used to analyze building operations?

Chapter 12 Review and Resources

Appendix

LEED for New Construction v2.2
Registered Project Checklist

Project Name:
Project Address:

Yes	?	No			
			Sustainable Sites		**14** Points

Y			Prereq 1	**Construction Activity Pollution Prevention**	Required
			Credit 1	**Site Selection**	1
			Credit 2	**Development Density & Community Connectivity**	1
			Credit 3	**Brownfield Redevelopment**	1
			Credit 4.1	**Alternative Transportation**, Public Transportation Access	1
			Credit 4.2	**Alternative Transportation**, Bicycle Storage & Changing Rooms	1
			Credit 4.3	**Alternative Transportation**, Low-Emitting & Fuel-Efficient Vehicles	1
			Credit 4.4	**Alternative Transportation**, Parking Capacity	1
			Credit 5.1	**Site Development,** Protect or Restore Habitat	1
			Credit 5.2	**Site Development,** Maximize Open Space	1
			Credit 6.1	**Stormwater Design,** Quantity Control	1
			Credit 6.2	**Stormwater Design,** Quality Control	1
			Credit 7.1	**Heat Island Effect,** Non-Roof	1
			Credit 7.2	**Heat Island Effect,** Roof	1
			Credit 8	**Light Pollution Reduction**	1

Yes	?	No			
			Water Efficiency		**5** Points

			Credit 1.1	**Water Efficient Landscaping**, Reduce by 50%	1
			Credit 1.2	**Water Efficient Landscaping**, No Potable Use or No Irrigation	1
			Credit 2	**Innovative Wastewater Technologies**	1
			Credit 3.1	**Water Use Reduction**, 20% Reduction	1
			Credit 3.2	**Water Use Reduction**, 30% Reduction	1

			Energy & Atmosphere		**17** Points

Y			Prereq 1	**Fundamental Commissioning of the Building Energy Systems**	Required
Y			Prereq 2	**Minimum Energy Performance**	Required
Y			Prereq 3	**Fundamental Refrigerant Management**	Required

*Note for EAc1: All LEED for New Construction projects registered after June 26th, 2007 are required to achieve at least two (2) points under EAc1.

			Credit 1	**Optimize Energy Performance**	1 to 10
				10.5% New Buildings or 3.5% Existing Building Renovations	1
				14% New Buildings or 7% Existing Building Renovations	2
				17.5% New Buildings or 10.5% Existing Building Renovations	3
				21% New Buildings or 14% Existing Building Renovations	4
				24.5% New Buildings or 17.5% Existing Building Renovations	5
				28% New Buildings or 21% Existing Building Renovations	6
				31.5% New Buildings or 24.5% Existing Building Renovations	7
				35% New Buildings or 28% Existing Building Renovations	8
				38.5% New Buildings or 31.5% Existing Building Renovations	9
				42% New Buildings or 35% Existing Building Renovations	10
			Credit 2	**On-Site Renewable Energy**	1 to 3
				2.5% Renewable Energy	1
				7.5% Renewable Energy	2
				12.5% Renewable Energy	3
			Credit 3	**Enhanced Commissioning**	1
			Credit 4	**Enhanced Refrigerant Management**	1
			Credit 5	**Measurement & Verification**	1
			Credit 6	**Green Power**	1

continued…

Yes ? No

			Materials & Resources	**13** Points

Y		Prereq 1	**Storage & Collection of Recyclables**	Required
		Credit 1.1	**Building Reuse**, Maintain 75% of Existing Walls, Floors & Roof	1
		Credit 1.2	**Building Reuse**, Maintain 100% of Existing Walls, Floors & Roof	1
		Credit 1.3	**Building Reuse**, Maintain 50% of Interior Non-Structural Elements	1
		Credit 2.1	**Construction Waste Management**, Divert 50% from Disposal	1
		Credit 2.2	**Construction Waste Management**, Divert 75% from Disposal	1
		Credit 3.1	**Materials Reuse**, 5%	1
		Credit 3.2	**Materials Reuse**, 10%	1
		Credit 4.1	**Recycled Content**, 10% (post-consumer + ½ pre-consumer)	1
		Credit 4.2	**Recycled Content**, 20% (post-consumer + ½ pre-consumer)	1
		Credit 5.1	**Regional Materials**, 10% Extracted, Processed & Manufactured Regiona	1
		Credit 5.2	**Regional Materials**, 20% Extracted, Processed & Manufactured Regiona	1
		Credit 6	**Rapidly Renewable Materials**	1
		Credit 7	**Certified Wood**	1

Yes ? No

			Indoor Environmental Quality	**15** Points

Y		Prereq 1	**Minimum IAQ Performance**	Required
Y		Prereq 2	**Environmental Tobacco Smoke** (ETS) **Control**	Required
		Credit 1	**Outdoor Air Delivery Monitoring**	1
		Credit 2	**Increased Ventilation**	1
		Credit 3.1	**Construction IAQ Management Plan**, During Construction	1
		Credit 3.2	**Construction IAQ Management Plan**, Before Occupancy	1
		Credit 4.1	**Low-Emitting Materials**, Adhesives & Sealants	1
		Credit 4.2	**Low-Emitting Materials**, Paints & Coatings	1
		Credit 4.3	**Low-Emitting Materials**, Carpet Systems	1
		Credit 4.4	**Low-Emitting Materials**, Composite Wood & Agrifiber Products	1
		Credit 5	**Indoor Chemical & Pollutant Source Control**	1
		Credit 6.1	**Controllability of Systems**, Lighting	1
		Credit 6.2	**Controllability of Systems**, Thermal Comfort	1
		Credit 7.1	**Thermal Comfort**, Design	1
		Credit 7.2	**Thermal Comfort**, Verification	1
		Credit 8.1	**Daylight & Views**, Daylight 75% of Spaces	1
		Credit 8.2	**Daylight & Views**, Views for 90% of Spaces	1

Yes ? No

			Innovation & Design Process	**5** Points

		Credit 1.1	**Innovation in Design**: Provide Specific Title	1
		Credit 1.2	**Innovation in Design**: Provide Specific Title	1
		Credit 1.3	**Innovation in Design**: Provide Specific Title	1
		Credit 1.4	**Innovation in Design**: Provide Specific Title	1
		Credit 2	**LEED® Accredited Professional**	1

Yes ? No

			Project Totals (pre-certification estimates)	**69** Points

Certified: 26-32 points, **Silver:** 33-38 points, **Gold:** 39-51 points, **Platinum:** 52-69 points

LEED for Existing Buildings v2.0
Registered Building Checklist

Project Name:
Project Address:

Yes	?	No			
			Sustainable Sites		**14** Points

Yes	?	No				
Y			Prereq 1	**Erosion & Sedimentation Control**		Required
Y			Prereq 2	**Age of Building**		Required
			Credit 1.1	**Plan for Green Site & Building Exterior Management -**4 specific actions		1
			Credit 1.2	**Plan for Green Site & Building Exterior Management -**8 specific actions		1
			Credit 2	**High Development Density Building & Area**		1
			Credit 3.1	**Alternative Transportation -** Public Transportation Access		1
			Credit 3.2	**Alternative Transportation -** Bicycle Storage & Changing Rooms		1
			Credit 3.3	**Alternative Transportation -** Alternative Fuel Vehicles		1
			Credit 3.4	**Alternative Transportation -** Car Pooling & Telecommuting		1
			Credit 4.1	**Reduced Site Disturbance -** Protect or Restore Open Space (50% of site area)		1
			Credit 4.2	**Reduced Site Disturbance -** Protect or Restore Open Space (75% of site area)		1
			Credit 5.1	**Stormwater Management -** 25% Rate and Quantity Reduction		1
			Credit 5.2	**Stormwater Management -** 50% Rate and Quantity Reduction		1
			Credit 6.1	**Heat Island Reduction -** Non-Roof		1
			Credit 6.2	**Heat Island Reduction -** Roof		1
			Credit 7	**Light Pollution Reduction**		1

Yes	?	No			
			Water Efficiency		**5** Points

Yes	?	No				
Y			Prereq 1	**Minimum Water Efficiency**		Required
Y			Prereq 2	**Discharge Water Compliance**		Required
			Credit 1.1	**Water Efficient Landscaping -** Reduce Potable Water Use by 50%		1
			Credit 1.2	**Water Efficient Landscaping -** Reduce Potable Water Use by 95%		1
			Credit 2	**Innovative Wastewater Technologies**		1
			Credit 3.1	**Water Use Reduction -** 10% Reduction		1
			Credit 3.2	**Water Use Reduction -** 20% Reduction		1

Yes	?	No			
			Energy & Atmosphere		**23** Points

Yes	?	No				
Y			Prereq 1	**Existing Building Commissioning**		Required
Y			Prereq 2	**Minimum Energy Performance -** Energy Star 60		Required
Y			Prereq 3	**Ozone Protection**		Required

*Note for EAc1: All LEED for Existing Buildings projects registered after June 26th, 2007 are required to achieve at least two (2) points under EAc1.

Yes	?	No				
			Credit 1	**Optimize Energy Performance**		1 to 10
					Energy Star Rating - 63	1
					Energy Star Rating - 67	2
					Energy Star Rating - 71	3
					Energy Star Rating - 75	4
					Energy Star Rating - 79	5
					Energy Star Rating - 83	6
					Energy Star Rating - 87	7
					Energy Star Rating - 91	8
					Energy Star Rating - 95	9
					Energy Star Rating - 99	10
			Credit 2.1	**Renewable Energy -** On-site 3% / Off-site 15%		1
			Credit 2.2	**Renewable Energy -** On-site 6% / Off-site 30%		1
			Credit 2.3	**Renewable Energy -** On-site 9% / Off-site 45%		1
			Credit 2.4	**Renewable Energy -** On-site 12% / Off-site 60%		1
			Credit 3.1	**Building Operation & Maintenance -** Staff Education		1
			Credit 3.2	**Building Operation & Maintenance -** Building Systems Maintenance		1
			Credit 3.3	**Building Operation & Maintenance -** Building Systems Monitoring		1
			Credit 4	**Additional Ozone Protection**		1
			Credit 5.1	**Performance Measurement -** Enhanced Metering (4 specific actions)		1
			Credit 5.2	**Performance Measurement -** Enhanced Metering (8 specific actions)		1

continued…

	Credit 5.3	**Performance Measurement -** Enhanced Metering (12 specific actions)	1
	Credit 5.4	**Performance Measurement** - Emission Reduction Reporting	1
	Credit 6	**Documenting Sustainable Building Cost Impacts**	1

Yes ? No

	Materials & Resources		**16** Points

Y	Prereq 1.1	**Source Reduction & Waste Management -** Waste Stream Audit	Required
Y	Prereq 1.2	**Source Reduction & Waste Management -** Storage & Collection	Required
Y	Prereq 2	**Toxic Material Source Reduction -** Reduced Mercury in Light Bulbs	Required
	Credit 1.1	**Construction, Demolition & Renovation Waste Management -** Divert 50%	1
	Credit 1.2	**Construction, Demolition & Renovation Waste Management -** Divert 75%	1
	Credit 2.1	**Optimize Use of Alternative Materials -** 10% of Total Purchases	1
	Credit 2.2	**Optimize Use of Alternative Materials -** 20% of Total Purchases	1
	Credit 2.3	**Optimize Use of Alternative Materials -** 30% of Total Purchases	1
	Credit 2.4	**Optimize Use of Alternative Materials -** 40% of Total Purchases	1
	Credit 2.5	**Optimize Use of Alternative Materials -** 50% of Total Purchases	1
	Credit 3.1	**Optimize Use of IAQ Compliant Products -** 45% of Annual Purchases	1
	Credit 3.2	**Optimize Use of IAQ Compliant Products -** 90% of Annual Purchases	1
	Credit 4.1	**Sustainable Cleaning Products & Materials -** 30% of Annual Purchases	1
	Credit 4.2	**Sustainable Cleaning Products & Materials -** 60% of Annual Purchases	1
	Credit 4.3	**Sustainable Cleaning Products & Materials -** 90% of Annual Purchases	1
	Credit 5.1	**Occupant Recycling -** Recycle 30% of the Total Waste Stream	1
	Credit 5.2	**Occupant Recycling -** Recycle 40% of the Total Waste Stream	1
	Credit 5.3	**Occupant Recycling -** Recycle 50% of the Total Waste Stream	1
	Credit 6	**Additional Toxic Material Source Reduction -** Reduced Mercury in Light Bulbs	1

Yes ? No

	Indoor Environmental Quality		**22** Points

Y	Prereq 1	**Outside Air Introduction & Exhaust Systems**	Required
Y	Prereq 2	**Environmental Tobacco Smoke (ETS) Control**	Required
Y	Prereq 3	**Asbestos Removal or Encapsulation**	Required
Y	Prereq 4	**PCB Removal**	Required
	Credit 1	**Outside Air Delivery Monitoring**	1
	Credit 2	**Increased Ventilation**	1
	Credit 3	**Construction IAQ Management Plan**	1
	Credit 4.1	**Documenting Productivity Impacts -** Absenteeism & Healthcare Cost Impacts	1
	Credit 4.2	**Documenting Productivity Impacts -** Other Productivity Impacts	1
	Credit 5.1	**Indoor Chemical & Pollutant Source Control -** Reduce Particulates in Air System	1
	Credit 5.2	**Indoor Chemical & Pollutant Source Control -** Isolation of High Volume Copy/Print/Fax	1
	Credit 6.1	**Controllability of Systems -** Lighting	1
	Credit 6.2	**Controllability of Systems -** Temperature & Ventilation	1
	Credit 7.1	**Thermal Comfort -** Compliance	1
	Credit 7.2	**Thermal Comfort -** Permanent Monitoring System	1
	Credit 8.1	**Daylight & Views -** Daylight for 50% of Spaces	1
	Credit 8.2	**Daylight & Views -** Daylight for 75% of Spaces	1
	Credit 8.3	**Daylight & Views -** Views for 45% of Spaces	1
	Credit 8.4	**Daylight & Views -** Views for 90% of Spaces	1
	Credit 9	**Contemporary IAQ Practice**	1
	Credit 10.1	**Green Cleaning -** Entryway Systems	1
	Credit 10.2	**Green Cleaning -** Isolation of Janitorial Closets	1
	Credit 10.3	**Green Cleaning -** Low Environmental Impact Cleaning Policy	1
	Credit 10.4	**Green Cleaning -** Low Environmental Impact Pest Management Policy	1
	Credit 10.5	**Green Cleaning -** Low Environmental Impact Pest Management Policy	1
	Credit 10.6	**Green Cleaning -** Low Environmental Impact Cleaning Equipment Policy	1

Yes ? No

	Innovation & Design Process		**5** Points

	Credit 1.1	**Innovation in Upgrades, Operation & Maintenance**	1
	Credit 1.2	**Innovation in Upgrades, Operation & Maintenance**	1
	Credit 1.3	**Innovation in Upgrades, Operation & Maintenance**	1
	Credit 1.4	**Innovation in Upgrades, Operation & Maintenance**	1
	Credit 2	**LEED™ Accredited Professional**	1

Yes ? No

	Project Totals (pre-certification estimates)	**85** Points

Certified: 32-39 points, **Silver:** 40-7 points, **Gold:** 48-63 points, **Platinum:** 64-85 points

NUMBERING SYSTEMS

A numbering system is a unique way to represent numerical quantities with a certain range of symbols. The same number can be represented in different ways by different numbering systems. **See Figure A-1.** Every numbering system is defined by a base that determines the total number of symbols used to represent quantities.

Numbering Systems				
Decimal	Binary	Binary Coded Decimal	Octal	Hexadecimal
1	1	0001	1	1
2	10	0010	2	2
3	11	0011	3	3
4	100	0100	4	4
5	101	0101	5	5
6	110	0110	6	6
7	111	0111	7	7
8	1000	1000	10	8
9	1001	1001	11	9
10	1010	0001 0000	12	A
11	1011	0001 0001	13	B
12	1100	0001 0010	14	C
13	1101	0001 0011	15	D
14	1110	0001 0100	16	E
15	1111	0001 0101	17	F
16	10000	0001 0110	20	`10
17	10001	0001 0111	21	11
18	10010	0001 1000	22	12
19	10011	0001 1001	23	13
20	10100	0010 0000	24	14
30	11110	0011 0000	36	1E
45	101101	0100 0101	55	2D
68	1000100	0110 1000	104	44
99	1100011	1001 1001	143	63
671	1010011111	0110 0111 0001	1237	29F
1876	11101010100	0001 1000 0111 0110	3524	754

Figure A-1

Decimal Numbers

The most common numbering system is decimal. A decimal number is a number expressed in a base of 10. The symbols used in this system are the digits 0, 1, 2, 3, 4, 5, 6, 7, 8, and 9.

A place value, or weight, is assigned to each position that a number greater than 9 holds. **See Figure A-2.** The weighted value of each position can be expressed as the base (10 in this case) raised to the power of n, the position. For example, 10 to the power of 2, or 10^2, is 100, and 10 to the power of 3, or 10^3, is 1000. For the decimal system, then, the position weights from right to left are 1, 10, 100, 1000, etc. Multiplying each digit by the weighted value of its position and then summing the results gives the value of the equivalent decimal number. For example, the number 9876 can be expressed as $(9 \times 1000) + (8 \times 100) + (7 \times 10) + (6 \times 1)$. This method is used with all bases to convert numbers to the decimal numbering system.

Decimal numbers are familiar from everyday use. However, in computing applications, the functions of processors and memory lend themselves better to arranging numbers in other numbering systems. Numbering systems common in computing applications include binary, octal, and hexadecimal.

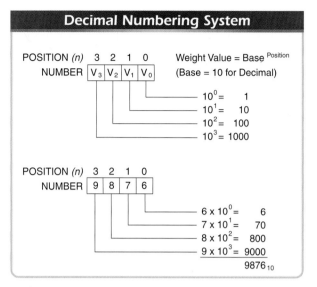

Figure A-2

Binary Numbers

A binary number is a number expressed in a base of 2. The symbols used in this system are 0 and 1. A bit is a binary digit, consisting of a 0 or a 1. Just as in the decimal system, the weighted value of each position is expressed as the base raised to the power of n, the position. **See Figure A-3.** For the binary numbering system, the weighted values of each position, from right to left, are 1, 2, 4, 8, 16, 32, 64, etc.

A binary number can be signed or unsigned. A signed binary number can be positive (no sign) or negative (−). The sign uses the left-most position of a binary number. The use of a signed binary number reduces the size of the largest decimal number represented. A 16-bit unsigned binary number is always positive and can represent a maximum decimal number of 65,535. A 16-bit signed binary number can represent a decimal range of −32,767 to +32,767.

Communications using binary numbers typically organize the information in groups of 4, 8, or 16 bits. **See Figure A-4.** Smaller numbers may not require all of the bits in a logical group to represent its value, but since a certain number of bits may be expected by the message receiver, leading zeroes are sometimes added. For example, the decimal number 18 can be expressed in binary as 10010, but if 8 bits were reserved in a communication protocol to represent this number, the transmitted signal would be 00010010.

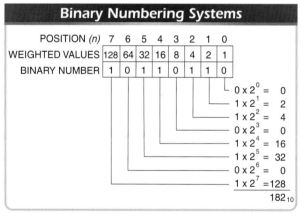

Figure A-3

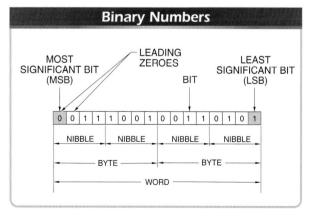

Figure A-4

A variation of the binary numbering system is binary-coded decimal (BCD). This system uses groups of four bits to represent each decimal digit. For example, the decimal number 87 is represented as binary 1000 0111. **See Figure A-5.** The first group of four binary digits (1000) represents decimal 8 and the second group (0111) represents 7. Spaces are typically added between the 4-bit groups to indicate each digit. Because the 4-bit groups for each decimal digit must be preserved in this scheme, leading zeroes are included as needed to each group.

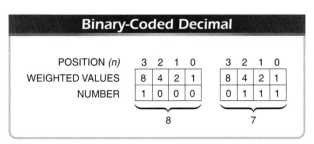

Figure A-5

It is interesting to compare a number in BCD to its representation in pure decimal. The binary number 87 is 1010111. BCD encoding requires more digits than traditional binary to represent the same number, but it is typically easier for people to convert to and from decimal digits.

While nearly all raw communication between computer-based devices is binary, the sequences for representing numbers can be very long and difficult for a person to interpret. This is not a common necessity, but sometimes troubleshooting communication problems can involve inspecting the raw message codes. In these cases, the information is often represented in octal or hexadecimal format, which shortens the sequences and makes the codes more recognizable.

Octal Numbers

An octal number is a number expressed in a base of 8. The symbols used in this system are the digits 0, 1, 2, 3, 4, 5, 6, and 7. The octal numbering system is often used to represent binary numbers using fewer digits, as each octal digit represents three bits in a binary system. According to the same position weighting scheme as other numbering systems, the weighted values of each position, from right to left, are 1, 8, 64, 512, etc. **See Figure A-6.**

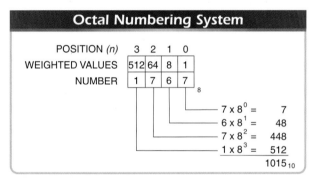

Figure A-6

Hexadecimal Numbers

A hexadecimal number is a number expressed in a base of 16. Since sixteen different symbols are required for this system, the digits 0 to 9 are insufficient. The letters A, B, C, D, E, and F are added for the numbers 10 through 15. The hexadecimal numbering system uses one digit to represent four bits in the binary numbering system. Just as in the decimal system, the weighted value of each position is expressed as the base raised to the power of n. **See Figure A-7.** The weighted values of each position, from right to left, are 1, 16, 256, 4096, etc.

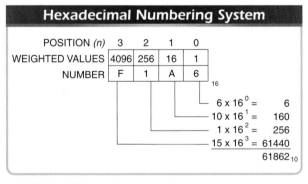

Figure A-7

Numbering System Designations

When there is the possibility of confusion between different numbering systems, subscripts or symbols are added to distinguish the base. These are not standardized, so there are several different designations that are commonly used. **See Figure A-8.** The most common may depend on the industry or field using the designations.

Decimal is often assumed to be the default numbering system, unless the context specifically explains otherwise. When it is necessary to designate decimal numbers, such as when they are used in combination with numbers in other systems, decimal numbers often use "dec" as a prefix or suffix. They may instead use a subscript "10" after the number.

Binary numbers are commonly indicated with "bin" or "b" as a prefix or suffix, subscript "2" after the number, or the symbol "%." The "%" has become a common convention because the symbol resembles the numeral one with two zeroes.

Common Numbering System Designations			
Decimal	Binary	Octal	Hexidecimal
37	0b100101	0o47	0x56A
37_{10}	100101_2	47_8	$56A_{16}$
37_{dec}	bin 100101	47_{oct}	$56A_{hex}$
	100101_b	o47	h56A
	%100101		56A#
			56Ah

Figure A-8

Octal numbers are commonly indicated with an "oct" subscript or an "o" prefix, such as "47_{oct}" or "o47." However, the "o" can be confused with 0 (zero). Hexadecimal numbers are commonly written with a "0x" or "h" prefix, "hex" subscript, or "#" or "h" suffix. The "0x" prefix is the most common.

Numbering System Conversions

It is often necessary to convert numbers between decimal and other bases. Computer programs or calculators can be used to automatically perform conversions, but the processes are actually very simple and involve only multiplication and addition. It is important to understand the concepts of numbering system conversions so that they can be done manually if necessary.

Converting Numbers in Other Bases to Decimal. The same method can be used with all numbering systems to convert numbers to the decimal numbering system. The weighted value of each position is expressed as the base raised to the power of n, the position. The weighted value is multiplied by the digit in that position, and all the products are added together, according to the following formula:

$$N_{10} = \ldots + (d_{3,b} \times b^3) + (d_{2,b} \times b^2) + (d_{1,b} \times b^1) + (d_{0,b} \times b^0)$$

where:

N_{10} = number represented in decimal

$d_{n,b}$ = nth-position digit of number to be converted to decimal

b = base of number to be converted to decimal

The decimal equivalent of a binary number can be calculated by multiplying each bit by the position weighting. For example, the decimal equivalent of the binary number 10110110 is calculated with the following formula:

$$N_{10} = (1 \times 2^7) + (0 \times 2^6) + (1 \times 2^5) + (1 \times 2^4) + (0 \times 2^3) + (1 \times 2^2) + (1 \times 2^1) + (0 \times 2^0)$$

$$N_{10} = (1 \times 128) + (0 \times 64) + (1 \times 32) + (1 \times 16) + (0 \times 8) + (1 \times 4) + (1 \times 2) + (0 \times 1)$$

$$N_{10} = 128 + 32 + 16 + 4 + 2$$

$$N_{10} = \mathbf{182}$$

The decimal equivalent of an octal number can be calculated by multiplying each octal digit by the position weighting. For example, the decimal equivalent of the octal number 1767 is calculated with the following formula:

$$N_{10} = (1 \times 8^3) + (7 \times 8^2) + (6 \times 8^1) + (7 \times 8^0)$$

$$N_{10} = (1 \times 512) + (7 \times 64) + (6 \times 8) + (7 \times 1)$$

$$N_{10} = 512 + 448 + 48 + 7$$

$$N_{10} = \mathbf{1015}$$

The decimal equivalent of a hexadecimal number can be calculated by multiplying each hexadecimal digit by the position weighting. For example, the decimal equivalent of the hexadecimal number F1A6 is calculated with the following formula:

$$N_{10} = (15 \times 16^3) + (1 \times 16^2) + (10 \times 16^1) + (6 \times 16^0)$$

$$N_{10} = (15 \times 4096) + (1 \times 256) + (10 \times 16) + (6 \times 1)$$

$$N_{10} = 61{,}440 + 256 + 160 + 6$$

$$N_{10} = \mathbf{61{,}862}$$

Converting Numbers in Decimal to Other Bases. To convert a decimal number to its equivalent in any base, a series of divisions is performed. The conversion process starts by dividing the decimal number by the base. **See Figure A-9.** If there is a remainder, it is placed in the least significant digit (LSD) right-most position of the new base number. If there is no remainder, a 0 is placed in the LSD position. The result of the division is then brought down, and the process is repeated until the final result of the successive divisions is 0. This methodology is a bit cumbersome, but it is the easiest conversion method to understand and use.

For example, for the binary equivalent of the decimal number 35, the first step is to divide the 35 by 2. The result is 17 with a remainder of 1, which becomes the LSD. The next step is to divide the 17 by 2, resulting in 8 with a remainder of 1. This 1 is the next LSD. Then the 8 is divided by 2, getting 4 with a remainder of 0. This 0 is the next LSD. Next, divide the 4 by 2, getting 2 with a remainder of 0. This 0 is the next LSD. The next step is to divide the 2 by 2, getting 1 with a remainder of 0. This 0 is the next LSD. Lastly, the 1 is divided by 2, getting 0 with a remainder of 1. This is the final digit. Therefore, the number 35_{10} is equivalent to 100011_2.

The same method can be used to convert decimal numbers to any other base, except that the decimal number is divided by that base instead of by 2. For example, for the hexadecimal equivalent of 1355_{10}, it is divided by 16, getting 84 with a remainder of 11. Decimal 11 is equivalent to hexadecimal B. Therefore, 0xB is the LSD. The 84 is divided by 16, getting 5 with a remainder of 4. This is the next LSD. The next step is to divide 5 by 16, getting 0 with a remainder of 5. This is the last digit. Therefore, the number 1355_{10} is equivalent to 0x54B.

Decimal Conversion	
Division	**Remainder**
$35 \div 2 = 17$	1
$17 \div 2 = 8$	1
$8 \div 2 = 4$	0
$4 \div 2 = 2$	0
$2 \div 2 = 1$	0
$1 \div 2 = 0$	1

$$35_{10} = 100011_2$$

Division	**Remainder**
$1355 \div 16 = 84$	11
$84 \div 16 = 5$	4
$5 \div 16 = 0$	5

$$1355_{10} = 54B_{hex}$$

Figure A-9

HVAC Symbols

Equipment Symbols	Ductwork	Heating Piping
EXPOSED RADIATOR	DUCT (1ST FIGURE, WIDTH; 2ND FIGURE DEPTH) 12 X 20	HIGH-PRESSURE STEAM ——HPS——
RECESSED RADIATOR	DIRECTION OF FLOW →	MEDIUM-PRESSURE STEAM ——MPS——
FLUSH ENCLOSED RADIATOR	FLEXIBLE CONNECTION	LOW-PRESSURE STEAM ——LPS——
PROJECTING ENCLOSED RADIATOR	DUCTWORK WITH ACOUSTICAL LINING	HIGH-PRESSURE RETURN ——HPR——
UNIT HEATER (PROPELLER)—PLAN	FIRE DAMPER WITH ACCESS DOOR FD AD	MEDIUM-PRESSURE RETURN ——MPR——
UNIT HEATER (CENTRIFUGAL)—PLAN	MANUAL VOLUME DAMPER VD	LOW-PRESSURE RETURN ——LPR——
UNIT VENTILATOR—PLAN	AUTOMATIC VOLUME DAMPER	BOILER BLOW OFF —— BD ——
STEAM	EXHAUST, RETURN OR OUTSIDE AIR DUCT—SECTION 20 X 12	CONDENSATE OR VACUUM PUMP DISCHARGE ——VPD——
DUPLEX STRAINER	SUPPLY DUCT—SECTION 20 X 12	FEEDWATER PUMP DISCHARGE ——PPD——
PRESSURE-REDUCING VALVE	CEILING DIFFUSER SUPPLY OUTLET 20" DIA CD 1000 CFM	MAKEUP WATER —— MU ——
AIR LINE VALVE	CEILING DIFFUSER SUPPLY OUTLET 20 X 12 CD 700 CFM	AIR RELIEF LINE —— V ——
STRAINER	LINEAR DIFFUSER 96 X 6-LD 400 CFM	FUEL OIL SUCTION ——FOS——
THERMOMETER	FLOOR REGISTER 20 X 12 FR 700 CFM	FUEL OIL RETURN ——FOR——
PRESSURE GAUGE AND COCK	TURNING VANES	FUEL OIL VENT ——FOV——
RELIEF VALVE	FAN AND MOTOR WITH BELT GUARD	COMPRESSED AIR —— A ——
AUTOMATIC 3-WAY VALVE		HOT WATER HEATING SUPPLY ——HW——
AUTOMATIC 2-WAY VALVE	LOUVER OPENING 20 X 12-L 700 CFM	HOT WATER HEATING RETURN ——HWR——
SOLENOID VALVE		

Air Conditioning Piping
REFRIGERANT LIQUID —— RL ——
REFRIGERANT DISCHARGE —— RD ——
REFRIGERANT SUCTION —— RS ——
CONDENSER WATER SUPPLY ——CWS——
CONDENSER WATER RETURN ——CWR——
CHILLED WATER SUPPLY ——CHWS——
CHILLED WATER RETURN ——CHWR——
MAKEUP WATER —— MU ——
HUMIDIFICATION LINE —— H ——
DRAIN —— D ——

Instrument Tag Identification

	First Letter		Second Letter		
	Measured or Initiating Variable	Modifier	Readout or Passive Function	Output Function	Modifier
A	Analysis		Alarm		
B	Burner Flame		User's Choice	User's Choice	User's Choice
C	Conductivity (Electrical)			Control	
D	Density (Mass) or Specific Gravity	Differential			
E	Voltage (EMF)		Primary Element		
F	Flow Rate	Ratio (Fraction)			
G	Gaging (Dimensional)		Glass		
H	Hand (Manually Initiated)				High
I	Current (Electrical)		Indicate		
J	Power	Scan			
K	Time or Time Schedule			Control Station	
L	Level		Light (Pilot)		Low
M	Moisture or Humidity				Middle or Intermediate
N	User's Choice		User's Choice	User's Choice	User's Choice
O	User's Choice		Orifice (Restriction)		
P	Pressure or Vacuum		Point (Test Connection)		
Q	Quantity or Event	Integrate or Totalize			
R	Radioactivity, radiation		Record or Print		
S	Speed or Frequency	Safety		Switch	
T	Temperature			Transmit	
U	Multivariable		Multifunction	Multifunction	Multifunction
V	Viscosity, Vibration			Valve, Damper, or Louver	
W	Weight or Force		Well		
X	Unclassified		Unclassified	Unclassified	Unclassified
Y	Event or State			Relay or Compute	
Z	Position			Drive, Actuate, or Unclassified Final Control Element	

A

absolute humidity: The amount of water vapor in a particular volume of air.

absorbent: A fluid that has a strong attraction for another fluid.

absorber: An absorption refrigeration system component in which refrigerant is absorbed by the absorbent.

absorption refrigeration system: A nonmechanical refrigeration system that uses a fluid with the ability to absorb a vapor when it is cool and release a vapor when heated.

access control system: A system used to deny those without proper credentials access to a specific building, area, or room.

actuator: A device that accepts a control signal and causes a mechanical motion.

adaptive control algorithm: A control algorithm that automatically adjusts its response time based on environmental conditions.

adaptive start time control: A process that adjusts the actual start time for HVAC equipment based on the building temperature responses from previous days.

address space: The logical collection of all possible LAN addresses for a given MAC layer type.

air changes per hour (ACH): A measure of the number of times the entire volume of air within a building space is circulated through the HVAC system in one hour.

air cleaning: A method of reducing air pollution that includes electronic air cleaning, ultraviolet (UV) air cleaning, and the use of filters such as carbon filters.

air conditioner: A self-contained cooling unit for forced-air HVAC systems.

air conditioning: The process of cooling the air in building spaces to provide a comfortable temperature.

airflow station: A sensor that measures the velocity of the air in a duct system.

air-handler economizer: An HVAC unit that uses outside air for free cooling.

air-handling unit (AHU): A forced-air HVAC system device consisting of some combination of fans, ductwork, filters, dampers, heating coils, cooling coils, humidifiers, dehumidifiers, sensors, and controls to condition and distribute supply air.

air velocity: The speed at which air moves from one point to another.

alarm monitoring: The detection and notification of abnormal building conditions.

algorithm: A sequence of instructions for producing the optimal result to a problem.

alternate scheduling: The programming of more than one unique time schedule per year.

analog signal: A signal that has a continuous range of possible values between two points.

appliance: A plumbing fixture that performs a special function and is controlled and/or energized by motors, heating elements, or pressure-sensing or temperature-sensing elements.

application layer: The OSI Model layer that provides communication services between application programs.

architect: A professional who designs and draws plans for a structure.

authority having jurisdiction (AHJ): The organization, office, or individual responsible for approving the equipment and materials used for building automation installation.

average value (V_{avg}) of a sine wave: The mathematical mean of all instantaneous voltage values in the sine wave.

averaging control: A control strategy that calculates an average value from multiple inputs, which is then used in control decisions.

B

BACnet XML: A proposed XML-based model for conveying control information between BACnet and other protocols.

bandwidth: The maximum rate at which bits can be conveyed by a signaling method over a certain media type.

bimetallic element: A sensing device that consists of two different metals jointed together that expand and contract at different rates with temperature change.

bipolar junction transistor (BJT): A transistor that controls the flow of current through the emitter (E) and collector (C) with a properly biased base (B).

boiler: A closed metal container that heats water to produce steam or hot water.

boost pump: A pump in a water supply system used to increase the pressure of the water while it is flowing to fixtures.

branch: A water distribution pipe that routes a water supply horizontally to fixtures or other pipes at approximately the same level.

branch circuit: The circuit in a power distribution system between the final overcurrent protective device and the associated end-use points, such as receptacles or loads.

bridge: A network device that joins two LANs at the data link and physical layers.

bridge rectifier: A circuit containing four diodes that permits both halves of the input AC sine wave to pass.

British thermal unit (Btu): The amount of heat energy required to raise the temperature of 1 lb of water 1°F.

broadcast: The transmission of a message intended for all nodes on the network.

building automation: The control of the energy-using and resource-using devices in a building for optimization of building system operations.

building automation system (BAS): A system that uses a distributed system of microprocessor-based controllers to automate any combination of building systems.

building drain: The lowest part of a drainage system, and it receives the discharge from all drainage pipes in the building and conveys it to the building sewer.

building main: A water distribution pipe that is the principal pipe artery supplying water to the entire building.

building sewer: The part of a drainage system that connects the building drain to the sanitary sewer.

bus topology: A linear arrangement of networked nodes with two specific endpoints.

C

cable: Two or more conductors grouped together within a common protective cover and used to connect individual components.

calibration: The adjustment of control parameters to the optimal values for the desired control response. Also known as tuning.

carbon dioxide (CO_2) sensor: A sensor that detects the concentration of CO_2 in air.

carbon filter: A type of air cleaner in which an activated carbon medium is used to remove most odors, gases, smoke, and smog from the air by means of an adsorption process.

cavitation: The process in which microscopic gas bubbles expand in a vacuum and suddenly implode when entering a pressurized area.

centralized generation: An electrical distribution system in which electricity is distributed through a utility grid from a central generating station to millions of customers.

centrifugal pump: A pump with a rotating impeller that uses centrifugal force to move water.

check valve: A valve that permits fluid flow in only one direction and closes automatically to prevent backflow (flow in a reverse direction).

chiller: A refrigeration system that cools water.

circuit breaker: An overcurrent protective device with a mechanism that automatically opens a switch in a circuit when an overcurrent condition occurs.

circulation: The continuous movement of air through a building and its HVAC system.

closed-loop control: Control in which feedback occurs between the controller, sensor, and controlled device.

closed-loop control system: A control system in which the result of an output is fed back into a controller as an input.

coaxial cable: A two-conductor cable in which one conductor runs along the central axis of the cable and the second conductor is formed by a braided wrap.

collapsed architecture: A protocol stack that excludes layers that are not needed for the application.

collision: The interaction of two messages on the same network media that can cause data corruption and errors.

combustion: The chemical reaction that occurs when oxygen (O) reacts with the hydrogen (H) and carbon (C) present in a fuel at ignition temperature.

comfort: The condition of a person not being able to sense a difference between themselves and the surrounding air.

common mode noise: Electrical noise produced between the ground and hot lines or the ground and neutral lines.

condensation: The formation of liquid (condensate) as moisture or other vapor cools below its dew point.

condenser: A heat exchanger that removes heat from high-pressure refrigerant vapor.

conditioned air: Indoor air that has been given desirable qualities by the HVAC system.

conduction: The transfer of heat from molecule to molecule through a material.

conductor: A material that has little resistance and permits electrons to move through it easily.

constant-air-volume AHU: An AHU that provides a steady supply of air and varies the heating, cooling, or other conditioning functions as necessary to maintain the desired setpoints within a building zone.

Construction Specifications Institute (CSI): An organization that develops standardized construction specifications.

consulting-specifying engineer: A building automation professional that designs the building automation system from the owner's list of desired features.

contactor: A heavy-duty relay for switching circuits with high-power loads.

contract documents: Documents produced by the consulting-specifying engineer for use by a contractor to bid a project and consist of construction drawings and a book of contract specifications.

control device: A building automation device that monitors or changes system variables, makes control decisions, or interfaces with other types of systems.

controller: A device that makes decisions to change some aspect of a system based on sensor information and internal programming.

control logic: The portion of controller software that produces the calculated outputs based on the inputs.

control loop: The continuous repetition of control logic decisions.

control point: The actual value that a control system experiences at any given time.

control signal: A changing characteristic used to communicate building automation information between control devices.

control strategy: A method for optimizing the control of building system equipment.

convection: The transfer of heat from warm to cool regions of a fluid from the circulation of currents.

cooling coil: A heat exchanger that transfers heat from the air surrounding or flowing through it.

cooling tower: A device that uses evaporation and airflow to cool water.

copper-clad aluminum: A conductor that has copper bonded to an aluminum core.

cord: A group of two or more conductors in one cover that is used to deliver power to a load by means of a plug.

current (*I*): A measure of the flow of charged particles flowing in a circuit and is measured in amperes (A).

D

daily multiple time period scheduling: The programming of time-based control functions for atypical periods of building occupancy.

daisy chain: A wiring implementation of bus topology that connects each node to its neighbor on either side.

damper: A set of adjustable metal blades used to control the amount of airflow between two spaces.

data link layer: The OSI Model layer that provides the rules for accessing the communication medium, uniquely identifying (addressing) each node, and detecting errors produced by electrical noise or other problems.

data trending: The recording of past building equipment operating information.

data warehouse: A dedicated storage area for electronic data.

deadband: The range between two setpoints in which no control action takes place.

dehumidifier: A device that removes moisture from the air by causing the moisture to condense.

derivative control algorithm: A control algorithm in which the output is determined by the instantaneous rate of change of a variable.

dew point: The air temperature below which moisture begins to condense.

differential pressure: The difference between two pressures.

differential pressure switch: A switch that activates at a differential pressure either above or below a certain value.

digital signal: A signal that has only two possible states.

diode: A semiconductor device that allows current to flow in one direction only.

direct digital control (DDC) system: A control system in which the building automation system controller is wired directly to controlled devices and can turn them on or off or start a motion.

direct expansion (DX) system: A system in which the refrigerant expands directly inside a coil in the main airstream itself to affect the cooling of the air.

distributed generation: An electrical distribution system in which many small power-generating systems create electrical power near the point of consumption.

distributed generator interconnection relay: A specialized relay that monitors both a primary power source and a secondary power source for the purpose of paralleling systems.

doping: The addition of material to a base element to alter the crystal structure of the element.

dry-bulb economizer: A type of economizer that operates strictly in proportion to the outside-air temperature, with no reference to humidity values.

dry-bulb temperature: The temperature of air measured by a thermometer freely exposed to the air but shielded from radiation and moisture.

dust-holding capacity: A filter's ability to hold dust without seriously reducing the filter's efficiency.

dust spot efficiency: A filter rating that measures a filter's ability to remove large particles from the air that tend to soil building interiors.

duty cycling control: A supervisory control strategy that reduces electrical demand by turning off certain HVAC loads temporarily.

dynamic pressure drop: The pressure drop in a duct fitting or transition caused by air turbulence as the air flows through the fitting or transition.

E

economizing: A cooling strategy that adds cool outside air to the supply air.

electrical demand: The amount of electrical power drawn by a load at a specific moment.

electrical demand control: A supervisory control strategy designed to reduce a building's overall electrical demand.

electrical service: The electrical power supply to a building or structure.

electrical system: A combination of electrical devices and components connected by conductors that distributes and controls the flow of electricity from its source to a point of use.

electric control system: A control system in which the power supply is line voltage (120 VAC or 220 VAC) or 24 VAC from a step-down transformer that is wired into the building power supply.

electric heating element: A device that consists of wire coils that become hot when energized.

electricity: The energy resulting from the flow of electrons through a conductor.

electricity consumption: The total amount of electricity used during a billing period.

electric motor: A device that converts electrical energy into rotating mechanical energy.

electronic air cleaner: A type of air cleaner in which particles in the air are positively charged in an ionizing section, and then the air moves to a negatively charged collecting area (i.e., plates) to separate the particles.

electronic control system: A control system in which the power supply is 24 VDC or less.

elevator system: A conveying system used for transporting people and/or materials vertically between floors in a building.

end switch: A switch that indicates the fully actuated damper positions.

energy harvesting: A strategy where a device obtains power from its surrounding environment.

enterprise system: A software and networking solution for managing business operations.

enthalpy: The total heat content of a substance.

enthalpy economizer: A type of economizer that uses temperature and humidity levels of the outside air to control operation.

equipment grounding conductor (EGC): A conductor that provides a low-impedance path from electrical equipment and enclosures to a grounding system.

estimation start time control: A process that calculates the actual start time for HVAC equipment based on building temperature data and a thermal recovery coefficient.

evaporation: The process of a liquid changing to a vapor by absorbing heat.

evaporator: A heat exchanger that adds heat to low-pressure refrigerant liquid.

exhaust air: The air that is ejected from a forced-air HVAC system. Also known as relief air.

Extensible Markup Language (XML): A general-purpose specification for annotating text with information on structure and organization.

F

fan: A mechanical device with spinning blades that move air.

feedback: The measurement of the results of a control action by a sensor.

fiber optics: A form of signaling based on light pulses to convey signals.

field-effect transistor (FET): A transistor that controls the flow of current through a drain (D) and source (S) with a properly biased gate (G).

filter: A type of air cleaner in which a mechanical device is used to remove particles from the air.

filter airflow resistance: The pressure drop across a filter at a given velocity.

filtration: The process of removing particulate matter from air.

fire protection system: A building system for protecting the safety of building occupants during a fire.

firewall: A router-type device that allows or blocks the passage of packets depending on a set of rules for restricting access.

fitting: A device used to connect two lengths of pipe.

fixture: A receptacle or device that is connected to the water distribution system, demands a supply of potable water, and discharges the waste directly or indirectly into the sanitary drainage system.

fixture branch: A water supply pipe that extends between a water distribution pipe and fixture supply pipe.

fixture drain: A drainage pipe that extends from the fixture trap to the junction of the next drainage pipe.

fixture supply pipe: A water supply pipe that connects the fixture to the fixture branch.

fixture trim: A set of water supply and drainage fittings installed on a fixture or appliance to control the water flowing into a fixture and the wastewater flowing from the fixture to the sanitary drainage system.

flat network architecture: A network configuration in which control devices are arranged in a peer-to-peer way.

flow meter: A device used to measure the flow rate and/or total flow of fluid flowing through a pipe.

flow pressure: The water pressure in a water supply pipe near a fixture while it is wide open and flowing.

flow rate: The volume of water passing a point at a particular moment.

flow switch: A switch with a vane that moves from the force exerted by the water or air flowing within a duct or pipe.

foot valve: A check valve installed at the bottom of the suction line on a suction-lift pump that keeps the suction line primed when the pump shuts down.

forced-air HVAC system: A system that distributes conditioned air throughout a building in order to maintain the desired conditions.

forward bias: The condition of a diode while it conducts current.

frame: A packet surrounded by additional data to facilitate its successful transmission and reception by delineating the start and end (or length) of the packet.

free topology: An arrangement of networked nodes that does not require a specific structure and may include any combination of buses, stars, rings, and meshes.

frequency: The number of AC waveforms per interval of time.

fuel oil: A petroleum-based product made from crude oil.

full-duplex communication: A system in which data signals can flow in both directions simultaneously.

full-wave rectifier: A circuit containing two diodes and a center-tapped transformer that permits both halves of the input AC sine wave to pass.

full-way valve: A valve designed to operate in only the fully open or fully closed positions. Also known as a shutoff valve.

furnace: A self-contained heating unit for forced-air HVAC systems.

fuse: An overcurrent protective device with a fusible link that melts and opens a circuit when an overcurrent condition occurs.

G

gateway: A network device that translates transmitted information between different protocols.

gateway network architecture: A network configuration in which a gateway is used to integrate separate control systems based on different protocols.

generator: An absorption refrigeration system component that vaporizes and separates the refrigerant from the absorbent.

geoexchange heat pump: *See* geothermal heat pump.

geothermal heat pump: A heat pump that uses the ground below the frost line as the heat source and heat sink. Also known as a geoexchange heat pump.

globe valve: A valve designed to control water flow to allow for throttling, mixing, and diverting of the water.

grain: A unit of measure that equals $1/7000$ lb.

grid: The network of conductors, substations, and equipment of a utility that distributes electricity from a central generation point to consumers.

grounded conductor: A current-carrying conductor that has been intentionally grounded.

grounding: The intentional connection of all exposed non-current-carrying metal parts to the earth.

grounding electrode conductor (GEC): A conductor that connects a grounding system to a buried grounding electrode.

H

half-duplex communication: A system in which data signals can flow in both directions but only one direction at a time.

half-wave rectifier: A circuit containing one diode that allows only half of the input AC sine wave to pass.

harmonic: Voltage or current at a frequency that is an integer (whole number) multiple (2nd, 3rd, 4th, etc.) of the fundamental frequency.

head: The difference in water pressure between points at different elevations.

heat exchanger: A device that transfers heat from one fluid to another fluid without allowing the fluids to mix.

heating, ventilating, and air conditioning (HVAC) system: A building system that controls the indoor climate of a building.

heating coil: A heat exchanger that transfers heat to the air surrounding or flowing through it.

heating value: The amount of British thermal units (Btu) per pound or gallon of fuel.

heat pump: A mechanical compression refrigeration system that moves heat from one area to another area.

heat transfer: The movement of heat from one material to another.

hierarchical network architecture: A network configuration in which control devices are arranged in a tiered network and have limited interaction with other control devices.

high-limit control: A control strategy that makes system adjustments necessary to maintain a control point below a certain value.

high/low signal select: A control strategy in which the building automation system selects the highest or lowest values from among multiple inputs for use in the control decisions.

high-priority load: A load that is important to the operation of a building and is shed last when electrical demand goes up.

holiday and vacation scheduling: The programming of time-based control functions during holidays and vacations.

hot water loop: A closed circuit of hot water supply distribution pipes, including the water heater, through which hot water is circulated.

hub: A multiport repeater that operates at the physical layer of an OSI Model and repeats messages from one port onto all of its other ports.

human-machine interface (HMI): An interface terminal that allows an individual to access and respond to building automation system information.

humidifier: A device that adds moisture to the air by causing water to evaporate into the air.

humidistat: A switch that activates at humidity levels either above or below a setpoint.

humidity: The amount of moisture present in the air.

humidity ratio (W): The ratio of the mass (weight) of the moisture in a quantity of air to the mass of the air and moisture together.

humidity sensor: *See* hygrometer.

hunting: An oscillation of output resulting from feedback that changes from positive to negative.

hybrid control system: A control system that uses multiple control technologies.

hydronic system: A system that distributes water throughout a building as the heat-transfer medium for heating and cooling systems.

hydropneumatic tank: A water tank with an air-filled bladder that raises the pressure of water as it is pumped into the tank.

hygrometer: A device that measures the amount of moisture in the air. Also known as a humidity sensor.

hygroscopic element: A material that changes its physical or electrical characteristics as the humidity changes.

I

ignition temperature: The intensity of heat required to start a chemical reaction.

impeller: A bladed, spinning hub of a fan or a pump that forces fluid to the perimeter of a housing.

integral control algorithm: A control algorithm in which the output is determined by the sum of the offset over time.

integrated circuit: An electronic device in which all components (transistors, diodes, and resistors) are contained in a single package or chip.

intermediary framework: A complete software and/or hardware solution for integrating multiple network-based protocol systems through a centralized interface.

internetwork: A network that involves the interaction between LANs through routers at the network layer.

L

latency: The time delay involved in the transmission of data on a network.

latent heat: Heat identified by a change of state and no temperature change.

lateral service: An electrical service in which service-entrance conductors are run underground from the utility service to the building.

lead/lag control: A control strategy that alternates the operation of two or more similar pieces of equipment in the same system.

level switch: A switch that activates at liquid levels either above or below a certain setpoint.

life safety control: A supervisory control strategy for life safety issues such as fire detection and suppression.

lift check valve: A check valve with a disk that moves vertically.

light-emitting diode (LED): A diode designed to produce light when forward biased.

lighting system: A building system that provides artificial light for indoor areas.

liquid chiller: A system that uses a liquid (normally water) to cool building spaces.

local area network (LAN): The infrastructure for data communication within a limited geographic region, such as a building or a portion of a building.

logical segment: A combination of multiple segments that are joined together with network devices that do not change the fundamental behavior of a LAN.

low-limit control: A control strategy that makes system adjustments necessary to maintain a control point above a certain value.

low-priority load: A load that is shed first for electrical demand control.

M

MAC address: A node address that is based on the addressing scheme of the associated data link layer protocol.

MAC layer: A sublayer of the OSI Model that combines functions of the physical and data link layers to provide a complete interface to the communications medium.

magnetic motor starter: A specialized contactor used for switching electrical power to a motor and includes overload protection.

main bonding jumper (MBJ): A connection at service equipment that connects an equipment grounding conductor, grounding electrode conductor, and grounded conductor (neutral conductor).

mapping: The process of making an association between comparable concepts in a gateway.

mechanical seal: An assembly of mechanical parts installed around a pump shaft that prevents leakage of the pumped liquid along the shaft.

media type: The specification of the characteristics and/or arrangement of the physical conductors or electromagnetic frequencies used for digital communication.

mesh topology: An interconnected arrangement of networked nodes.

micron: A unit of measure equal to one-millionth of a meter or 0.000039″.

microwave sensor: A sensor that activates when it senses changes in the reflected microwave energy caused by people moving within its field-of-view.

mixed air: The blend of return air and outside air that is combined inside an AHU and goes on to be conditioned.

multicast: The transmission of a message intended for multiple nodes, which are all assigned to the same multicast group.

N

natural gas: A colorless, odorless fossil fuel.

network architecture: The physical design of a communication network, including the network devices and how they connect segments together to form more complex networks.

network layer: The OSI Model layer that provides for the interconnection of multiple LAN types (MAC layers) into a larger network.

network management tool: *See* network tool.

network tool: A software application that runs on a computer connected to a network and is used to make changes to the operation of the nodes on a network. Also known as a network management tool.

node: A computer-based device that communicates with other similar devices on a shared network.

N-type material: Material created by doping a region of a crystal with atoms from an element that has more electrons in its outer shell than the crystal.

O

occupancy sensor: A sensor that detects whether an area is occupied by people.

octet: A sequence of eight bits.

offset: The difference between the value of a control point and its corresponding setpoint.

open building information exchange (oBIX): An XML-based model for conveying control information between any building automation protocol, including enterprise and proprietary protocols.

open-loop control: Control in which no feedback occurs between the controller, sensor, and controlled device.

open-loop control system: A control system in which decisions are made based only on the current state of the system and a model of how it should work.

open protocol: A standardized communication and network protocol that is published for use by any device manufacturer.

Open Systems Interconnection (OSI) Model: A standard description of the various layers of data communication commonly used in computer-based networks.

optimum start control: A supervisory control strategy in which the HVAC load is turned on as late as possible to achieve the indoor environment setpoints by the beginning of building occupancy.

optimum stop control: A supervisory control strategy in which the HVAC load is turned off as early as possible to maintain the proper building space temperature until the end of building occupancy.

outlet: An end-use point in the power distribution system.

outside air: Fresh air from outside a building that is incorporated into a forced-air HVAC system.

overcurrent: Electrical current in excess of the equipment limit, total amperage load of the circuit, or conductor or equipment rating.

overcurrent protective device (OCPD): A device that prevents conductors or devices from reaching excessively high temperatures from high currents by opening the circuit.

overhead service: An electrical service in which service-entrance conductors are run from the utility pole through the air and to the building.

overload: A small-magnitude overcurrent that, over a period of time, may trip the fuse or circuit breaker.

P

packet: A collection of data message information to be conveyed.

packing: An assembly of compressible sealing material installed around a pump shaft that prevents the pumped liquid from leaking along the shaft.

panelboard: A wall-mounted power distribution cabinet containing overcurrent protective devices for lighting, appliance, or power distribution branch circuits.

passive infrared (PIR) sensor: A sensor that activates when it senses the heat energy of people within its field-of-view.

peak value (V_{max}) of a sine wave: The maximum instantaneous value of a sine wave of either positive or negative alternation.

permanent holiday: A holiday that remains on the same date each year.

photodarlington transistor: A transistor that consists of a phototransistor and a standard NPN transistor in a single package.

photodiode: An electronic device that changes resistance or switches on when exposed to light.

phototransistor: An NPN transistor that has a large, thin base region that is switched on when exposed to a light source.

physical layer: The OSI Model layer that provides for signaling (the transmission of a stream of bits) over a communication channel.

plumbing system: A system of pipes, fittings, and fixtures within a building that conveys a water supply and removes wastewater and waterborne waste.

pneumatic control system: A control system in which compressed air is used to provide power and control signals for the control system.

pole: A set of contacts that belong to a single circuit.

port: 1. A virtual data connection used by nodes to exchange data directly with certain application programs on other nodes. **2.** An opening in a valve that allows a connection to a pipe.

positive-displacement pump: A pump that creates flow by trapping a certain amount of water and then forcing that water through a discharge outlet.

potable water: Water that is free from impurities that could cause disease or other harmful health conditions.

power factor (*PF*): The ratio of true power used in an AC circuit to apparent power delivered to the circuit.

powerline signaling: A communications technology that encodes data onto the alternating current signals in existing power wiring.

power quality: A measure of how closely the power in an electrical system matches the nominal (ideal) characteristics.

predictive maintenance: The monitoring of wear conditions and equipment characteristics in comparison to a predetermined tolerance to predict possible malfunctions or failures.

presentation layer: The OSI Model layer that provides transformation of the syntax of the data exchanged between application layer entities.

pressure gauge: A pressure-sensing device that indicates the pressure of a fluid on a numeric scale.

pressure loss due to friction: The loss of water pressure resulting from the resistance between water and the interior surface of a pipe or fitting.

pressure sensor: A sensor that measures the pressure exerted by a fluid, such as air or water.

pressure switch: A switch that activates at pressures either above or below a certain value.

preventive maintenance: Scheduled inspection and work that is required to maintain equipment in peak operating condition.

priming: The process of overcoming suction lift and getting liquid to a pump inlet.

priming pump: A vacuum pump that ejects air from the suction line of a larger suction-lift pump installation.

project closeout information: A set of documents produced by the controls contractor for the owner's use while operating the building.

proportional control algorithm: A control algorithm in which the output is in direct response to the amount of offset in the system.

proportional/integral (PI) control: The combination of proportional and integral control algorithms.

proportional/integral/derivative (PID) control: The combination of proportional, integral, and derivative algorithms.

proprietary protocol: A communication and network protocol that is developed and used by only one device manufacturer.

protocol stack: A combination of OSI layers and the specific protocols that perform the functions in each layer. Also known as a protocol suite.

protocol suite: *See* protocol stack.

psychrometrics: The scientific study of the properties of air and water vapor and the relationships between them.

P-type material: Material created by doping a region of a crystal with atoms from an element that has fewer electrons in its outer shell than the crystal.

pulse meter: A meter that outputs a pulse for every predetermined amount of flow in a circuit or pipe.

pump: A device that moves water through a piping system.

R

radiation: The transfer of heat between nontouching objects through radiant energy (electromagnetic waves).

radio frequency signaling: Communications technology that encodes data onto carrier waves in the radio frequency range.

rainwater leader: A vertical drainage pipe that conveys rainwater from a drain to the building storm drain or to another point of disposal.

reactive power (*VAR*): Power supplied to a reactive load.

receptacle: An outlet for the temporary connection of corded electrical equipment.

rectifier: A device that changes AC voltage into DC voltage.

refrigerant: A fluid that is used for transferring heat.

register: A cover for the opening of ductwork into a building space.

relative humidity: The ratio of the amount of water vapor in the air to the maximum moisture capacity of the air at a certain temperature.

relay: An electrical switch that is actuated by a separate electrical circuit.

relief air: *See* exhaust air.

repeater: A network device that amplifies and repeats electrical signals and provides a simple way to extend the length of a segment.

reset control: A control strategy in which a primary setpoint is adjusted automatically as another value (the reset variable) changes.

reset schedule: A chart that describes the setpoint changes in a pneumatic control system.

resistance temperature detector (RTD): A temperature-sensing element made from a metal or alloy conductor with an electrical resistance that changes with temperature.

restored load: A shed load that has been turned on after electrical demand control.

return air: The air from within a building space that is drawn back into a forced-air HVAC system to be exhausted or reconditioned.

reverse bias: The condition of a diode when it acts as an insulator.

ring topology: A closed-loop arrangement of networked nodes.

riser: A water distribution pipe that routes a water supply vertically one full story or more.

rooftop unit: An HVAC package unit that provides heating and cooling to a building space but is mounted in an enclosure on the roof.

rotating priority load shedding: An electrical demand control strategy in which the order of loads to be shed is changed with each high electrical demand condition.

router: A network device that joins two or more LANs together at the network layer and manages the transmission of messages between them.

routing: The process of determining the path between LANs that is required to deliver a message.

S

sanitary drainage system: A plumbing system that conveys wastewater and waterborne waste from plumbing fixtures and appliances to a sanitary sewer.

sanitary sewer: A sewer that carries sewage but does not convey rainwater, surface water, groundwater, and similar nonpolluting wastes.

scheduled control: A supervisory control strategy in which the date and time are used to determine the desired operation of a load or system.

schedule linking: The association of loads within the building automation system that are always used during the same time.

security system: A building system that protects against intruders, theft, and vandalism.

segment: A portion of a network in which all of the nodes share common wiring.

segmentation: A protocol mechanism that controls the orderly transmission of large data in small pieces.

self-contained control system: A control system that does not require an external power supply.

semiconductor: A material in which electrical conductivity is between that of a conductor (high conductivity) and that of an insulator (low conductivity).

sensor: A device that measures the value of a controlled variable, such as temperature, pressure, or humidity, and transmits a signal that conveys this information to a controller.

session layer: The OSI Model layer that provides mechanisms to manage a long series of messages that constitute a dialog.

setback: The unoccupied heating or cooling setpoint.

setpoint: The desired value to be maintained by a system.

setpoint control: A control strategy that maintains a setpoint in a system.

setpoint schedule: A description of the amount a reset variable resets the primary setpoint.

seven-day scheduling: The programming of time-based control functions that are unique for each day of the week.

sewage: Any liquid waste containing animal or vegetable matter in suspension or solution and/or chemicals in solution.

sewer gas: The mixture of vapors, odors, and gases found in sewers.

shed load: An electric load that has been turned off for electrical demand control.

shed table: A table that prioritizes the order in which electrical loads are turned off.

short circuit: A is a circuit in which current takes a low-impedance path around the normal path of current flow.

shutoff valve: *See* full-way valve.

signal: The conveyance of information.

signaling: The use of electrical, optical, and radio frequency changes in order to convey data between two or more nodes.

simplex communication: A system in which data signals can flow in only one direction.

smoke detector: An initiating device that is activated by the presence of smoke particles.

solar energy: Energy (radiant heat) transmitted from the sun.

solenoid: A device that converts electrical energy into a linear mechanical force.

solenoid valve: A full-way valve that is actuated by an electromagnet.

source control: A method of reducing air pollution by identifying strategies to reduce the origin of pollutants.

stack: A vertical drainage pipe that extends through one or more floors.

stale air: Air with high concentrations of carbon dioxide and/or other vapor pollutants.

star topology: A radial arrangement of networked nodes.

static pressure: The air pressure in a duct that pushes outward against the sides of the duct.

static pressure drop: The decrease in air pressure caused by friction between the air moving through a duct and the internal surfaces of the duct.

storm sewer: A sewer used for conveying groundwater, rainwater, surface water, or similar nonpolluting wastes.

stormwater drainage system: A plumbing system that conveys precipitation collecting on a surface to a storm sewer or other place of disposal.

supervisory control strategy: A method for controlling certain overall functions of a building automation system.

supply air: The newly conditioned mixed air that is distributed from a forced-air HVAC system to a building space.

swing check valve: A check valve with a hinged disk.

switch: 1. A multiport bridge that can forward messages selectively to one of its other ports based on the destination address. **2.** A device that isolates an electrical circuit from a power source.

switchboard: The last point on the power distribution system for the power company and the beginning of the power distribution system for the electrician of the property.

switching: The complete interruption or resumption of electrical power to a device.

system-powered control system: A control system in which the duct pressure developed by the fan system is used as the power supply.

T

temperature: The measurement of the intensity of the heat of a substance.

temperature sensor: A device that measures a property related to temperature.

temperature stratification: An undesirable variation of air temperature between the top and bottom of a space.

temporary scheduling: The programming of time-based control functions for a one-time temporary schedule.

terminal unit: The end point in an HVAC distribution system where the conditioned medium (air, water, or steam) is added to or directly influences the environment of the conditioned building space.

terminator: A resistor-capacitor circuit connected at one or more points on a communication network to absorb signals, avoiding signal reflections.

thermal recovery coefficient: The ratio of a temperature change to the length of time it takes to obtain that change.

thermistor: A temperature-sensing element made from a semiconductor with an electrical resistance that changes with temperature.

thermocouple: A temperature-sensing element consisting of two dissimilar metal wires joined at the sensing end that generate a small voltage that varies in proportion to the temperature at the hot junction.

thermodynamics: The science of thermal energy (heat) and how it transforms to and from other forms of energy.

thermostat: A switch that activates at temperatures either above or below a certain setpoint.

thermostatic mixing valve: A valve that mixes hot and cold water in proportion to achieve a desired temperature.

thermowell: A watertight and thermally conductive casing for immersion temperature sensors that mounts the sensing element inside the pipe, vessel, or fixture containing the water to be measured.

three-way valve: A valve with three ports that can control water flow between them.

throttling valve: A valve designed to control water flow rate by partially opening or closing.

throughput: The actual rate at which bits are transmitted over a certain media at a specific time.

throw: A position that a switch can adopt.

thyristor: A solid-state switching device that switches current on by a quick pulse of control current.

time-based control: A control strategy in which the time of day is used to determine the desired operation of a load.

timed override: A control function in which occupants temporarily change a zone from an UNOCCUPIED to OCCUPIED state.

topology: The shape of the wiring structure of a communications network.

total flow: The volume of water that passes a point during a specific time interval.

TP/FT: Signaling technology that is only used with LonTalk devices where the type of signaling is a differential Manchester encoded signal for serially transmitted data.

TP/XF: A twisted-pair technology that is used only with LonTalk devices where the signal is a transformer-isolated differential Manchester encoded signal for serially transmitted data.

transceiver: A hardware component that provides the means for nodes to send and receive messages over a network.

transducer: A device that converts one form of energy into another form of energy.

transfer switch: A switch that allows an electrical system to be switched between two power sources.

transient holiday: A holiday that changes its date each year.

transient voltage: A temporary, undesirable voltage in an electrical circuit, ranging from a few volts to several thousand volts and lasting from a few microseconds up to a few milliseconds.

transistor: A three-terminal semiconductor device that controls current according to the amount of voltage applied to the base.

transport layer: The OSI Model layer that manages the end-to-end delivery of messages across multiple LAN types.

transverse mode noise: Electrical noise produced between the hot and neutral lines.

triac: A thyristor used to switch AC.

true power (P_T): The actual power used in an electrical circuit.

tuning: *See* calibration.

twisted-pair cable: A multiple-conductor cable in which pairs of individually insulated conductors are twisted together.

U

ultrasonic sensor: A sensor that activates when it senses changes in the reflected sound waves caused by people moving within its detection area.

ultraviolet (UV) air cleaner: A type of air cleaner that kills biological contaminants using a specific light wavelength.

uninterruptible power supply (UPS): An electrical device that provides stable and reliable power, even during fluctuations or failures of the primary power source.

utility: A company that generates and/or distributes electricity to consumers in a certain region or state.

V

vacuum tube: A device that switches or amplifies electronic signals.

valve: A fitting that regulates the flow of water within a piping system.

variable: Some changing characteristic in a system.

variable-air-volume AHU: An AHU that provides air at a constant air temperature but varies the amount of supply air in order to maintain the desired setpoints within a building zone.

variable-air-volume (VAV) terminal box: A device located at a building zone that provides heating and airflow as needed in order to maintain the desired setpoints within the building zone.

variable-frequency drive (VFD): A motor controller that is used to change the speed of an AC motor by changing the frequency of the supply voltage.

velocity pressure: The air pressure in a duct that is measured parallel to the direction of airflow.

ventilation: The process of introducing fresh outside air into a building.

vent piping system: A plumbing system that provides for the circulation of air in a sanitary drainage system.

virtual point: A control point that exists only in software.

viscosity: The ability of a liquid to resist flow.

voice-data-video (VDV) system: A building system used for the transmission of information.

voltage: The difference in electrical potential between two points in an electrical circuit.

vortex damper: A pie-shaped damper at the inlet of a centrifugal fan that reduces the ability of the fan to grip and move air.

W

water column (WC): The pressure exerted by a square inch of a column of water.

water heater: A plumbing appliance used to heat water for the plumbing system's hot water supply.

water supply fixture unit (wsfu): An estimate of a plumbing fixture's water demand based on its operation.

water supply system: A plumbing system that supplies and distributes potable water to points of use within a building.

wet-bulb temperature: The temperature of air measured by a thermometer that has its bulb kept in contact with moisture.

wireless mesh network: A wireless network where the devices on the network can communicate directly with each other to relay signals.

X

X10 technology: A control protocol that uses power line signals to communicate between devices.

Z

zener diode: A diode designed to operate in a reverse-biased mode without being damaged.

zone: An area within a building that shares the same HVAC requirements.

Index

transport layers, 84
transverse mode noise, 122
triacs
    defined, 54, *55*
    heating elements and, 185, *186*
    testing, *73*, 73
troubleshooting, 56–57
tuning (calibration), 35–36, *36*
turbines, 101
twisted-pair cable, *93*, 93

U

ultrasonic sensors, 183
ultraviolet (UV) air cleaners, *168*, 168
underfloor air distribution systems, *307*, 307
uninterruptible power supplies (UPSs), 130–133
unit heater control, 221–222, *222*
utilities, 100

V

vacuum tubes, *46*, 46
valves
    HVAC systems and, *195*, 195
    plumbing systems and, 234, *248*, 248–251, *249*, *250*
variable-air-volume AHUs, *154*, *155*, 155
variable-air-volume terminal boxes, *155*, 155
variable-frequency drives (VFDs), 128–129, *129*, 191
variables, 21
VDV (voice-data-video) systems, 19
velocity pressure, 166
ventilation, *172*, 172, *197*
vent piping systems, 236
VFDs (variable-frequency drives), 191
virtual points, 27
viscosity, 142
voice-data-video (VDV) systems, 19
voltage, *119*, 119–120, *120*
vortex dampers, 191

W

water, 147–148
water chiller control, 226–227, *228*
water column (WC), 166
water condition sensors, 241–245
water distribution control devices, 194–195

water flow, 240–241, *241*
water heaters, 238
water levels, *241*, 241
water meters, *243*, 243
water pressure, 238–240
water supply characteristics, 237–241
water supply fixture units (wsfu), 241
water supply systems, 233–235, 234
water temperature, 237–238
WC (water column), 166
wet-bulb temperature, *163*, 163
wind turbines, 101
wireless building automation systems, *64*, 64–65
wireless mesh networks, *65*, 65
wireless networking technologies, 302–303, *303*
wireless networking trends, 305–306
wireless technologies, 97
wires, 111

X

XML (extensible markup language), *263*, 263–265, *264*
X10 technology, 11, *12*

Z

zener diodes, *49*, 49
zones, *158*, 158–161